1993

EXCURSIONS IN CALCULUS

AN INTERPLAY OF THE
CONTINUOUS AND THE DISCRETE

ROBERT M. YOUNG

D1636830

THE
DOLCIANI MATHEMATICAL EXPOSITIONS

Published by
THE MATHEMATICAL ASSOCIATION OF AMERICA

———

The Dolciani Mathematical Expositions

NUMBER THIRTEEN

EXCURSIONS IN CALCULUS
AN INTERPLAY OF THE
CONTINUOUS AND THE DISCRETE

ROBERT M. YOUNG

Published and Distributed by
THE MATHEMATICAL ASSOCIATION OF AMERICA

©*1992 by*
The Mathematical Association of America (Incorporated)
Library of Congress Catalog Card Number 91-67385

Complete Set ISBN 0-88385-300-0
Vol. 13 ISBN 0-88385-317-5

Printed in the United States of America

Current printing (last digit):
10 9 8 7 6 5 4 3 2 1

The DOLCIANI MATHEMATICAL EXPOSITIONS series of the Mathematical Association of America was established through a generous gift to the Association from Mary P. Dolciani, Professor of Mathematics at Hunter College of the City University of New York. In making the gift, Professor Dolciani, herself an exceptionally talented and successful expositor of mathematics, had the purpose of furthering the ideal of excellence in mathematical exposition.

The Association, for its part, was delighted to accept the gracious gesture initiating the revolving fund for this series from one who has served the Association with distinction, both as a member of the Committee on Publications and as a member of the Board of Governors. It was with genuine pleasure that the Board chose to name the series in her honor.

The books in the series are selected for their lucid expository style and stimulating mathematical content. Typically, they contain an ample supply of exercises, many with accompanying solutions. They are intended to be sufficiently elementary for the undergraduate and even the mathematically inclined high-school student to understand and enjoy, but also to be interesting and sometimes challenging to the more advanced mathematician.

DOLCIANI MATHEMATICAL EXPOSITIONS

In memory of my father

PREFACE

Without the concepts, methods and results found and developed by previous generations right down to Greek antiquity one cannot understand either the aims or the achievements of mathematics in the last fifty years.

—Hermann Weyl

Dissatisfaction with elementary calculus is growing. Is it because the subject has changed so little since Euler's great treatise of 1748, the *Introductio in analysin infinitorum*; or, as Paul Halmos has suggested, is it because it has changed so much?

> Teachers of elementary mathematics in the U.S.A. frequently complain that all calculus books are bad. That is a case in point. Calculus books are bad because there is no such subject as calculus; it is not a subject because it is many subjects. What we call calculus nowadays is the union of a dab of logic and set theory, some axiomatic theory of complete ordered fields, analytic geometry and topology, the latter in both the "general" sense (limits and continuous functions) and the algebraic sense (orientation), real-variable theory properly so called (differentiation), the combinatoric symbol manipulation called formal integration, the first steps of low-dimensional measure theory, some differential geometry, the first steps of the classical analysis of the trigonometric, exponential, and logarithmic functions, and, depending on the space available and the personal inclination of the author, some cook-book differential equations, elementary mechanics, and a small assortment of applied mathematics. Any one of these is hard to write a good book on; the mixture is impossible.[1]

But if Halmos is right, then the rich diversity of the calculus is at once its greatest allure. For "Mathematical science," said David Hilbert, "is in my opinion an indivisible whole, an organism whose vitality is conditioned upon the connection of its parts."[2] Besides, what better goal to pursue than one that remains elusive?

Excursions in Calculus is one possible supplement to a more traditional calculus course. It is a collection of six essays, a potpourri of topics and techniques that have fascinated me in the teaching of elementary calculus at Oberlin College over the past twenty years. Its underlying theme is the elegant interplay that exists between the two main currents of mathematics, the continuous and the discrete. Accordingly, such fundamental notions from discrete mathematics as induction, recursion, combinatorics, and the algorithmic point of view as a unifying principle are prominently featured.

Throughout, I have been guided by the precepts of George Pólya, whose craft of discovery has helped shape the teaching and learning of so many of

---

[1] P. R. Halmos, How to write mathematics, *L'Enseignement Mathématique,* 16 (1970) 125.

[2] David Hilbert, Mathematical Problems, *Bulletin of the American Mathematical Society,* 8 (1901 –1902) 478.

us. The proliferation of new knowledge conspires against self-discovery, but it remains true that "the worst way to teach is to talk," to which one might well add, *and the worst way to learn is to listen.* Wherever possible, I have tried to let the masters do the teaching.

A word about the contents. In the sixth Josiah Willard Gibbs Lecture presented in New York City in 1928, G. H. Hardy said:

> The elementary theory of numbers should be one of the very best subjects for early mathematical instruction. It demands very little previous knowledge; its subject matter is tangible and familiar; the processes of reasoning which it employs are simple, general and few; and it is unique among the mathematical sciences in its appeal to natural human curiosity. A month's intelligent instruction in the theory of numbers ought to be twice as instructive, twice as useful, and at least ten times as entertaining as the same amount of "calculus for engineers."[3]

The present work attests to the author's belief that Hardy was right.

A final word about the problems. Gathered from a great many sources they are an integral part of the text. Most are challenging, few are routine, hopefully all are interesting. Some of the problems remain unsolved. These have been included not so much as a challenge, but as a way of indicating what ought to be known but currently is not. At the same time, many a talented student has succeeded in uncovering what the professional mathematician has not. The way to excite good students is to give them hard problems. (A complete listing of Sources for Solutions is included at the end of the book.)

I am indebted to a great many people for their encouragement and support.

To Ralph Boas, who read every word of the manuscript, made many of them better, and prevented me from making many foolish mistakes.

To Doug Dickson, Mike Henle, Joe Malkevitch, Joe Roberts, and Jonathan Scherer for their sharp criticism and advice.

To Chris Gernon and Mok Oh for their invaluable assistance in preparing the computer-generated figures.

And, most especially, to Linda Miller, whose unwavering support and forbearance helped make this work possible.

---

[3] G. H. Hardy, An Introduction to the Theory of Numbers, *Bulletin of the American Mathematical Society,* 35 (1929) 818.

CONTENTS

"Cheshire Puss," she began, rather timidly, . . . "Would you tell me, please, which way I ought to go from here?"

"That depends a good deal on where you want to get to," said the Cat.

"I don't much care where—" said Alice.

"Then it doesn't matter which way you go," said the Cat.

—Lewis Carroll, *Alice's Adventures in Wonderland*

                —listen:there's a hell
of a good universe next door;let's go

            —e. e. cummings, "pity this busy monster,manunkind,"

CHAPTER 1

INFINITE ASCENT, INFINITE DESCENT: THE PRINCIPLE OF MATHEMATICAL INDUCTION

A well-known scientist (some say it was Bertrand Russell) once gave a public lecture on astronomy. He described how the earth orbits around the sun and how the sun, in turn, orbits around the center of a vast collection of stars called our galaxy. At the end of the lecture, a little old lady at the back of the room got up and said: "What you have told us is rubbish. The world is really a flat plate supported on the back of a giant tortoise." The scientist gave a superior smile before replying, "What is the tortoise standing on?" "You're very clever, young man, very clever," said the old lady. "But it's turtles all the way down!"

—Stephen W. Hawking, *A Brief History of Time*

1. Patterns

> But as if a magic lantern threw the nerves in patterns on a screen . . . .
> —T. S. Eliot, "The Love Song of J. Alfred Prufrock"

The positive integers,

$$1, 2, 3, 4, 5, \ldots,$$

the ordinary counting numbers: Has anything so profound ever come in simpler garb?

We tend to forget the advanced stage of intellectual development needed to perceive the abstract character of a number. And yet, as Bertrand Russell wrote, "It must have required many ages to realize that a brace of pheasants and a couple of days were both instances of the number two."

We choose to believe in the existence of all the positive integers 1,2,3,. . ., as if the mere presence of three dots were enough to bridge the gap between the finite and the infinite. Mathematics needs the infinite and yet its existence is an act of faith.

We hope that arithmetic is sound, that the building blocks of mathematics are unflawed, that they contain no inherent contradictions. And yet, in 1931, Kurt Gödel presented mathematics with the astonishing and melancholy conclusion that it is impossible to establish the internal logical consistency of a very large class of deductive systems—among them elementary arithmetic—"unless one adopts principles of reasoning so complex that their internal consistency is as open to doubt as that of the systems themselves" [311]. Hermann Weyl has said, "God exists because arithmetic *is* consistent—the Devil exists because we can't prove it!"

The integers. Once shrouded in mysticism by the ancients, who believed that whole numbers were the essence of the universe, the integers have remained the primary source of all of mathematics. From Pythagoras (circa 550 B.C.) to Euclid (circa 300 B.C.), to Diophantus of Alexandria (circa A.D. 275), to Pierre de Fermat (1601–1665), who initiated the modern theory of numbers, to Euler (1707–1783) and Gauss (1777–1855), interest in the integers has never waned. "All things are number," intoned Pythagoras; and, more than two thousand years later, Leopold Kronecker (1823–1891) reaffirmed the foundation on which most of mathematics now rests: "God created the integers; everything else is the work of man."

Having been given the "miraculous jar" of the positive integers, the first mathematicians set about to uncover the patterns among them.

Hardy said it best: "A mathematician, like a painter or a poet, is a maker of patterns." Whether the patterns are simple or elaborate, transparent or concealed, whether they are creations of the mind, or, in the Platonic sense, that which the mind has but uncovered—whatever their origin, they often emerge as the result of observation and experimentation. That the same scientific methods that have proved so successful in dealing with physical objects should prove just as successful in the study of their abstract mathematical counterparts is a source of no small wonder. Observation and experimentation—and today, of course, the computer facilitates both.

Accordingly, a mathematician interested in the theory of numbers begins by observing and experimenting with the positive integers.

Example 1. Summing the Odd Integers. Let us suppose, for example, that by some chance we come across the relations

$$1 = 1^2, \quad 1 + 3 = 2^2, \quad 1 + 3 + 5 = 3^2,$$
$$1 + 3 + 5 + 7 = 4^2, \quad 1 + 3 + 5 + 7 + 9 = 5^2$$

obtained by adding together the first few odd numbers. Even on the basis of such modest evidence, we can hardly refrain from formulating the conjecture that *for every natural number n, the sum of the first n odd numbers is n^2*, or

$$1 + 3 + 5 + \cdots + (2n - 1) = n^2.$$

Is the conjecture true—not just for the values $n = 1, 2, 3, 4, 5$ which suggested it, but for every value of n? While additional evidence might reinforce our belief, it could never confirm it—we simply cannot carry out infinitely many verifications. Besides, even if we verify that a proposition is true for every number up to a million or a million million, we are no nearer to establishing that it is true always.

Consider for example the following disarmingly simple proposition about the primes. It was first formulated by a Russian civil servant, Christian Goldbach, in 1742 in a letter to the great Swiss mathematician Leonhard Euler.

Example 2. The Goldbach Conjecture. Recall that a positive integer greater than 1 whose only natural divisors are itself and 1 is said to be a *prime number,* or simply a *prime*. Thus 2 is the first prime and the first few primes are

$$2, \; 3, \; 5, \; 7, \; 11, \; 13, \; 17, \; 19, \ldots.$$

The number 1 is special; its only natural divisor is itself, but because it behaves so differently from other numbers, it is usually considered not to be a prime.

Prime numbers, as the reader is undoubtedly aware, are the fundamental building blocks from which all the natural numbers can be constructed. *Every positive integer greater than 1 can be written as a product of primes, and, apart from the order of the factors, in one way only.* This is the content of the *fundamental theorem of arithmetic* (for a proof, see Section 3). While multiplication is the natural operation to perform on the primes, it is not unreasonable to ask what can be accomplished through addition. In the simplest case, *what numbers can be obtained as a sum of two odd primes?* Such a sum must of course be even, and Goldbach proposed, solely on the basis of empirical evidence, that every even number greater than 4 can be represented in this way.

Testing the conjecture for the first few even numbers, we find

$$6 = 3 + 3$$
$$8 = 3 + 5$$
$$10 = 3 + 7$$
$$12 = 5 + 7$$
$$14 = 3 + 11$$
$$16 = 3 + 13$$
$$18 = 5 + 13$$
$$20 = 3 + 17$$
$$22 = 3 + 19$$
$$24 = 5 + 19$$
$$26 = 3 + 23$$
$$28 = 5 + 23$$
$$30 = 7 + 23.$$

We could, of course, proceed further—into the far reaches of the number system—and we should find overwhelming evidence in support of the conjecture. Computer searches up to 20 billion have never revealed an exception [386]. Nevertheless, no finite amount of evidence can ever prove conclusive— the next even number may yet violate the rule—and after two and a half centuries the general proposition remains elusive.

Example 3. A Question of Divisibility.   As a final illustration, let us consider a pattern that is far more recondite—but at the same time far more practicable—

than those previously mentioned. It concerns the divisibility of numbers of the form

$$2^n - 1$$

where $n = 1, 2, 3, \ldots$. These numbers have played an important role in the continuing search for prime numbers, and they have consistently produced the largest known primes.

In experimenting with numbers of this form, we may happen to observe that they are sometimes divisible by one more than their exponent. For example,

$$2^2 - 1 \text{ is divisible by } 3$$

$$2^4 - 1 \text{ is divisible by } 5$$

$$2^6 - 1 \text{ is divisible by } 7.$$

What can be said about the other even exponents? Do they behave similarly? (Of course, it is not necessary to consider odd exponents, for then the numbers $2^n - 1$ and $n + 1$ are a fortiori of opposite parity, i.e., one even and the other odd.) When we come to the next even exponent we see that the pattern fails: $2^8 - 1$ is not divisible by 9. Continuing beyond $n = 8$, we find that $2^{10} - 1$ is divisible by 11, $2^{12} - 1$ is divisible by 13, while $2^{14} - 1$ is not divisible by 15.

What rule can possibly underlie these isolated examples? Is it possible to tell in advance whether or not $2^n - 1$ is divisible by $n + 1$ without laboriously going through the implied division? Gathering more evidence, we examine all exponents up to $n = 30$ and find that $2^n - 1$ is divisible by $n + 1$ for each of the values

$$n + 1 = 3, 5, 7, 11, 13, 17, 19, 23, 29, 31$$

and for these values only. The pattern is now unmistakable—the values listed above are all primes. In fact, they are all the odd primes up to 31. We are thereby led to conjecture that $2^n - 1$ *is exactly divisible by* $n + 1$ *if and only if* $n + 1$ *is an odd prime.*

Notice that our conjecture is made up of two separate assertions. One states that the ratio

$$\frac{2^n - 1}{n + 1}$$

is always a whole number whenever the denominator is an odd prime. The other half, the converse, asserts that this ratio can never be a whole number if the denominator is composite (that is, neither 1 nor a prime). Does the evidence support both statements equally? Do you feel as confident about the validity of this conjecture as you did about the previous ones? Would it strengthen your conviction to know that Leibniz himself believed it?

We press further. We test the next thirty exponents, and then the thirty after that, and so on until all values of n up to 300 have been checked and the results compared with our conjecture. In each case we find agreement. By now $2^n - 1$ has nearly one hundred decimal digits, so that the necessary divisions must be performed indirectly by means of "modular arithmetic". (The reader who is unfamiliar with this important algebraic tool may wish to consult the Appendix now.)

Each new verification of our conjecture reinforces our belief in its validity. This time, however, we have been misled; we have stumbled upon a half-truth. It is true that the relation

$$2^n - 1 \text{ is divisible by } n + 1 \tag{1}$$

must always hold whenever $n + 1$ is an odd prime, but it may sometimes hold when $n + 1$ is composite. The first assertion is a special case of Fermat's little theorem, one of the cornerstones of the modern theory of numbers. In the next chapter, we shall provide a proof. Its converse, however, is false. In fact, there are infinitely many composite numbers $n + 1$ for which (1) holds, but the smallest of these is $341 = 11 \times 31$.[1] Little wonder that even Leibniz was in error![2]

The lesson is clear—inductive inference may prove faulty, and nothing short of a rigorous proof can ever suffice.[3]

---

[1] To see that $2^{340} - 1$ is divisible by 341, observe that

$$2^5 \equiv 1 \pmod{31} \quad \text{and} \quad 2^5 \equiv -1 \pmod{11},$$

from which it follows that $2^{340} \equiv 1$ with respect to each of the moduli 31 and 11, and hence also with respect to their product. This first and smallest counterexample was discovered by Pierre Sarrus in 1819.

[2] Dickson, *History of the Theory of Numbers* [135], volume 1, p. 91. Dickson also asserts—and many Western writers have copied him—that Chinese mathematicians claimed 25 centuries ago that $2^n - 2$ is divisible by n if and only if n is a prime. The statement appears to be false. In his monumental treatise *Science and Civilisation in China,* volume 3: *Mathematics and the Sciences of the Heavens and the Earth,* p. 54, footnote d, Needham writes: "We have not been able to trace this with certainty to its origin, but the source of the statement . . . is a mysterious remark made by J. H. Jeans in a note of 1897 (written while he was still an undergraduate) . . ." Needham suggests that Jeans's remark may have been due to an error in translation.

[3] For a delightful discussion of patterns that seem to appear when we look at small numbers, see Richard Guy, "The Strong Law of Small Numbers" [192] and the sequel, "The Second Strong Law of Small Numbers" [193].

PROBLEMS

A mathematician's work is mostly a tangle of guesswork, analogy, wishful thinking and frustration, and proof, far from being the core of discovery, is more often than not a way of making sure that our minds are not playing tricks.

—Gian-Carlo Rota

Familiarity with numbers, acquired by innate faculty sharpened by assiduous practice, does give insight into the profounder theorems of algebra and analysis.

—Aitken

It is a great nuisance that knowledge can only be acquired by hard work.

—W. Somerset Maugham

1. Verify the following relations which were observed by Galileo (1615) and used in his work on freely falling bodies:

$$\frac{1}{3} = \frac{1+3}{5+7} = \frac{1+3+5}{7+9+11} = \cdots.$$

2. Suppose that the odd numbers are grouped in the following pattern:

$$
\begin{array}{ccccc}
 & & & & 1 \\
 & & & 3 & 5 \\
 & & 7 & 9 & 11 \\
 & 13 & 15 & 17 & 19 \\
21 & 23 & 25 & 27 & 29
\end{array}
$$

Formulate a conjecture about the sums of the entries in each row, express it in suitable mathematical notation, and prove it. The result is attributed to Nicomachus (circa A.D. 100).

3. Palindrome Conjecture (unsolved). To any integer add its reverse. Then add the sum's reverse to the sum. Continue until a sum is palindromic (the same from right to left as from left to right). For example:

$$38 + 83 = 121$$

$$139 + 931 = 1070, \quad 1070 + 0701 = 1771.$$

It has been conjectured that every integer will eventually produce a palindrome.

4. (Ramanujan) Compute the value of

$$\sqrt{1 + 2\sqrt{1 + 3\sqrt{1 + 4\sqrt{1 + \cdots}}}} \ .$$

5. Observe that

$$\frac{16}{64} = \frac{1}{4}$$

and that "cancelling the 6's" produces the correct result. There are exactly three similar patterns with numbers less than 100 (aside from trivial variations obtained by inverting the fraction). Find them. (Boas calls it "a frivolous piece of mathematics but one with serious overtones" [43].)

6. Plutarch writes of the Pythagorean number mysticism: "The Pythagoreans also have a horror for the number 17. For 17 lies exactly halfway between 16, which is a square, and the number 18, which is the double of a square, these two being the only two numbers representing areas, for which the perimeter (of the rectangle) equals the area." (Van der Waerden, *Science Awakening* [426], p. 96) Verify that 16 and 18 are indeed the only two natural numbers with this property.

7. Consider the infinite ladder of whole numbers:

a	b
1	1
2	3
5	7
12	17
29	41
⋮	⋮

a. Discern the rule by which the numbers on the ladder are formed.

b. Show that on each rung of the ladder,

$$b^2 - 2a^2 = \pm 1.$$

c. Conclude that, as one descends the ladder, the ratios

$$\frac{1}{1}, \frac{3}{2}, \frac{7}{5}, \frac{17}{12}, \frac{41}{29}, \dots$$

approach $\sqrt{2}$, and that these approximations improve on each successive rung.

Remark. Concealed within the ladder is another elegant pattern concerning Pythagorean triangles (right-triangles with integral sides). Every other rung of the ladder has two odd entries. If we express b as the sum of two consecutive integers, say m and $m + 1$, then

$$m^2 + (m + 1)^2 = a^2.$$

For example, on the fifth rung, $b = 41 = 20 + 21$, so that $20^2 + 21^2 = 29^2$. All Pythagorean triangles whose legs differ by 1 are obtained in this way (see [385], pp. 42–46). For yet another concealed pattern, see chapter 2, §2, problem 8.

8. Observe that

$$1 + 2 = 3$$

$$4 + 5 + 6 = 7 + 8$$

$$9 + 10 + 11 + 12 = 13 + 14 + 15$$

$$16 + 17 + 18 + 19 + 20 = 21 + 22 + 23 + 24.$$

Guess the general law suggested by these examples, express it in suitable mathematical notation, and prove it.

9. Consider the table

$$1 = 1$$

$$2 + 3 + 4 = 1 + 8$$

$$5 + 6 + 7 + 8 + 9 = 8 + 27$$

$$10 + 11 + 12 + 13 + 14 + 15 + 16 = 27 + 64.$$

Guess the general law suggested by these examples, express it in suitable mathematical notation, and prove it.

10. Observe that

$$1 = 1$$
$$1 - 4 = -(1 + 2)$$
$$1 - 4 + 9 = 1 + 2 + 3$$
$$1 - 4 + 9 - 16 = -(1 + 2 + 3 + 4).$$

Guess the general law suggested by these examples, express it in suitable mathematical notation, and prove it.

11. Prove that every natural number greater than 11 is the sum of two composite numbers.

12. Find three distinct natural numbers such that the sum of their reciprocals is an integer. (The solution is unique.)

    Remark. R. L. Graham ("A theorem on partitions," *Journal of the Australian Mathematical Society* 4 (1963), pp. 435–441) has shown that every integer greater than 77 can be decomposed into a sum of integers whose reciprocals add up to unity (77 does not possess this property). For example, $78 = 2 + 6 + 8 + 10 + 12 + 40$ and

$$\frac{1}{2} + \frac{1}{6} + \frac{1}{8} + \frac{1}{10} + \frac{1}{12} + \frac{1}{40} = 1.$$

13. The sum

$$1 + \frac{1}{2} + \frac{1}{3} + \cdots + \frac{1}{n}$$

is never an integer if $n > 1$. More generally, the sum of two or more consecutive terms of the harmonic series is never an integer.

14. The following scheme furnishes a simple geometric interpretation of the *sieve of Eratosthenes.* Between every point $1, 1/2, 1/3, \ldots$ on the y-axis and every point $2, 3, 4, \ldots$ on the x-axis, we draw a straight line. Show that the set of x-coordinates at which these lines meet the horizontal line $y = -1$ is precisely the set of composite numbers.

15. Find a simple formula for the sums

$$1^3 + 2^3 + 3^3 + \cdots + n^3.$$

16. Notice that

$$3^2 + 4^2 = 5^2$$
$$3^3 + 4^3 + 5^3 = 6^3.$$

Find all values of n for which

$$3^n + 4^n + 5^n + \cdots + (n+2)^n = (n+3)^n.$$

17. Simplify the following sums:
   a. $1 \cdot 1! + 2 \cdot 2! + 3 \cdot 3! + \cdots + n \cdot n!$
   b. $1/2! + 2/3! + 3/4! + \cdots + n/(n+1)!$

18. Prove or disprove: All the numbers in the sequence

$$31, 331, 3331, 33331, \ldots$$

are primes.

19. Observe that the numbers

$$3! - 2! + 1! = 5$$
$$4! - 3! + 2! - 1! = 19$$
$$5! - 4! + 3! - 2! + 1! = 101$$
$$6! - 5! + 4! - 3! + 2! - 1! = 619$$
$$7! - 6! + 5! - 4! + 3! - 2! + 1! = 4421$$

are all primes. Does this pattern continue?

20. True or false: Every number of the sequence

$$49, 4489, 444889, 44448889, \ldots$$

is a perfect square.

21. The $3x+1$ Problem (unsolved).   Starting with any natural number, divide it by 2 if it is even and multiply it by 3 and add 1 if it is odd. Repeat the process continually. For example, the sequence obtained by starting with 17 runs

$$17, 52, 26, 13, 40, 20, 10, 5, 16, 8, 4, 2, 1.$$

The $3x + 1$ problem asserts that every natural number produces a sequence
that eventually ends with 4, 2, 1. (This has been verified by computer for all
natural numbers up to 1,000,000,000.) Try it for $x = 27$. How many iterations
of the algorithm are needed?

22.   Any number belonging to the sequence

$$11, 111, 1111, 11111, \ldots$$

is called a **repunit** (a contraction of repeated unit). Prove that no repunit can
be a square.

23.   Into how many regions do n lines divide the plane? (Assume that the lines
are in "general position," i.e., no two are parallel and no three are concurrent.)

24.   Find the number of regions determined in the inside of a circle by joining
n points on the circumference in all possible ways by straight lines, no three
of which are concurrent inside the circle. It is easy to calculate that the correct numbers for $n = 1, 2, 3, 4, 5$ are 1, 2, 4, 8, 16, respectively. Formulate a
conjecture for general n, and test it for $n = 6$.

25.   **Brocard's Problem (unsolved).**   When is $n! + 1$ a square? The only
known solutions are $n = 4, 5, 7$.

26.   Find all natural numbers which cannot be expressed as the difference of
two natural squares.

27.   Prove that for every natural number n:

$$n^3 - n \text{ is divisible by 3}$$
$$n^5 - n \text{ is divisible by 5}$$
$$n^7 - n \text{ is divisible by 7.}$$

Formulate a conjecture for all odd exponents and test it for the exponent 9.

28.   Define a sequence $a_0, a_1, a_2, \ldots$ by setting $a_0 = a_1 = 1$ and

$$a_{n+1} = \frac{a_0^2 + a_1^2 + \cdots + a_n^2}{n} \qquad \text{for } n \geq 1.$$

The first few terms are found to be $1, 1, 2, 3, 5, 10, 28, \ldots$. Is a_n always an integer?

29. Write down the multiples of $\sqrt{2}$, ignoring fractional parts, and underneath write the numbers missing from the first sequence:

1	2	4	5	7	8	9	11	12
3	6	10	13	17	20	23	27	30.

Is there a pattern?

30. The 13th of the month is more likely to be Friday than any one of the other days of the week.

2. Proof by Induction

Let us return to the problem of computing the sum of the first n odd numbers. The formula

$$1 + 3 + 5 + \cdots + (2n - 1) = n^2 \qquad (1)$$

comprises a whole sequence of assertions, one for every natural number n. Rather than focus attention on any particular value of n, we shall investigate instead the transition from any one value to the next. Specifically, we shall show that *every true assertion is followed by another true assertion.* It will then follow that if the first assertion is true, so are all the others. For the truth of the first will guarantee the truth of the second, and the truth of the second will then guarantee that of the third, and so on. It is this fundamentally important method of proof which is called the *principle of mathematical induction.*[4]

---

[4] Although discovery of the method is usually attributed to Pascal, it appears that the first person to apply mathematical induction to rigorous proofs was the Italian scientist Francesco Maurolico in his *Arithmetica* of 1575. (See, for example, Giovanni Vacca, "Maurolycus, the first discoverer of the principle of mathematical induction," *Bulletin of the Amer. Math. Society* 16 (1909–1910), 70–73, and also W. H. Bussey, "The origin of mathematical induction," *Amer. Math. Monthly* 24 (1917), 199–207.) Further improvements were made in the early 17th century by Pierre de Fermat. Pascal in one of his letters acknowledged Maurolico's introduction of the method and used it himself in his *Traité du triangle arithmétique* (1665), in which he presents what is now called Pascal's triangle (see chapter 2, §2). The phrase "mathematical induction" was apparently coined by De Morgan in the early 19th century. Of course the method is implicit even in Euclid's proof of the infinitude of the primes. For he shows (see chapter 2, §1) that if there are n primes, then there must be $n+1$ primes; and since there is a first prime, the number of primes must be infinite. For further discussion of mathematical induction, see the great works of George Pólya, *Mathematical Discovery* [341] and *Mathematics and Plausible Reasoning* [338], [339].

Let us use it to verify formula (1). Here, the transition from one case to the next is particularly simple. For suppose that the equation

$$1 + 3 + 5 + \cdots + (2n - 1) = n^2$$

is known to be true for a certain natural number n. Adding $2n + 1$, the next odd number, to both sides, we find

$$1 + 3 + 5 + \cdots + (2n + 1) = n^2 + 2n + 1$$

which is indeed $(n + 1)^2$, the next square number. This means that if our conjecture is true for a certain integer n, it remains necessarily true for the next integer $n + 1$. But we know that the conjecture is true for $n = 1$, so by the principle of mathematical induction, it must be true for all integers: true for 1, therefore, also for 2; true for 2, therefore, also for 3; and so on. We have succeeded in proving the conjecture in full generality.

Students who are not yet well-versed in this method of proof are often skeptical of its validity. Their objection: "You have *assumed* the proposition that is to be proved." Of course, we have done no such thing. Infinitely many propositions are to be proved; all that the principle of induction allows us to do is to suppose, when proving any one case, that the previous case has already been established. It is, in essence, a reformulation of the way in which the positive integers are ordered.

We would be remiss in not pointing out that formula (1) can be established without the use of mathematical induction, and, as is often the case, the inductive proof supplies the method. Let us begin, accordingly, with the formula

$$(k + 1)^2 = k^2 + 2k + 1$$

which describes the transition from one square number to the next. Rewriting this as

$$(k + 1)^2 - k^2 = 2k + 1$$

and then choosing successively $k = 0, 1, 2, \ldots, n - 1$, we obtain

$$
\begin{aligned}
1^2 - \ \ 0^2 \ &= \ \ 1 \\
2^2 - \ \ 1^2 \ &= \ \ 3 \\
3^2 - \ \ 2^2 \ &= \ \ 5 \\
4^2 - \ \ 3^2 \ &= \ \ 7 \\
&\cdots \\
n^2 - (n - 1)^2 &= 2n - 1.
\end{aligned}
$$

Adding the results—notice that the sum on the left "telescopes"—we find

$$n^2 = 1 + 3 + 5 + \cdots + (2n - 1).$$

Careful scrutiny of the previous proof leads to yet another—the simplest and most illuminating of all—a proof based on a single picture:

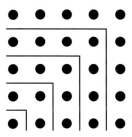

FIGURE 1
The sum of the first n odd numbers is n^2.

This elegant scheme, known already to the Pythagoreans, is nothing more than a geometric depiction of the identity $(k + 1)^2 = k^2 + 2k + 1$.

Finally, let us mention an extension of the method of induction which is frequently useful. Given an infinite sequence of propositions

$$P_1, P_2, P_3, \ldots$$

in order to demonstrate that all of them are true, it is enough to know two things:

1. P_1 is true.

2. For all n, if $P_1, P_2, P_3, \ldots, P_n$ are true, then so is P_{n+1}.

Thus, in trying to prove P_{n+1} we are entitled to assume that all previous propositions, not just P_n, have already been established. The reader will have no difficulty in verifying this simple variant.

PROBLEMS

1. Numbers of the form

$$1 + 2 + 3 + \cdots + n$$

where n is any natural number, are known as **triangular numbers**. The name derives from their geometric interpretation as triangular arrays of dots:

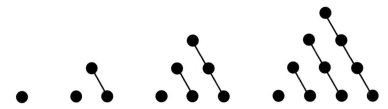

FIGURE 2
The first four triangular numbers.

In the nth figure there are n dots aligned along each side of the equilateral triangular shape and the total number of dots in this figure is the nth triangular number.

    a. Show that the nth triangular number is equal to $n(n + 1)/2$.
    b. If T_n denotes the nth triangular number, then

$$T_n^2 - T_{n-1}^2 = n^3.$$

Use this to show that

$$1^3 + 2^3 + \cdots + n^3 = T_n^2$$
$$= \left[\frac{n(n + 1)}{2}\right]^2.$$

2. Show that all the numbers $21, 2211, 222111, \ldots$ are triangular.

3. There are infinitely many numbers that are simultaneously triangular and square.
    Hint. If T is such a number, then $4T$ and $8T + 1$ are both squares.

4. Establish the following formula for the sum of the first n squares:

$$1^2 + 2^2 + 3^2 + \cdots + n^2 = n(n+1)(2n+1)/6.$$

5. Find a simple formula for

$$1^2 + 3^2 + 5^2 + \cdots + (2n-1)^2.$$

6. Calculate the following sums:
   a. $1/1 \cdot 2 + 1/2 \cdot 3 + 1/3 \cdot 4 + \cdots + 1/(n-1)n$
   b. $1/1 \cdot 2 \cdot 3 + 1/2 \cdot 3 \cdot 4 + 1/3 \cdot 4 \cdot 5 + \cdots + 1/(n-2)(n-1)n$
   c. $1/1 \cdot 2 \cdot 3 \cdot 4 + 1/2 \cdot 3 \cdot 4 \cdot 5 + 1/3 \cdot 4 \cdot 5 \cdot 6 + \cdots + 1/(n-3)(n-2)(n-1)n.$

7. Establish the following formulas:
   a. $1 \cdot 2 + 2 \cdot 3 + 3 \cdot 4 + \cdots + n(n+1) = n(n+1)(n+2)/3$
   b. $1 \cdot 2 \cdot 3 + 2 \cdot 3 \cdot 4 + 3 \cdot 4 \cdot 5 + \cdots + n(n+1)(n+2) = n(n+1)(n+2)(n+3)/4.$

8. Consider the table

$$3^2 + 4^2 = 5^2$$

$$10^2 + 11^2 + 12^2 = 13^2 + 14^2$$

$$21^2 + 22^2 + 23^2 + 24^2 = 25^2 + 26^2 + 27^2.$$

Guess the general law suggested by these examples, express it in suitable mathematical notation, and prove it. (Compare this pattern with problem 8 of the previous section).

9. Show that

$$\sqrt[3]{3} > \sqrt[4]{4} > \sqrt[5]{5} > \cdots.$$

What is the limit of these numbers?

10. On computing square roots by successive approximations. The following method for approximating square roots is often attributed to Heron of Alexandria (circa 100 B.C.), but it was known to the Babylonians hundreds of years earlier. To compute $\sqrt{a}$ first choose an initial approximation $x_1 > 0$ and then compute the successive approximations according to the formula

$$x_{n+1} = \frac{1}{2}\left(x_n + \frac{a}{x_n}\right), \qquad n = 1, 2, \ldots.$$

Notice that when the nth approximation to $\sqrt{a}$ has been found, $\sqrt{a} = \sqrt{x_n(a/x_n)}$, so that $\sqrt{a}$ is the geometric mean of x_n and a/x_n. As an approximation to this geometric mean, we take x_{n+1} to be the arithmetic mean of these two numbers.

    a.   By considering the hyperbola $xy = a$, show that if x_n is any approximation to $\sqrt{a}$ then x_{n+1} is a better one.

    b.   Show that, for any initial approximation x_1, the sequence $\{x_n\}$ always converges to $\sqrt{a}$.

    Hint. By comparing the errors $\epsilon_n = x_n - \sqrt{a}$ and $\epsilon_{n+1} = x_{n+1} - \sqrt{a}$ of two successive approximations, show that the absolute values of the successive errors are at least halved at each step after the first.

    c.   Show that when Newton's method for approximating the zeros of a function is applied to $f(x) = x^2 - a$, the result is Heron's method.

    d.   If we take $x_1 = 1$ as a first approximation to $\sqrt{2}$, we obtain the successive approximations

$$1, \frac{3}{2}, \frac{17}{12}, \frac{577}{408}, \ldots$$

How is this sequence related to the corresponding sequence of approximations given in problem 7 (section 1)?

11.   If $0 \le k \le n$, the **binomial coefficient** $\binom{n}{k}$ is defined by

$$\binom{n}{k} = \frac{n!}{k!(n-k)!},$$

with the understanding that $0! = 1$.

    a.   Prove that

$$\binom{n+1}{k} = \binom{n}{k-1} + \binom{n}{k}.$$

(The proof does not require an induction argument.) This relation gives rise to the following configuration, known as **Pascal's triangle**—a number not on one of the sides is the sum of the two numbers above it.

$$
\begin{array}{ccccccccc}
 & & & & 1 & & & & \\
 & & & 1 & & 1 & & & \\
 & & 1 & & 2 & & 1 & & \\
 & 1 & & 3 & & 3 & & 1 & \\
1 & & 4 & & 6 & & 4 & & 1 \\
\end{array}
$$

$$1 \quad\quad 5 \quad\quad 10 \quad\quad 10 \quad\quad 5 \quad\quad 1$$

$$\ldots$$

The binomial coefficient $\binom{n}{k}$ is the $(k+1)$st number in the $(n+1)$st row.

b. Show that $\binom{n}{k}$ is the number of sets of exactly k integers chosen from the numbers $1, 2, 3, \ldots, n$.

c. Prove the **binomial theorem**: *If x and y are any numbers and n is a natural number, then*

$$(x+y)^n = \sum_{k=0}^{n} \binom{n}{k} x^{n-k} y^k.$$

In the next chapter we shall investigate some of the far-reaching consequences of this important algebraic identity.

12. The product of n consecutive integers is divisible by $n!$.

13. Show that every positive integral power of $\sqrt{2} - 1$ is of the form $\sqrt{m} - \sqrt{m-1}$.

14. Prove that for any integer $k > 1$ and any positive integer n, n^k is the sum of n consecutive odd integers.

15. Derive the following estimate for $n!$:

$$\left(\frac{n}{3}\right)^n < n! < \left(\frac{n}{2}\right)^n$$

for $n \geq 6$. (The lower bound holds for all n.) This estimate is crude. A precise estimate is given by *Stirling's formula* (see chapter 5, §1, problem 14).

16. a. Show that if a_n denotes the length of the side of a regular n-gon inscribed in the unit circle (circle of radius 1), then the side of the $2n$-gon is

$$a_{2n} = \sqrt{2 - \sqrt{4 - a_n^2}}.$$

b. Deduce Viète's famous infinite product expansion

$$\frac{2}{\pi} = \frac{\sqrt{2}}{2} \cdot \frac{\sqrt{2 + \sqrt{2}}}{2} \cdot \frac{\sqrt{2 + \sqrt{2 + \sqrt{2}}}}{2} \cdots.$$

c. A sequence is defined recursively by taking the first two terms to be 0 and 1 and each successive term to be alternately the arithmetic and geometric mean of the two immediately preceding terms. Thus the sequence begins

$$0, \ 1, \ 1/2, \ 1/\sqrt{2}, \ldots.$$

Does the sequence approach a limit, and, if so, what is its value?

17. Consider the sequence

$$\left(1+\frac{1}{2}\right)^2, \left(1+\frac{1}{3}\right)^3, \left(1+\frac{1}{4}\right)^4, \ldots$$

    a. Show that each term in the sequence is smaller than the next one.

    b. Show that each term in the sequence is smaller than 3.

The limit of these simple fractions is the fundamental constant e. Correct to five decimal places,

$$e \approx 2.71828.$$

18. Prove *Bernoulli's inequality*: *If* $x > -1$, *then*

$$(1+x)^n \geq 1 + nx.$$

19. Prove *Hermite's identity*: *If* x *is real and* $n = 1, 2, 3, \ldots$, *then*

$$[x] + [x + 1/n] + [x + 2/n] + \cdots + [x + (n-1)/n] = [nx].$$

(Here, $[x]$ denotes the "bracket function," i.e., the largest integer not exceeding x.)

20. The following paradoxical argument is attributed to Tarski. It purports to prove by mathematical induction that *any n natural numbers are equal*! For $n = 1$ the statement is obviously true. It remains to pass from n to $n + 1$. For the purpose of illustration, let us pass from 3 to 4, and leave the general case to the reader. Suppose then that any four numbers a, b, c, and d have been given. By assumption ($n = 3$), a, b, and c are equal and, likewise, b, c, and d are equal. Consequently, all four numbers are equal. Explain the paradox.

3. Applications

The principle of mathematical induction finds applications throughout all of mathematics. Some additional examples will further attest to the power and scope of the method.

a. The Fundamental Theorem of Arithmetic

It is a relatively simple matter to show that every natural number, except 1, is expressible as a product of primes.[5] For example

$$666 = 2 \times 3 \times 3 \times 37.$$

It is not nearly so simple to show that such a representation is always unique, except for possible rearrangement of the factors. This is the content of the fundamental theorem of arithmetic.

Theorem.   *Apart from the order of the factors, any natural number greater than 1 can be represented as a product of primes in one way only.*

Proof.   The theorem is evidently true for the numbers 2 and 3, which are themselves primes. Proceeding by induction, we suppose that every natural number smaller than N (and greater than 1) has a unique representation as a product of primes, and we show that N must also. Suppose to the contrary that N had two different representations, say

$$N = p_1 p_2 \cdots p_n = q_1 q_2 \cdots q_m \tag{1}$$

where the prime factors in each product may be assumed to be in ascending order of magnitude, that is,

$$p_1 \leq p_2 \leq \cdots \leq p_n \quad \text{and} \quad q_1 \leq q_2 \leq \cdots \leq q_m.$$

Observe to begin with that every p must be different from every q, for if $p_i = q_j$ for some i and j, then we could cancel these factors in (1), thereby obtaining a number smaller than N with two different prime factorizations. Suppose, without loss of generality, that $p_1 < q_1$, and consider the number

$$M = p_1 q_2 \cdots q_m < q_1 q_2 \cdots q_m = N.$$

Since M and N are both divisible by p_1, so is their difference

$$N^* = N - M.$$

Therefore N^* possesses a prime factorization that includes p_1. On the other hand, we have

$$N^* = (q_1 - p_1) q_2 \cdots q_m$$

---

[5] It is convenient to regard a prime number as a *product* of primes, where the product has only one factor, namely the number itself.

and in this representation *not a single factor is divisible by* p_1. Consequently, N^* also has a prime factorization in which p_1 does not appear, and thus possesses two different prime factorizations. But $N^* < N$, and we have thereby contradicted the induction hypothesis. The contradiction shows that N can have at most one representation as a product of primes, and the proof is complete.

b. The Tower of Hanoi

In 1883 there appeared an ingenious puzzle known as the Tower of Hanoi.[6] It consisted of three pegs fastened to a stand together with eight circular rings, each of a different size. Initially, all of the rings are placed on one peg, the largest at the bottom and the radii of successive rings decreasing as we ascend. This arrangement is called the *tower* (see Figure 3). The problem is to transfer the tower to another peg, following these rules—only one ring at a time must be moved, and no ring may be placed atop a smaller one.

The following year, De Parville gave this fanciful account of the origin of the puzzle.[7]

> In the great temple at Benares, beneath the dome which marks the center of the world, there rests a brass plate in which are fixed three diamond needles, each a cubit high and as thick as the body of a bee. On one of these needles, at the creation, God placed sixty-four disks of pure gold, the largest disk resting on the brass plate, and the others getting smaller and smaller up to the top one. This is the Tower of Bramah. Day and night unceasingly the priests transfer the disks from one diamond needle to another according to the fixed and immutable laws of Bramah, which require that the priest on duty must not move more than one disk at a time and that he must place this disk on a needle so that there is no smaller disk below it. When the sixty-four disks shall have been transferred from the needle on which at the creation God placed them to one of the other needles, tower, temple, and Brahmins alike will crumble into dust, and with a thunderclap the world will vanish.

---

[6] Brought out by a certain "N. Claus (de Siam)," an anagram of "Lucas d'Amiens" or Edouard Lucas (1842–1891), a distinguished mathematician of his time. For a general survey of the puzzle, see [222].

[7] *La Nature*, 12 (1884) 285–286. This translation appears in Ball and Coxeter, *Mathematical Recreations and Essays* [21], p. 317.

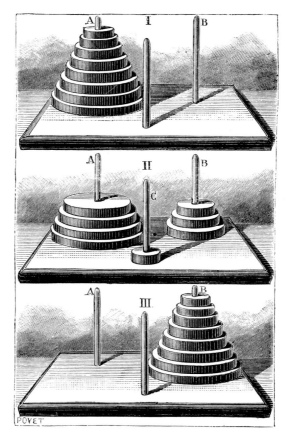

FIGURE 3
The Tower of Hanoi
From E. Lucas, *Récréations Mathématiques,* volume III, p. 56.

Whether the initial configuration contains eight disks, eight times eight disks, or any number of disks, the puzzle can always be solved. This we shall establish by induction. Suppose then that there are n disks on the peg A. We shall effect the transfer in three stages.

   i. By the induction hypothesis we are entitled to assume that the top $n - 1$ disks can be transferred, one by one, to the peg B, leaving the peg C vacant.

  ii. The bottom disk may now be moved to the peg C.

148,782

iii. Reversing the first process, we may now transfer all $n - 1$ disks from B to C.

The solution given by this 3-stage process is an example of a *recursive procedure*—a procedure which invokes itself. Thus, to solve the n-disk problem we must first solve the $(n - 1)$-disk problem, and to solve that we must go back to the $(n - 2)$-disk problem, and so on, until eventually we arrive at the trivial problem of transferring a single disk. Such a procedure is easily implemented in many high-level computer languages (such as Pascal), in which a procedure, or subroutine, is permitted to invoke itself.

While recursive programs have the merits of brevity and elegance, they require large amounts of storage—a great many intermediate solutions must be saved. An *iterative* program, on the other hand, is one that carries out a repetitive task by means of a simple loop rather than through a succession of recursions. For the Tower of Hanoi problem, there is such a simple iterative scheme: *Every other move consists of a transfer of the smallest disk from one peg to another, the pegs being taken in cyclic order.* (For another solution, based on binary numbers, see Computer Recreations, *Scientific American,* November 1984, pp. 19–28.)

Let us determine how many individual moves must be made. If it requires P_n transfers of single disks to move a tower of n disks, then the 3-step process described above shows that

$$P_n = 2\,P_{n-1} + 1.$$

This is an example of a *recursive formula* (the terms *difference equation* or *recurrence relation* are also used)—to find P_n we must go back to P_{n-1}. Now, with 2 disks, it requires 3 transfers, i.e., $2^2 - 1$ transfers; hence with 3 disks the number of transfers required will be $2(2^2 - 1) + 1$, that is, $2^3 - 1$. Proceeding in this way, we see that with a tower of n disks it will require $2^n - 1$ transfers of single disks to effect the complete transfer. Thus, the eight disks of the puzzle will require 255 single transfers, while the number of transfers which the Brahmins must make is $2^{64} - 1 = 18{,}446{,}744{,}073{,}709{,}551{,}615$. At the rate of one transfer every second, all day and every day, without errors, this works out to more than 500 billion years! Five hundred billion years is considerably longer than the life expectancy of the sun—the prophecy is correct.

c. The Euclidean Algorithm

If m and n are positive integers, their **greatest common divisor**, denoted by $\gcd(m, n)$, is the largest integer that exactly divides both m and n. According to the fundamental theorem of arithmetic, we may factor m and n into primes and thereby obtain their greatest common divisor. Thus, for example, if $m = 210$ and $n = 330$, then

$$m = 2 \cdot 3 \cdot 5 \cdot 7$$

$$n = 2 \cdot 3 \cdot 5 \cdot 11$$

so that

$$\gcd(m, n) = 2 \cdot 3 \cdot 5 = 30.$$

While useful for theoretical purposes, this method is of little practical value— at present, there is no means for finding the prime factors of an integer very rapidly. Fortunately, there is an efficient procedure for calculating the greatest common divisor of two integers without factoring them. It is the **Euclidean algorithm**,[8] which Euclid chose as the very first step in his development of the theory of numbers (Book 7, Propositions 1 and 2 of the *Elements* (circa 300 B.C.)). After more than two thousand years, it remains one of the fundamental algorithms of arithmetic.

Euclidean Algorithm.   Given two positive integers m and n, find their greatest common divisor.

Step 1:   Divide m by n and let r be the remainder. (We will have $0 \leq r < n$.)

Step 2:   If $r = 0$ the algorithm terminates; n is the answer.

Step 3:   Set $m \leftarrow n, n \leftarrow r$, and go back to step 1.

The arrow ($\leftarrow$) in step 3 is the fundamental *replacement* operation (also called *assignment* or *substitution*); "$m \leftarrow n$" means the value of variable m is to be replaced by the current value of variable n. When the algorithm begins, the values of m and n are the numbers originally given; but these values will, in general, change during the execution of the algorithm.

Observe that the order of the actions in step 3 is important; "set $m \leftarrow n$, $n \leftarrow r$" is entirely different from "set $n \leftarrow r, m \leftarrow n$."

The algorithm begins with step 1 and proceeds sequentially. In step 2, if $r \neq 0$ then no action is taken and the algorithm automatically proceeds to step

---

[8] Scholars believe that the algorithm was probably not invented by Euclid, but was known up to 200 years earlier; it was almost certainly known to Eudoxus (circa 375 B.C.)—cf. K. von Fritz, "The discovery of incommensurability by Hippasus of Metapontum," *Ann. Math.* (2) 46 (1945), pp. 242–264. It is based on the simple observation that any common divisor of two positive integers also divides their difference. The form of the algorithm given here is from Knuth, *The Art of Computer Programming* [263], volume 1, p. 2.

3. The algorithm must always terminate after a finite number of steps. For the value of r *diminishes* each time that step 1 is encountered, so that ultimately we must arrive at $r = 0$.

As an illustration, let us find the greatest common divisor of the first pair of *amicable numbers*,[9] $m = 220$ and $n = 284$. The six iterations of the algorithm are conveniently summarized in the table below.

Iteration	m	n	Remainder r
1	220	284	220
2	284	220	64
3	220	64	28
4	64	28	8
5	28	8	4
6	8	4	0

Beginning at step 1, we see that the quotient is zero and the remainder is 220. Thus $r \leftarrow 220$. At step 2, r is not zero, so that no action is taken. We proceed to step 3 and set $m \leftarrow 284$ and $n \leftarrow 220$. (Notice that whenever $m < n$ initially, the quotient in step 1 is always zero and the algorithm merely interchanges m and n.)

Returning to step 1 for the second iteration, we find that $284/220 = 1 + 64/220$, so that $r \leftarrow 64$. Once again, step 2 is inapplicable, and at step 3 we set $m \leftarrow 220$ and $n \leftarrow 64$. The next cycle sets $r \leftarrow 28$ and ultimately $m \leftarrow 64$ and $n \leftarrow 28$. Two more iterations yield $m \leftarrow 8$ and $n \leftarrow 4$. Finally, when 8 is divided by 4, the remainder is zero, and the algorithm terminates at step 2. The greatest common divisor of 220 and 284 is 4.

Let us now prove that the algorithm is valid for all values of m and n. For this purpose we let P_n be the statement "The algorithm works for n and all integers m." If m is a multiple of n, then $r = 0$ and the algorithm terminates the first time that step 2 is executed: clearly n is the greatest common divisor of m and n. This case always occurs when $n = 1$. Proceeding by induction, we assume that $P_1, P_2, \ldots, P_{n-1}$ are true and we verify that P_n is also true.

If m is not a multiple of n, then after step 1 we have

$$m = qn + r \qquad \text{where } 0 < r < n$$

---

[9] Two unequal natural numbers are said to be *amicable* if each is the sum of the divisors of the other except the number itself. The smallest pair, 220 and 284, was known to Pythagoras (Dickson [135], volume 1, p. 38). Since then more than a thousand pairs of amicable numbers have been found, but it remains unknown whether there are infinitely many such pairs. For the history and discovery of these intriguing numbers see the articles of Lee and Madachy [275].

(r is not zero because m is not a multiple of n). This shows that any number that divides both m and n must divide $m - qn = r$, and any number that divides both n and r must divide $qn + r = m$. Thus, the pairs m, n and n, r have the same common divisors, and, in particular, they must have the same greatest common divisor.

The algorithm now proceeds to step 3, where it sets $m \leftarrow n$, $n \leftarrow r$. Thus, the value of n decreases the next time step 1 is encountered, and the induction hypothesis guarantees that the final value of n is the gcd of n and r. Since the pairs m, n and n, r have the same gcd, the proof is complete.

Careful examination of the argument above will reveal a general method applicable to proving the validity of any algorithm. For this together with further details about Euclid's algorithm, including other ways to calculate the greatest common divisor, see Knuth, *The Art of Computer Programming* [263], volume 1, section 1.2.1 and volume 2, section 4.5.

d. Viète's Formula for Pi

$$\frac{2}{\pi} = \frac{\sqrt{2}}{2} \frac{\sqrt{2 + \sqrt{2}}}{2} \frac{\sqrt{2 + \sqrt{2 + \sqrt{2}}}}{2} \cdots$$

This remarkable representation for π, making use only of the number 2 and repeated square roots, was discovered by the French mathematician François Viète (1540–1603) at the end of the 16th century. It was the first numerically exact expression for π and also the first to express π as an infinite product. (The symbolism on the right side of the formula means that we are to compute the product of the first n terms and then take the limit of these "partial products" as $n \to \infty$.)

During the next century, the calculus, and, in particular, the theory of infinite series and infinite products, would provide new and more powerful techniques for computation, freeing the number π from its geometric origins and clarifying its fundamental role throughout all of mathematical analysis. Toward that end, Viète's formula helped point the way.

Viète based his derivation on the recurrence relation that exists between the areas of regular polygons of 2^n and 2^{n+1} sides inscribed in a circle of unit radius. The simplest proof, however, is based on the beautiful formula due to Euler

$$\frac{\sin \theta}{\theta} = \cos \frac{\theta}{2} \cos \frac{\theta}{4} \cos \frac{\theta}{8} \cdots$$

which includes that of Viète in the special case $\theta = \pi/2$.

To prove Euler's formula, we start from the elementary trigonometric identity

$$\sin 2\theta = 2 \sin \theta \cos \theta.$$

When this identity is applied repeatedly, we obtain

$$\begin{aligned}
\sin x &= 2 \sin \frac{x}{2} \cos \frac{x}{2} \\
&= 4 \sin \frac{x}{4} \cos \frac{x}{2} \cos \frac{x}{4} \\
&= 8 \sin \frac{x}{8} \cos \frac{x}{2} \cos \frac{x}{4} \cos \frac{x}{8} \\
&\ \ \vdots \\
&= 2^n \sin \frac{x}{2^n} \cos \frac{x}{2} \cos \frac{x}{4} \cdots \cos \frac{x}{2^n}.
\end{aligned} \tag{2}$$

But from elementary calculus we know that $\lim_{\theta \to 0} \frac{\sin \theta}{\theta} = 1$, and therefore

$$\lim_{n \to \infty} 2^n \sin \frac{x}{2^n} = x. \tag{3}$$

By combining (2) and (3) we arrive at Euler's identity.

e. The Principle of Inclusion and Exclusion

Among the many important methods of indirect counting, we single out one that is both elementary and far-reaching. Suppose we are given n finite sets $S_1, S_2, \ldots, S_n$ and we wish to determine the number of elements in their union[10]

$$n(S_1 \cup S_2 \cup \cdots \cup S_n).$$

(The number of elements in any set S will be denoted by the symbol $n(S)$.) For this purpose it is not enough to know how many elements belong to the individual sets; we must be given complete information concerning all possible overlaps. Let us assume, therefore, that for every pair (i, j), every triple (i, j, k), etc., we know the number of elements in $S_i \cap S_j$, or $S_i \cap S_j \cap S_k$, etc., and try to discover a formula for $n(S_1 \cup S_2 \cup \cdots \cup S_n)$.

---

[10] It is worth emphasizing that any two objects which appear as members of the same set are different. In other words, an object may belong, or not belong, to a given set, but it cannot "more than belong." Thus, for example, the set $\{1\} \cup \{1\} = \{1\}$ contains one element, not two. By contrast, an object may belong repeatedly to a sequence, such as the sequence $(1, 1/2, 1, 1/3, 1, 1/4, \ldots)$ in which 1 appears infinitely often.

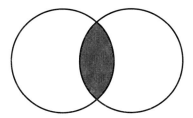

FIGURE 4

When there are only two sets the solution is practically trivial:

$$n(S_1 \cup S_2) = n(S_1) + n(S_2) - n(S_1 \cap S_2), \tag{4}$$

since the elements common to S_1 and S_2 have been counted twice in the sum $n(S_1)+n(S_2)$ and must therefore be subtracted from this sum in order to obtain the correct count for $n(S_1 \cup S_2)$.

When there are three sets S_1, S_2, S_3, the formula is more interesting. From (4) we have

$$n(S_1 \cup S_2 \cup S_3) = n[(S_1 \cup S_2) \cup S_3]$$
$$= n(S_1 \cup S_2) + n(S_3) - n[(S_1 \cup S_2) \cap S_3]. \tag{5}$$

From the algebra of sets we know that $(S_1 \cup S_2) \cap S_3 = (S_1 \cap S_3) \cup (S_2 \cap S_3)$, and therefore

$$n[(S_1 \cup S_2) \cap S_3] = n(S_1 \cap S_3) + n(S_2 \cap S_3) - n(S_1 \cap S_2 \cap S_3). \tag{6}$$

Combining (4), (5), and (6), we obtain the desired formula:

$$n(S_1 \cup S_2 \cup S_3) = n(S_1) + n(S_2) + n(S_3) - n(S_1 \cap S_2) - n(S_1 \cap S_3)$$
$$- n(S_2 \cap S_3) + n(S_1 \cap S_2 \cap S_3). \tag{7}$$

A pattern has begun to emerge. The general formula for the union of n sets is given by the following theorem.

The Principle of Inclusion and Exclusion.   *If $S_1, S_2, \ldots, S_n$ are finite sets, then*

$$n(S_1 \cup S_2 \cup \cdots \cup S_n) = \sum n(S_i) - \sum n(S_i \cap S_j)$$
$$+ \sum n(S_i \cap S_j \cap S_k) - \cdots \pm n(S_1 \cap S_2 \cap \cdots \cap S_n).$$

The subscripts $i, j, k, \ldots$ range between 1 and n; their order is irrelevant, but for uniqueness we shall always write them in increasing order. Two subscripts are never equal. Thus, the formula tells us that we must first add the numbers of elements in each set S_i, then subtract the numbers of elements in each pair of sets $S_i \cap S_j$, then add the numbers in each triple $S_i \cap S_j \cap S_k$, and so on. Notice that every possible combination of sets occurs exactly once in the sums. The first sum contains n terms, the second sum $\binom{n}{2}$ terms, the third $\binom{n}{3}$, and so on. The last sum reduces to the single term $n(S_1 \cap S_2 \cap \cdots \cap S_n)$, which is the number of elements contained in all n sets.

This formula may be established by mathematical induction in precisely the same way that we derived (7) from (4). There is, however, a more elegant solution based on the binomial theorem and a simple counting argument. If an element x belongs to exactly k of the sets $S_1, S_2, \ldots, S_n$ then its contribution to the first sum in the formula will be $\binom{k}{1}$; its contribution to the second sum will be $\binom{k}{2}$; to the third sum $\binom{k}{3}$; etc. Therefore, its total contribution to the right side of the formula will be

$$\binom{k}{1} - \binom{k}{2} + \binom{k}{3} - \cdots \pm \binom{k}{k}.$$

We need now only recall the binomial expansion of $(1-1)^k$ to see that the expression above is always equal to 1. Thus no matter how many sets x belongs to, it will be counted exactly once on both sides of the formula. On the other hand, elements that do not belong to the union $S_1 \cup S_2 \cup \cdots \cup S_n$ are not counted at all by any term in the formula. Thus the formula is correct.

Example. *A Matching Game.* There is an interesting game, dating back to the 18th century, which the French call *jeu de rencontre*[11] (or, matching game). It has many variants and a surprising outcome.

In one formulation, two identical decks of cards, each numbered $1, 2, \ldots,$ N, are put into random order and matched against each other. Whenever a card occupies the same position in both decks we say there is a *match*. A match may occur at any one of the N places or at several places simultaneously. The problem is to determine the likelihood that there will be at least one match. In one amusing variant, the two decks are represented by a set of N customers in a restaurant together with the hats they have deposited in the cloakroom. When they leave a playful attendant returns the hats at random. A match occurs if a person gets the correct hat. The reader may care to speculate on how

---

[11] The game was discussed by Montmort and N. Bernoulli under the name of *Treize* (see Todhunter, *A History of the Mathematical Theory of Probability from the Time of Pascal to that of Laplace* [416], p. 91), but it was Euler who provided the general solution (see his memoir entitled "Calcul de la Probabilité dans le Jeu de Rencontre" (1753) [*Opera Omnia*, series prima, vol. 7, p. 11]). The description of the game given here is adapted from Feller, *An Introduction to Probability Theory and Its Applications* [156], volume 1, p. 100.

the probability of a match depends on N: Is the probability of a match of hats at a dinner party for ten significantly different from the corresponding probability at a banquet for one thousand? We shall find, quite surprisingly, that the probability is practically independent of N and roughly 2/3.

Let us suppose, without loss of generality, that the cards have been numbered in such a way that one deck appears in its natural order. There are $N!$ possible permutations of the second deck and we must determine how many of these contain a match. Let S_i be the set of those permutations in which the ith card stays in its original position. The remaining $N - 1$ cards may then be in any order, and so $n(S_i) = (N - 1)!$. Similarly, for every pair (i, j) we have $n(S_i \cap S_j) = (N - 2)!$, and so on. Therefore, by the principle of inclusion and exclusion, the total number of permutations containing at least one match is given by

$$N(N - 1)! - \binom{N}{2}(N - 2)! + \binom{N}{3}(N - 3)! - \cdots \pm \binom{N}{N}0!$$

$$= N! \left(1 - \frac{1}{2!} + \frac{1}{3!} - \cdots \pm \frac{1}{N!}\right).$$

Assuming that all permutations are equally likely, the required probability is then

$$P = 1 - \frac{1}{2!} + \frac{1}{3!} - \cdots \pm \frac{1}{N!}.$$

Notice that $1 - P$ is precisely the Nth partial sum of the Taylor expansion

$$e^{-1} = 1 - 1 + \frac{1}{2!} - \frac{1}{3!} + \cdots,$$

and hence

$$P \approx 1 - e^{-1} = .63212\ldots.$$

The convergence of P to its limit is extremely rapid; in fact, it follows from the elementary theory of alternating series that, for any value of N, P differs from $1 - e^{-1}$ by at most $1/(N + 1)!$. The degree of approximation is illustrated by the following table:

$N =$	3	4	5	6	7
$P =$	.66667	.62500	.63333	.63194	.63214

Thus even when N is small the probabilities of a match are approximately the same. Notice that the successive approximations do not converge monotonically, but alternate between being a bit larger and a bit smaller than the limit.

PROBLEMS

1. Let us consider the system S of numbers

$$1, 5, 9, 13, 17, 21, 25, 29, \ldots$$

consisting of all numbers of the form $4x + 1$. The product of any two such numbers is again of the same form. We define a "prime" in S to be a number in this system (other than 1) which cannot be properly factored *within the system.* Thus the first few primes in S are 5, 9, 13, 17, 21, and the first number in S which is not a prime is 25. Show that every number in S, other than 1, can be written as a product of primes in S. Show by an example that such a representation is not necessarily unique. What does this tell us about the logical structure of any proof of the fundamental theorem of arithmetic? (This important illustration was given by Hilbert.)

2. The **Fermat numbers** are defined by

$$F_n = 2^{2^n} + 1 \qquad n = 0, 1, 2, \ldots.$$

  a. F_n is prime if $n = 0, 1, 2, 3, 4$. What about $n = 5$?
  b.   Find a simple formula for the product of the first n Fermat numbers

$$F_0 F_1 F_2 \cdots F_{n-1}.$$

  c.   The Fermat numbers are pairwise *relatively prime* (this means that the greatest common divisor of any two of them is 1). Corollary: The number of primes is infinite.

3. Bertrand's postulate states that for all natural numbers $n > 1$ there is a prime between n and $2n$. (Bertrand verified this empirically for every natural number from 2 to 3,000,000. He then conjectured that the result holds for all $n > 1$.) Considering unity as a prime, show that every positive integer can be written as a sum of distinct primes. (If $n > 6$, then it is not necessary to consider unity as a prime.)

4. Any number which is not a power of 2 can be represented as the sum of at least 2 consecutive positive integers, but such a representation is impossible for powers of 2.

5. Use Euclid's algorithm to find a simple formula for

$$\gcd(2^m - 1, 2^n - 1)$$

where m and n are positive integers.

6. **a.** If d is the greatest common divisor of m and n, then there exist integers a and b such that

$$d = am + bn.$$

Find an extension of the Euclidean algorithm which determines a and b as well as d.

    b. Show that a and b are not unique.

    c. Find a and b such that

$$\gcd(220, 284) = 220a + 284b.$$

7. The **Fibonacci numbers** are defined recursively by the formulas

$$f_1 = 1, \quad f_2 = 1$$

$$f_n = f_{n-1} + f_{n-2} \qquad \text{for } n \geq 3.$$

The sequence begins with the numbers

$$1, 1, 2, 3, 5, 8, 13, 21, \ldots,$$

each term past the second being the sum of the two preceding terms. This famous sequence will be investigated in detail in chapter 3. For now, we show its intimate connection with the Euclidean algorithm.

    a. The Euclidean algorithm requires exactly n division steps to show that

$$\gcd(f_{n+1}, f_{n+2}) = 1.$$

    b. (G. Lamé, 1845) Let a and b be positive integers, with $a < b$, such that Euclid's algorithm applied to a and b requires exactly n division steps. Then $a \geq f_{n+1}$ and $b \geq f_{n+2}$. Accordingly, if b is the smallest number satisfying these conditions, then

$$a = f_{n+1} \quad \text{and} \quad b = f_{n+2}.$$

8. **a.** Prove the following theorem of E. Lucas (1876): If f_m and f_n are Fibonacci numbers, then

$$\gcd(f_m, f_n) = f_{\gcd(m,n)}.$$

b.   Use Lucas's theorem to give another proof of the infinitude of the primes.

9.   Two lattice points in the plane are said to be *mutually visible* if there is no other lattice point on the line segment that joins them. Show that (a, b) and (m, n) are mutually visible if and only if $a - m$ and $b - n$ are relatively prime.

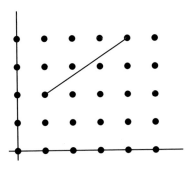

FIGURE 5

10.   The sieve of Eratosthenes asserts that, in order to find all the primes up to a prescribed limit N, we need only determine those primes up to $\sqrt{N}$ and then strike out their multiples. For example, to determine all the primes up to 100, it is sufficient to delete from the sequence $2, 3, 4, \ldots, 100$ all proper multiples of 2, 3, 5, and 7. By combining the sieve of Eratosthenes with the principle of inclusion and exclusion, show how we can calculate $\pi(N)$, the number of primes up to N, without knowing these primes individually. Show, for example, that $\pi(100) = 25$, $\pi(1000) = 168$, and $\pi(10000) = 1229$.

11.   The important arithmetical function $\phi(n)$, called the **Euler ϕ-function,** is defined to be the number of positive integers less than n that are relatively prime to n. By convention $\phi(1) = 1$. For example, $\phi(36) = 12$ because there are 12 numbers less than 36 that are relatively prime to 36, namely,

$$1, 5, 7, 11, 13, 17, 19, 23, 25, 29, 31, 35.$$

Using the principle of inclusion and exclusion, show that if $p_1, p_2, \ldots, p_k$ are the distinct prime factors of n, then

$$\phi(n) = n \left(1 - \frac{1}{p_1}\right) \left(1 - \frac{1}{p_2}\right) \cdots \left(1 - \frac{1}{p_k}\right).$$

For example, the prime factors of 36 are 2 and 3, so that $\phi(36) = 36 \left(\frac{1}{2}\right) \left(\frac{2}{3}\right) = 12$.

12.   Show that $\phi(666) = 6 \cdot 6 \cdot 6$.

>   Here is wisdom. Let him that hath understanding count the number of the beast: for it is the number of a man; and his number is 666.
>
>   From the book of *Revelations*

13.   Evaluate the determinant

$$\begin{vmatrix} \gcd(1,1) & \gcd(1,2) & \cdots & \gcd(1,n) \\ \gcd(2,1) & \gcd(2,2) & \cdots & \gcd(2,n) \\ \cdots & \cdots & \cdots & \cdots \\ \gcd(n,1) & \gcd(n,2) & \cdots & \gcd(n,n) \end{vmatrix}.$$

14.   Derangements.   A permutation of n elements in which no element stays in its original position is called a **derangement**. The solution of the matching game shows that the number D_n of derangements is given by the formula

$$D_n = n! \left[1 - \frac{1}{1!} + \frac{1}{2!} - \cdots + (-1)^n \frac{1}{n!}\right].$$

These numbers are called **subfactorials**. The following table gives their values for the first 12 natural numbers:

n	D_n	n	D_n	n	D_n
1	0	5	44	9	133 496
2	1	6	265	10	1 334 961
3	2	7	1 854	11	14 684 570
4	9	8	14 833	12	176 214 841

   a.   Derive the recurrence relation

$$D_n = (n-1)[D_{n-1} + D_{n-2}].$$

(For a purely combinatorial proof, see Vilenkin, *Combinatorics* [430], p. 53.) What is the corresponding relation for ordinary factorials?

   b.   Derive the simpler recurrence relation

$$D_n = n \, D_{n-1} + (-1)^n.$$

   c.   Show that D_n is the integer nearest $n!/e$.

15. On the average, how many matches can be expected in the matching game? The answer is independent of the number of objects being permuted. (For an application to ESP experiments, see Vilenkin, *Combinatorics* [430], p. 49.)

4. Infinite Descent

> I discovered at last a most singular method.
> —Fermat, in a letter to Huygens (1659)

a. The Method

In the seventeenth century, the great French mathematician Pierre de Fermat introduced a famous method of proof which he called the method of "infinite descent" and which he claimed to have used in all of his number-theoretical discoveries. It is based on the obvious assertion that *a decreasing sequence of natural numbers cannot continue indefinitely.* With it Fermat was able to establish some of the most remarkable properties of the integers.

Suppose, as is the case with mathematical induction, that we are given infinitely many statements $P_1, P_2, P_3, \ldots$, one for each natural number n, and that we wish to prove that all of them are true. This can be accomplished by showing that *every false statement is preceded by another false statement.* More precisely: once we assume hypothetically that any given statement P_n is false, then another statement P_m can be found, with $m < n$, which is also false. Notice that we assume only that m is *some* value smaller than n, and not necessarily the next smallest. Since n was arbitrary, repetition of the argument leads to yet another value $k < m$ for which P_k is also false, and again a third value still smaller, and so on ad infinitum. But this is impossible, since there are only finitely many natural numbers less than n, and we conclude by a reductio ad absurdum that all of the statements must be true.

The method of infinite descent is simply another form of the principle of proof by mathematical induction (the latter might very well be called the method of *infinite ascent*).[12] If we can show, without restriction, that every false statement is preceded by another false statement, then it follows a priori that

---

[12] Both methods of proof are equivalent to the well-ordering principle, which asserts that every nonempty subset of the natural numbers has a smallest element. For an excellent discussion of well-ordering, see Wilder, *Introduction to the Foundations of Mathematics* [451], chapter V.

the first statement cannot possibly be false (for it is preceded by no statement at all!). The second statement must then also be true, for its falsity would imply that of the first. In similar fashion, the third statement is found to be true since it is preceded by only true statements. And so on ad infinitum.

By inverting the logic we have gained an advantage. The transition from n to $n + 1$, which is the hallmark of an inductive proof, is replaced by a transition from n to some smaller value, but not necessarily $n - 1$, which might prove far too restrictive. Furthermore, in a proof "by descent" not a single proposition need ever be established directly; the final contradiction is enough to show that all propositions are true.[13]

Example. *The Irrationality of* $\sqrt{2}$. If legend is accurate, then the exultation that may have led Pythagoras to sacrifice a hecatomb, when he discovered the remarkable theorem that bears his name, must have been short-lived. For the Pythagorean theorem reveals at once the existence of irrational magnitudes—$\sqrt{2}$, in the simplest case, the diagonal of a unit square. This discovery shattered the Pythagorean doctrine that all phenomena in the universe could be reduced to whole numbers or their ratios.

The simplest way of showing that there can be no rational number whose square is 2 is by a reductio ad absurdum. Suppose, if possible, that there were a positive fraction $\frac{a}{b}$ such that $(\frac{a}{b})^2 = 2$, or

$$a^2 = 2b^2.$$

Then a^2 is even and hence a must also be even, since the square of an odd number is odd. Thus we can write $a = 2c$, so that $a^2 = 4c^2 = 2b^2$ and

$$b^2 = 2c^2.$$

But b is clearly smaller than a, and hence we have found another fraction, $\frac{b}{c}$, equal to $\sqrt{2}$ and having a smaller numerator. This begins the infinite descent and so leads to a contradiction.

In the usual proof, one begins with the added assumption that the fraction $\frac{a}{b}$ is in its lowest terms and arrives, in the same way, at the contradiction that a and b are both even. But this too is a proof by descent, somewhat hidden, for the proof that a fraction can indeed always be put in reduced form depends on

---

[13] It is worth pointing out that the contradiction needed to complete the proof by reductio ad absurdum can be obtained in another way, without exhibiting an infinitely descending sequence of positive integers. In fact, it is realized as soon as we arrive, after a finite number of steps, at the statement "P_1 is false," which, as we have already seen, cannot be true. Accordingly, when viewed in this way, Fermat's method is really a proof by *finite descent.*

the fact that a decreasing sequence of whole numbers must finally come to an end.

There is another interesting proof in which the hypothetical relation $a^2 = 2b^2$ is written in the equivalent form $(2b-a)^2 = 2(a-b)^2$, and so $(2b-a)/(a-b)$ is another fraction whose square is 2. But clearly $b < a < 2b$, and so $0 < a - b < b$. Hence we have found another fraction equal to $\frac{a}{b}$ and having a smaller denominator. The contradiction now follows either by descent or by assuming at the outset that $\frac{a}{b}$ is in lowest terms.

Thus the system of rational numbers is too incomplete, too full of gaps, to provide a solution for the equation $x^2 = 2$. "The material is too coarse for such finer purposes."[14] It is therefore instructive to see how the Pythagoreans succeeded in approximating $\sqrt{2}$ by means of a rapidly converging sequence of rational numbers.

The process can be described in terms of an "infinite ladder" of whole numbers, as pictured below.

a	b
1	1
2	3
5	7
12	17
29	41
⋮	⋮

On the first rung of the ladder, both entries are taken to be unity. Subsequent entries are determined by a simple recursive rule: *If the numbers on a given rung are a and b, then those on the next rung are a + b and 2a + b.*

Assertion. *As one descends the ladder,*

$$\frac{b}{a} \to \sqrt{2}$$

the approximations improving on each successive rung.

The assertion follows at once from the identity

$$b^2 - 2a^2 = \pm 1 \tag{1}$$

---

[14] Konrad Knopp, *Theory and Application of Infinite Series* [262], p. 24.

whose validity on every rung of the ladder is easily demonstrated by induction. Indeed, the relation

$$(2a + b)^2 - 2(a + b)^2 = 2a^2 - b^2,$$

which is valid for all a and b, shows that if (1) holds on a particular rung then it also holds on the next rung, but with the opposite sign. But if we set $a = b = 1$, then (1) holds on the first rung; hence, by induction, it holds on every rung.

In this way, the Pythagoreans were able to approach the simplest of all irrational numbers, the one that precipitated the first great crisis in Greek mathematics and "brought to the fore a difficulty that preoccupied all the Greeks, namely, the relation of the discrete to the continuous."[15]

Commenting on these early investigations, B. L. Van der Waerden writes in his classic work *Science Awakening* [426], p. 127:

> The problem of approximating to the ratio of the diagonal and the side [of a square] by means of rational numbers was proposed and solved by the Babylonians. But the Pythagoreans carried this old problem infinitely farther than the Babylonians. They found a whole set of approximations of indefinitely increasing accuracy; moreover they developed a scientific theory concerning these approximations and they proved the general proposition by complete induction. Again and again it becomes apparent that there were excellent number-theoreticians in the Pythagorean school.

b. Pythagorean Triangles

Although the theory of numbers was initiated by the ancient Greeks, it is Fermat who was the first mathematician to discover truly deep properties of the integers. His point of departure was the famous ancient treatise on arithmetic, the *Arithmetica* of Diophantus (3rd century A.D.), and in his copy of that great work, Fermat recorded, as marginal notes, most of his number-theoretical discoveries. In only one case, however, did he ever provide a proof, and that was only a sketch. For nearly two centuries, the best mathematicians worked hard to prove his assertions. Finally, all were verified, with two exceptions—an erroneous formula purporting to produce only primes (which we shall take up in the next chapter) and a notorious statement, still unproved, which has come to be known as *Fermat's last theorem.*

---

[15] Morris Kline, *Mathematical Thought from Ancient to Modern Times* [260], p. 34.

Opposite problem 8 in Book II of Diophantus, which asks "To divide a square number into two squares," Fermat records, in Latin: "On the other hand, it is impossible for a cube to be written as a sum of two cubes or a fourth power to be written as a sum of two fourth powers or, in general for any number which is a power greater than the second to be written as a sum of two like powers. For this I have discovered a truly wonderful proof, but the margin is too small to contain it."[16] This tantalizingly simple statement is known as Fermat's last theorem. In symbols, for any integer $n > 2$ the equation

$$a^n + b^n = c^n$$

cannot be solved in positive integers a, b, c.

If Fermat really was in possession of a proof, then it must have been truly extraordinary, for despite the efforts of many of the greatest mathematicians since that time, the general proposition remains unproved.[17]

We shall give a proof of Fermat's conjecture only in the simplest case possible, the case $n = 4$, where the insolubility of the equation was proved by Fermat himself. In preparation, however, we must first return to the problem in Diophantus which inspired Fermat's last theorem.

To represent a square as the sum of two other squares is one of the oldest problems in mathematics. Symbolically, we wish to solve the equation

$$a^2 + b^2 = c^2 \tag{2}$$

---

[16] *Oeuvres de Fermat*, I, p. 291; French translation, III, p. 241.

[17] By contrast, the fate of Euler's conjectured generalization of Fermat's last theorem has proved far less auspicious. In 1778 Euler wrote:

> It has seemed to many geometers that this theorem [Fermat's last theorem] may be generalized. Just as there do not exist two cubes whose sum or difference is a cube, it is certain that it is impossible to exhibit three biquadrates whose sum is a biquadrate, but that at least four biquadrates are needed. ... In the same manner it would seem to be impossible to exhibit four fifth powers whose sum is a fifth power, and similarly for higher powers.

(See Dickson [135], volume II, p. 648.) Nearly two centuries later, however, a computer search by Lander and Parkin (*Mathematics of Computation*, 21 (1967), 101–103) uncovered the counterexample $27^5 + 84^5 + 110^5 + 135^5 = 144^5$. No other sum of four fifth powers equal to another fifth power has ever been found. Subsequently, Noam Elkies (*Mathematics of Computation*, 51 (1988), 825–835) showed that the equation

$$a^4 + b^4 + c^4 = d^4$$

has infinitely many nontrivial solutions. The smallest solution, $a = 95800$, $b = 217519$, $c = 414560$, $d = 422481$, was discovered by Roger Frye. In not a single instance has Euler's general conjecture ever been verified. For recent research on the problem, see Guy [191].

in positive integers a, b, c. The problem, by virtue of the Pythagorean theorem, amounts to finding all right triangles whose sides have integral lengths.[18] Because of this geometrical interpretation, any such triple of natural numbers a, b, c is called a **Pythagorean triple**, and the corresponding right triangle is called a **Pythagorean triangle**.

The most famous Pythagorean triple is of course 3, 4, 5 and knowledge of this particular example was widespread in the ancient world. In fact an Old Babylonian tablet has survived (Plimpton 322), dated between 1900 B.C. and 1600 B.C., which contains a list of fifteen Pythagorean triples, some of the numbers being quite large. This remarkable find[19] indicates that the ancient Babylonians may well have known the Pythagorean theorem at least a thousand years before Pythagoras lived. To Pythagoras himself is attributed the special solution

$$\frac{n^2 - 1}{2}, \ n, \ \frac{n^2 + 1}{2}$$

where n is an odd integer greater than 1. This solution produces infinitely many such triples, but not all of them. The first general solution of equation (2) in positive integers appears in Euclid's *Elements* (Book X, Lemma 1 to Proposition 29).

Theorem. *All the primitive Pythagorean triples are given by the formulas*

$$a = m^2 - n^2, \quad b = 2mn, \quad c = m^2 + n^2$$

where m and n are relatively prime positive integers, $m > n$, and one of them is even and the other odd (apart from the possibility of interchanging a and b).

(A Pythagorean triple is said to be *primitive* if the numbers a, b, c have no common factor greater than 1. It is easy to see that *every Pythagorean triple is a multiple of a primitive one.*)

To prove the theorem algebraically is entirely straightforward (see, e.g., [385], pp. 36f), yet the result often appears abstract. There is, however, a striking geometric argument which the reader may find both conceptually and technically illuminating.

---

[18] The assertion that (2) implies the existence of a right-angled triangle with sides a, b, c is actually the converse of the Pythagorean theorem.

[19] It was discovered by Neugebauer and Sachs and published in their *Mathematical Cuneiform Texts* (1945).

Primitive Pythagorean Triangles

Generating Numbers		Sides of the Triangle		
m	n	a	b	c
2	1	3	4	5
3	2	5	12	13
4	1	15	8	17
4	3	7	24	25
5	2	21	20	29
5	4	9	40	41
6	1	35	12	37
6	5	11	60	61
7	2	45	28	53
7	4	33	56	65
7	6	13	84	85
8	1	63	16	65
8	3	55	48	73
8	5	39	80	89
8	7	15	112	113

Starting with the equation $a^2 + b^2 = c^2$, we divide throughout by c^2 and then put

$$\frac{a}{c} = x, \quad \frac{b}{c} = y$$

thereby obtaining the new equation

$$x^2 + y^2 = 1. \tag{3}$$

It is now a matter of finding all the rational number pairs (x, y) that satisfy it, or, all the *rational points* that lie on the unit circle $x^2 + y^2 = 1$.

The set of all rational points in the plane fills the plane densely, and it is our goal to determine how this circle winds its way through that dense set.[20]

---

[20] By *dense* we mean that in every open disk, no matter how small, there are infinitely many rational points.

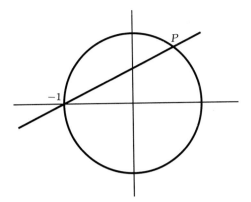

FIGURE 6

Let us consider a variable line L drawn through the point $(-1, 0)$. Assuming that L is not a tangent, it must intersect the circle at one other point P. We shall show that *P has rational coordinates if and only if L has rational slope*. The "only if" part is obvious. For the converse, write the equation of L in the form

$$y = \lambda(x + 1)$$

where the slope λ is assumed to be rational. Substituting in (3), we find

$$x^2 + \lambda^2(x + 1)^2 = 1$$

or

$$(1 + \lambda^2)x^2 + 2\lambda^2 x + \lambda^2 - 1 = 0$$

for the abscissas of the two points of intersection. We already know one solution $(x = -1)$ and the other is readily found; we have

$$x = \frac{1 - \lambda^2}{1 + \lambda^2} \tag{4}$$

whence

$$y = \frac{2\lambda}{1 + \lambda^2}. \tag{5}$$

Thus the coordinates of P are rational whenever the slope λ is rational.

These formulas constitute a particularly useful parameterization of the unit circle.[21] To every rational number λ there corresponds a rational point on the circle, and conversely, every rational point on the circle with the exception of $(-1, 0)$ is obtained in this way. Our construction shows in particular that the set of all rational points on the unit circle is *everywhere dense,* which means that on any arc of the circle, no matter how small, there are infinitely many such points. A similar remark holds for any conic, provided that the equation of the conic has rational coefficients, and provided also that there is at least one rational point on the curve. This, however, need not be the case; for example, the circle $x^2 + y^2 = 3$ contains not a single rational point (see problem 20).

Knowing the general solution of the equation $x^2 + y^2 = 1$ in rational numbers we can now determine the general solution of the equation

$$a^2 + b^2 = c^2$$

in integers. Let a, b, c be a given primitive solution and put

$$\lambda = \frac{n}{m}$$

where m and n are relatively prime positive integers. Then, by (4) and (5),

$$\frac{a}{c} = \frac{m^2 - n^2}{m^2 + n^2} \quad \text{and} \quad \frac{b}{c} = \frac{2mn}{m^2 + n^2}. \tag{6}$$

It is certainly tempting to argue that a, b, c must be the numbers $m^2 - n^2, 2mn, m^2 + n^2$, since these numbers form an obvious solution of equation (2). The problem, however, is that this solution may not be primitive (for example, the choice $\lambda = 1/3$ yields the solution $8, 6, 10$). If m and n happen to be of opposite parity, then all is well, for then the three numbers $m^2 - n^2, 2mn, m^2 + n^2$ can have no common factor greater than 1. Indeed, such a factor would have to be odd (since $m^2 - n^2$ is odd) and would have to divide $(m^2 + n^2) + (m^2 - n^2) = 2m^2$ as well as $(m^2 + n^2) - (m^2 - n^2) = 2n^2$, and this is impossible since m

---

[21] A simpler and more revealing derivation, which uses only rational operations, is obtained by writing the equation of the circle as

$$y^2 = 1 - x^2 = (1 - x)(1 + x),$$

which becomes, after dividing through by $(1 + x)^2$,

$$\left(\frac{y}{1 + x}\right)^2 = \frac{1 - x}{1 + x}.$$

If the point (x, y) also lies on L, then

$$\lambda^2 = \frac{1 - x}{1 + x}$$

and we can now solve for x as a rational function of λ.

and n are relatively prime. Therefore in this case we are justified in asserting that $a = m^2 - n^2$, $b = 2mn$, $c = m^2 + n^2$.

To avoid the difficulty when m and n are both odd, we resort to a common ploy: putting

$$p = \frac{m+n}{2}, \qquad q = \frac{m-n}{2},$$

we see that p and q are relatively prime, and also that they are of opposite parity, since $p + q$ is odd. Substituting for m and n in terms of p and q in (6), we obtain

$$\frac{a}{c} = \frac{2pq}{p^2 + q^2} \qquad \text{and} \qquad \frac{b}{c} = \frac{p^2 - q^2}{p^2 + q^2}.$$

These fractions are of the same form as before, except now a and b are interchanged and p and q take the place of m and n.

Thus we have proved that all primitive solutions of $a^2 + b^2 = c^2$ are given by the formulas of the theorem, apart from the possibility of interchanging a and b.

c. Fermat's Last Theorem: The Easy Case

Equipped with a complete characterization of Pythagorean triangles, we could now give a direct proof of Fermat's last theorem in the special case $n = 4$. However, we shall prefer a slightly circuitous route which leads to Fermat's one known proof and to a most remarkable fact about right triangles.[22]

Theorem. *The area of a Pythagorean triangle can never be a square.*

Proof. The proof is accomplished by showing that, if the area of such a triangle were a square, then there would be a smaller triangle with the same property, and so on ad infinitum, which is impossible. Suppose then that there are positive integers a, b, c such that

$$a^2 + b^2 = c^2$$

and for which the area of the triangle, $\frac{1}{2}ab$, is a square. There is no loss of generality in supposing that the given triangle is primitive (why?), so that a, b, c

---

[22] *Oeuvres de Fermat*, I, 340–341; French translation, III, 271–272. For an English translation, see Dickson [135], volume II, pp. 615–616.

take the form

$$a = m^2 - n^2, \qquad b = 2mn, \qquad c = m^2 + n^2$$

where m and n are relatively prime integers of opposite parity. (As before, by interchanging a and b, if necessary, we may suppose that a is odd.) Now the area of the triangle is just $mn(m - n)(m + n)$ and each factor is easily seen to be relatively prime to the other three. But *the only way for a product of relatively prime numbers to be a square is for each of them to be a square.* (This follows at once from the fundamental theorem of arithmetic.) Therefore $(m - n)(m + n)$ is a square, and if we call it p^2, then

$$p^2 + n^2 = m^2$$

where p and m are odd and n is even (remember that $a = p^2$ is odd). The new triangle with sides p, n, and m is also primitive and hence we can write

$$p = m_1^2 - n_1^2, \quad n = 2m_1 n_1, \quad m = m_1^2 + n_1^2$$

where the generators m_1 and n_1 are relatively prime and of opposite parity. But n is a square, and therefore either m_1 or n_1 must be an odd square while the other must be twice a square. Finally, m is also a square, say $m = u^2$, and so the third equation above becomes

$$m_1^2 + n_1^2 = u^2.$$

Thus m_1, n_1, and u are the sides of a Pythagorean triangle whose area $\frac{1}{2}m_1 n_1$ is a perfect square and whose hypotenuse is smaller than the hypotenuse of the original triangle, since

$$u = \sqrt{m} < m < m^2 < m^2 + n^2 = c.$$

This completes the proof "by descent."

Corollary. *The equation*

$$a^4 - b^4 = c^2$$

has no solutions in positive integers a, b, c.

Proof. If the equation were satisfied by three positive integers a, b, c, then the numbers

$$a^4 - b^4, \qquad 2a^2 b^2, \qquad a^4 + b^4$$

would form the sides of a right triangle, since

$$(a^4 - b^4)^2 + (2a^2b^2)^2 = (a^4 + b^4)^2,$$

and the area of the triangle would be

$$a^2b^2(a^4 - b^4) = a^2b^2c^2$$

which is a square. But the area of a Pythagorean triangle can never be a square, and the contradiction proves the corollary.

Corollary (Special Case of Fermat's Last Theorem). *The equation*

$$a^4 + b^4 = c^4$$

has no solutions in positive integers a, b, c.

The proof follows at once from the previous corollary.

Fermat's last theorem may be depicted geometrically in terms of the curves

$$x^n + y^n = 1$$

and the totality of all rational points in the x, y-plane. For $n = 3$ and $n = 4$ the curves have the general appearance indicated below, while for higher odd and even values of n we obtain curves similar in character to these. Fermat's theorem asserts that, despite the dense distribution of rational points in the plane, *these curves, in contrast to the unit circle, do not pass through a single rational point other than their intercepts on the coordinate axes.*

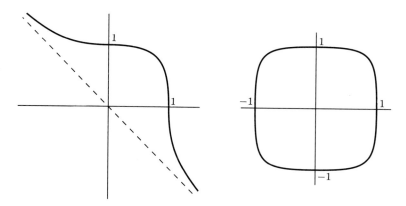

FIGURE 7

The continuing search for a general solution of Fermat's last theorem has given rise to important developments not only in number theory but throughout mathematics. Perhaps the most outstanding example is the work of Kummer (1810–1893), whose efforts in this area led him to create a new and extensive theory in higher algebra called *ideal theory*. Kummer described very general conditions for the insolubility of the Fermat equation, and almost all modern research on the problem has been based on his work. It is now known that Fermat's last theorem is true for all exponents up to 1,000,000 ([363], p. 164).

In a dramatic recent discovery, which makes use of deep results from algebraic geometry, Gerd Faltings (1983) has shown that for any $n > 2$, the equation $x^n + y^n = 1$ can have at most a finite number of rational solutions. Of course, Fermat's last theorem asserts that, trivial solutions aside, that finite number is zero.

d. The Loss of Certainty

> The science of the 20th century has a fascinating characteristic: it has been discovering its own limitations. The best known example of this is in quantum mechanics, where Heisenberg's uncertainty principle imposes limits on what can be measured. Perhaps even more disconcerting are the mathematical results demonstrated since 1930 concerning the impossibility of certain demonstrations, because they can assign a sort of limit to thought itself.
>
> —Jean-Paul Delahaye[23]

Is the Fermat conjecture solvable, is it capable of being either proved or disproved using the ordinary axioms of arithmetic? If the conjecture is false, then sooner or later a hypothetical computer must discover positive integers that satisfy the Diophantine equation $a^n + b^n = c^n$ for some n greater than 2. If, on the other hand, Fermat's last theorem is true, could a proof be inaccessible, beyond the scope of ordinary arithmetic? Could the axiomatic method, which has served mathematics so well throughout its history, be inadequate to answer all the questions that might reasonably be asked of it? The 20th century has revealed that such fundamental concerns about the ultimate foundations of logic and mathematics are far from groundless.

---

[23] "Chaitin's equation: An extension of Gödel's theorem," *Notices of the Amer. Math. Soc.* 36 (1989), 984–987.

At the Second International Congress of Mathematicians, held in Paris in 1900, David Hilbert, one of the leading mathematicians of the time, expressed his belief in the power of mathematical reasoning:

> ... every definite mathematical problem must necessarily be susceptible of an exact settlement, either in the form of an actual answer to the question asked, or by the proof of the impossibility of its solution and therewith the necessary failure of all attempts. ... However unapproachable these problems may seem to us and however helpless we stand before them, we have, nevertheless, the firm conviction that their solution must follow by a finite number of purely logical processes. ... We hear within us the perpetual call: There is the problem. Seek its solution. You can find it by pure reason, for in mathematics there is no *ignorabimus* [we shall not know].[24]

But Hilbert's optimism was unjustified. The crisis came in 1930, when the Austrian mathematician Kurt Gödel (1906–1978) established his great *incompleteness theorem*. In a subtle and profound intellectual accomplishment, Gödel showed that "any formal system that only demonstrates true arithmetical statements is necessarily incomplete, because certain true arithmetical statements cannot be proven within it." Thus, ordinary arithmetic contains propositions that are *formally undecidable,* incapable of being either proved or disproved within the system. Several years later, Alan Turing showed that Gödel's incompleteness theorem is equivalent to the assertion that there can be no general method for systematically deciding whether a computer program will ever halt, that is, whether it will ever cause the computer to stop running. The "halting problem" is undecidable.

Gödel's incompleteness theorem is, in essence, a denial of the law of the excluded middle. We are accustomed to thinking of statements as being either true or false, or more precisely, provable or disprovable. But Gödel has forced us to confront the possibility that some statements may very well be neither, that they may surpass the power of formal reasoning, that they may be undecidable.

More recently, it has been shown that undecidable propositions, once believed to be exceptional and pathological, are in fact very frequent, sometimes very simple, and reside in the most elementary branches of number theory. The work on Hilbert's tenth problem[25] has shown, for example, that Gödel's

---

[24] David Hilbert, "Mathematical Problems. Lecture Delivered before the International Congress of Mathematicians at Paris in 1900," *Proc. of Symp. in Pure Math.* XXVIII (1976), 1–34.

[25] In his address to the International Congress of 1900, Hilbert had asked for the design of an algorithm to determine whether an arbitrary prescribed polynomial Diophantine equation has

incompleteness theorem is equivalent to the assertion that one cannot always prove that a Diophantine equation has no solutions when this is the case. Perhaps the famous Fermat equation $a^n + b^n = c^n$ is of this sort. Will all efforts toward a solution have been in vain?

Mathematical reasoning is far more limited than anyone could have imagined. Mathematical truth is far less certain than anyone would have hoped.

> I wanted certainty in the kind of way in which people want religious faith. I thought that certainty is more likely to be found in mathematics than elsewhere. But I discovered that many mathematical demonstrations, which my teachers expected me to accept, were full of fallacies, and that, if certainty were indeed discoverable in mathematics, it would be in a new kind of mathematics, with more solid foundations than those that had hitherto been thought secure. But as the work proceeded, I was continually reminded of the fable about the elephant and the tortoise. Having constructed an elephant upon which the mathematical world could rest, I found the elephant tottering, and proceeded to construct a tortoise to keep the elephant from falling. But the tortoise was no more secure than the elephant, and after some twenty years of very arduous toil, I came to the conclusion that there was nothing more that *I* could do in the way of making mathematical knowledge indubitable.[26]

The underlying difficulty is—as it always has been—the nature of the infinite. "This concept, which created problems even for the Greeks in connection with irrational numbers and which they evaded in the method of exhaustion, has been a subject of contention ever since and prompted Weyl to remark that mathematics is the science of infinity."[27] Perhaps our intuition about the positive integers—as embodied in the principle of mathematical induction—is far from being as clear and precise as we had once thought. "The question for the

---

integer solutions. This is the tenth in the famous list of twenty-three major unsolved problems that Hilbert put forth for the new century. Subsequently, Alonzo Church in 1936, using his newly developed notion of recursive function, demonstrated *the absence of decision procedures for a large class of formal systems,* including ordinary arithmetic. It was not until 1970, however, that Hilbert's tenth problem was finally resolved. The problem is unsolvable—there is no algorithm that enables one to state, in advance, whether a Diophantine equation has solutions. The first proof was given by the 22-year-old Russian mathematician Yuri Matijasevich, building on earlier work of Martin Davis, Hilary Putnam, and Julia Robinson. (A complete account of the solution, requiring only a little number theory, is given in [121].)

[26] Bertrand Russell, "Reflections on My Eightieth Birthday," in *Portraits from Memory and Other Essays* (1956) p. 54.

[27] Morris Kline, *Mathematical Thought from Ancient to Modern Times* [260], p. 1209.

ultimate foundations and the ultimate meaning of mathematics remains open; we do not know in which direction it will find its final solution nor even whether a final objective answer can be expected at all. 'Mathematizing' may well be a creative activity of man, like language or music, of primary originality, whose historical decisions defy complete objective rationalization."[28]

The 20th century will only have added to the mystery and endless allure of the simplest of all mathematical objects.

PROBLEMS

1. Show that the sides of a Pythagorean triangle in which the hypotenuse exceeds the larger leg by 1 are given by Pythagoras's solution

$$\frac{n^2 - 1}{2}, \quad n, \quad \frac{n^2 + 1}{2}$$

where n is an odd integer greater than 1. Notice that these triples follow the simple pattern

$$3^2 = 9 = 4 + 5, \qquad 3^2 + 4^2 = 5^2$$
$$5^2 = 25 = 12 + 13, \qquad 5^2 + 12^2 = 13^2$$
$$7^2 = 49 = 24 + 25, \qquad 7^2 + 24^2 = 25^2$$
$$9^2 = 81 = 40 + 41, \qquad 9^2 + 40^2 = 41^2$$

and so on.

2. Prove that in order to find the generators m and n of a given primitive Pythagorean triple a, b, c, it is sufficient to represent the rational number $(a + c)/b$ in the form of the irreducible fraction m/n.

3. Find all the Pythagorean triangles whose perimeter equals their area.

4. Prove that the length of the radius of the circle inscribed in a Pythagorean triangle is always an integer.

5. Show that in any Pythagorean triangle, one of the sides is divisible by 3, one is divisible by 4, and one is divisible by 5.

---

[28] Hermann Weyl, *Obituary Notices of Fellows of the Royal Society* 4 (1944), 547–553. (See his *Gesammelte Abhandlungen*, IV, 121–129.)

6. The equation

$$3^n + 4^n = 5^n$$

has only one solution in natural numbers n. Show, more generally, that if $a^2 + b^2 = c^2$, then $a^n + b^n < c^n$ for all $n > 2$.

7. If n is a natural number greater than 2, then the equation

$$x^n + (x+1)^n = (x+2)^n$$

has no solution in natural numbers x.

8. The only Pythagorean triple consisting of consecutive natural numbers is 3, 4, 5.

9. Prove that the equation $x^2 + y^2 = z^2$ is not solvable in prime numbers.

10. Prove that there is no Pythagorean triangle whose legs are expressible as the squares of integers.

11. By gathering sufficient empirical evidence, formulate a conjecture about when a natural number is the hypotenuse of a Pythagorean triangle.

12. Show that there exists an infinite sequence of natural numbers $a_1, a_2, \ldots$ such that each of the numbers

$$a_1^2 + a_2^2 + \cdots + a_n^2 \qquad n = 1, 2, \ldots$$

is a square.
   Hint. Look for a sequence that begins with the progression

$$3^2 + 4^2 = 5^2$$
$$3^2 + 4^2 + 12^2 = 13^2$$
$$3^2 + 4^2 + 12^2 + 84^2 = 85^2.$$

13. Prove that the only solution of the equation

$$x^2 + y^2 + z^2 = 2xyz$$

in integers x, y, and z is $x = y = z = 0$.

14. Find all solutions of the equation

$$\frac{1}{x} + \frac{1}{y} = \frac{1}{z}$$

in natural numbers x, y, z.

15. (Unsolved) Is it possible to find a box whose sides and diagonals are all integers?

16. (Fermat) The sum and difference of two squares cannot both be squares.

17. (Fermat) The only integers x for which

$$1 + x + x^2 + x^3 + x^4$$

is a perfect square are $-1, 0, 3$.

18. Find four integral solutions of the equation

$$x^3 + y^3 + z^3 = 3.$$

It is not known whether there are any other solutions. (Mordell has compared the difficulty of this problem with that of determining whether the sequence 123456789 appears in the decimal expansion of π. It does! [349, p. 61])

19. (Unsolved) Is the equation

$$x^3 + y^3 + z^3 = 30$$

solvable in integers?

20. Show that there is no rational point on the circle $x^2 + y^2 = 3$.

21. (Continuation) Find a circle in the plane on which there lies precisely one rational point.

22. (Continuation) Find a circle in the plane on which there lie precisely two rational points.

23. (Continuation) If a circle contains three rational points, then it contains infinitely many rational points. In this case, the rational points form a dense subset of the circle.

24. Find all solutions of the equation $x^y = y^x$ in positive rational numbers x and y $(x \neq y)$. Prove that you have obtained all of them.

25. Find all the rational points on the folium of Descartes

$$x^3 + y^3 = xy.$$

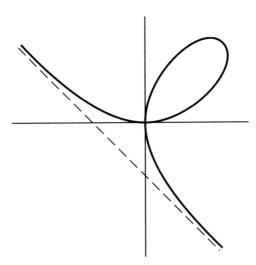

FIGURE 8
The folium of Descartes

CHAPTER 2

PATTERNS, POLYNOMIALS, AND PRIMES: THREE APPLICATIONS OF THE BINOMIAL THEOREM

About binomial theorems I'm teeming with a lot of news,
With many cheerful facts about the square on the hypotenuse.

—William S. Gilbert, *The Pirates of Penzance,* Act I

1. Disorder among the Primes

a. The Infinitude of the Primes

The most important class of positive integers—and, at the same time, the most elusive—is the class of primes:

$$2, 3, 5, 7, 11, 13, 17, \ldots.$$

These are the building blocks, the fundamental units from which all natural numbers can be built up through multiplication, but which themselves are incapable of being further decomposed. Like the "indivisibles" of the atomists, prime numbers are the stuff of which the world of numbers is fashioned.

The simplicity of the definition is misleading for prime numbers have turned out to be the object of the most profound mathematical inquiry. At the same time, they have led to some of the most intransigent problems, many of which lie well beyond the scope of current mathematical knowledge. Such, for example, is the famous Goldbach conjecture, which asserts that *every even number greater than 4 is the sum of two odd primes* (see chapter 1, §1). Goldbach proposed the problem in a letter to Euler of 1742, on the basis of empirical evidence. As we have seen, the evidence in favor of the conjecture is indeed overwhelming, but a rigorous proof has never been found. Perhaps this is the reason that number theory has always held so great an allure—that common observations "with the impress of simplicity upon them" can lead to patterns that lie so deep.

It is all the more remarkable that prime numbers should appear repeatedly and unexpectedly in so many number patterns. For the most striking feature of the sequence of primes is the very absence of any noticeable pattern or regularity. They seem to be scattered haphazardly among the natural numbers and no one can predict where the next one will appear. Certainly as one ascends to higher ranges of numbers the primes grow fewer, for a large number has more potential divisors than a smaller one. Thus, for example, the numbers of primes in the first five blocks of 100 numbers are

$$25, 21, 16, 16, 17$$

while those in the first five blocks of 100 above 100,000 are

$$6, 9, 8, 9, 8.$$

TABLE 1. PRIME NUMBERS

From D. N. Lehmer, List of prime numbers from 1 to 10,006,721, Carnegie Institution of Washington. Publication No. 165, Washington, D.C., 1914. Reprinted by permission of the publisher.

But whatever order the primes may exhibit in the large, their individual behavior seems totally chaotic. In this section we shall point out some of the reasons why.

The first question that must be asked about the class of primes, or about any class of numbers, is whether it contains infinitely many elements. Despite the diminishing frequency of the primes, the evidence suggests that they never come to an end, that there is not ultimately a last prime beyond which all numbers are composite. The matter was settled in antiquity, in the ninth book of Euclid's *Elements*: *There is no largest prime.* Euclid's proof remains a model of mathematical reasoning. After two thousand years it has lost none of its beauty or freshness.

Theorem (Euclid).　*There are infinitely many primes.*

Proof.　Let P be any prime and form the number

$$N = (2 \cdot 3 \cdot 5 \cdots P) + 1,$$

which is the product of all the primes up to P increased by one. Then N is certainly not divisible by any of the primes $2, 3, 5, \ldots, P$. Accordingly, either N is a prime (much greater than P) or all the prime factors of N lie between P and N. In either case, we have found a prime greater than P. No matter how large P may be, there is always a larger prime. Thus the number of primes is infinite.

To the many lines that have been written in praise of this remarkable proof, we can add little. Let us quote instead the following passage from Rademacher and Toeplitz[1]:

> This part of Euclid is quite remarkable, and it would be hard to name its most admirable feature. The problem itself is only of theoretical interest. It can be proposed, for its own sake, only by a person who has a certain inner feeling for mathematical thought. This feeling for mathematics and appreciation of the beauty of mathematics was very evident in the ancient Greeks, and they have handed it down to later civilizations. Also, this problem is one that most people would completely overlook. Even when it is brought to our attention it appears to be trivial and superfluous, and its real difficulties are not immediately apparent. Finally, we must admire the ingenious and simple way in which Euclid proves the theorem. The most natural way to try to

---

[1] *The Enjoyment of Mathematics* [353], p. 10.

prove the theorem is not Euclid's. It would be more natural to try to find the next prime number following any given prime. This has been attempted but has always ended in failure because of the extreme irregularity of the formation of the primes.

Euclid's proof circumvents the lack of a law of formation for the sequence of primes by looking for *some* prime beyond instead of for the *next* prime after P. For example, his proof gives 2311, not 13, as a prime past 11, and it gives 59 as one past 13. Frequently there are a great many primes between the one considered and the one given by the proof. This is not a sign of weakness of the proof, but rather it is evidence of the ingenuity of the Greeks in that they did not try to do more than was required.

We have established with stunning simplicity the existence of infinitely many objects of a certain type without knowing what they are or how to find them. There are, for example, prime numbers having at least one million digits, but we do not know any one of them. The largest known prime is

$$2^{756839} - 1,$$

and it contains only 227,832 digits. The proof that it is a prime was discovered in 1992 by David Slowinski and Paul Gage of Cray Research.[2]

It is, therefore, only natural to ask whether there exists a simple general formula for the nth prime number p_n. But a natural question is not always a reasonable one, and here the evidence suggests that such a formula, while perhaps possible, is highly improbable—the distribution of the primes is far too erratic. If p_n could be described by an elementary function, then so could all its differences $p_{n+1} - p_n$. While very little is known about the difference between two consecutive primes, what is known and what the evidence from the tables suggests is that its behavior is extremely irregular. To begin with, the tables indicate the presence of large gaps between successive primes. For example, the smallest consecutive primes whose difference is 100 are the numbers 396733 and 396833. It is easy to see that these large gaps must occur. In fact, for any positive integer N, all of the $N - 1$ numbers

$$N! + 2, N! + 3, N! + 4, \ldots, N! + N$$

are composite—the first number is divisible by 2, the second by 3, the third by 4, and so on. Since N was arbitrary, we can make the difference $p_{n+1} - p_n$ as large as we choose.

---

[2] See *The Amer. Math. Monthly,* 99 (1992) p. 360.

On the other hand, there is overwhelming evidence that the smallest possible differences between consecutive odd primes, namely, the numbers 2, 4, 6, ..., recur indefinitely. Pairs of primes such as

$$(11, 13) \quad \text{or} \quad (101, 103) \quad \text{or} \quad (1000000000061, 1000000000063)$$

which differ by 2 are called **twin primes**, and whether they go on indefinitely 2000 years of mathematical investigation has failed to show.

It is therefore highly unlikely that there should exist an elementary formula for the nth prime, and a similar remark applies to many other related problems, such as to find a general formula for the prime which immediately follows a given prime. A far more reasonable problem is *to find a simple arithmetical formula which yields prime numbers only*, although not necessarily all of them. More precisely, we would like to find an elementary function $f(n)$ such that $f(n) \rightarrow \infty$ and $f(n)$ is prime for every n, or for all n beyond a certain limit. But here too the evidence is not encouraging.

In 1640, Fermat wrote to Mersenne that he had discovered a formula that produces only primes,

$$F_n = 2^{2^n} + 1$$

although he confessed that he did not have a proof. The conjecture has not proved a happy one. The first five **Fermat numbers**, as the values of F_n are now known, are

$$F_0 = 3$$
$$F_1 = 5$$
$$F_2 = 17$$
$$F_3 = 257$$
$$F_4 = 65537$$

and indeed all are primes. But in 1732 Euler discovered the factorization

$$F_5 = 2^{32} + 1 = 641 \cdot 6700417;$$

hence F_5 is composite. In 1880 Landry proved that

$$F_6 = 2^{64} + 1 = 274177 \cdot 67280421310721.$$

Since then many additional Fermat numbers have been investigated, among them F_{16}, which was found to be composite in 1953 by J. L. Selfridge. The importance of this result lies in the fact that it disproves the conjecture that all of the numbers

$$2 + 1, \ 2^2 + 1, \ 2^{2^2} + 1, \ 2^{2^{2^2}} + 1, \ 2^{2^{2^{2^2}}} + 1, \ldots$$

are prime—indeed, F_{16} is the fifth term in the sequence; it has nearly twenty thousand digits.

Fermat maintained throughout his life that F_n is always a prime, one of his few mathematical statements that has ever been refuted. In fact, no one has ever found a Fermat prime beyond F_4. Nonetheless, in the 19th century, Fermat numbers became the object of renewed attention when Gauss proved that a regular polygon with p sides (p a prime) could be constructed by Euclidean methods (straight-edge and compass alone) if and only if p is a Fermat number. It was this remarkable discovery, which had eluded mathematicians since antiquity, that convinced Gauss to pursue a career in mathematics. When he made the discovery Gauss was only 18 years old!

Fermat numbers also satisfy a simple divisibility condition which gives rise to another proof of the infinitude of the primes.

Theorem. *Any two Fermat numbers are relatively prime.*

Proof. We begin by observing that

$$F_0 F_1 F_2 \cdots F_n = F_{n+1} - 2. \tag{1}$$

Indeed, we have only to multiply on the left by $2^{2^0} - 1$ to see that (1) follows almost at once.

Suppose now that F_m and F_n are distinct Fermat numbers, with $m < n$, and that each is divisible by d. Then d is also a divisor of

$$F_n - (F_0 \cdots F_m \cdots F_{n-1}) = 2.$$

Therefore d is either 1 or 2. But it cannot be 2 because F_n is odd. Therefore it must be 1, and F_m and F_n are relatively prime.

It follows at once that each Fermat number is divisible by an odd prime that does not divide any of the others, and hence there can be no largest prime. This elegant argument is due to Pólya (cf. [345], volume 2, p. 130).

b. Polynomials and Primes

There are a number of remarkable formulas which take on prime values for long sequences of consecutive integers. Such, for example, is the polynomial

$$P(x) = x^2 + x + 41,$$

attributed to Euler, which is prime for the forty(!) consecutive values $x = 0, 1, 2, \ldots, 39$:

$P(0) = 41$	$P(10) = 151$	$P(20) = 461$	$P(30) = 971$
$P(1) = 43$	$P(11) = 173$	$P(21) = 503$	$P(31) = 1033$
$P(2) = 47$	$P(12) = 197$	$P(22) = 547$	$P(32) = 1097$
$P(3) = 53$	$P(13) = 223$	$P(23) = 593$	$P(33) = 1163$
$P(4) = 61$	$P(14) = 251$	$P(24) = 641$	$P(34) = 1231$
$P(5) = 71$	$P(15) = 281$	$P(25) = 691$	$P(35) = 1301$
$P(6) = 83$	$P(16) = 313$	$P(26) = 743$	$P(36) = 1373$
$P(7) = 97$	$P(17) = 347$	$P(27) = 797$	$P(37) = 1447$
$P(8) = 113$	$P(18) = 383$	$P(28) = 853$	$P(38) = 1523$
$P(9) = 131$	$P(19) = 421$	$P(29) = 911$	$P(39) = 1601$

Notice that $P(x+1) = P(x) + 2(x+1)$, so that the values $P(0), P(1), \ldots$ are most easily calculated recursively by first starting with $P(0) = 41$ and then adding on successive terms of the arithmetic progression $2, 4, 6, 8, \ldots$. When $x = 40$, the formula fails: $P(40) = 1681 = (41)^2$. Notice also that $P(x - 1) = P(-x)$, and therefore $P(x)$ takes on these same prime values for $x = -1, -2, \ldots, -40$. Thus the polynomial

$$P(x - 41) = (x - 41)^2 + (x - 41) + 41$$

produces only primes for the eighty consecutive values $x = 1, 2, \ldots, 80$. There is no other quadratic known that yields a longer string of consecutive prime numbers.

Such examples, however striking, can only be regarded as curiosities for it is not difficult to show that no polynomial can take on prime values for all integral values of the argument.

Theorem (Goldbach).   *If P is a nonconstant polynomial, then the values*

$$P(1), P(2), P(3), \ldots$$

cannot all be primes.

In the simplest case, P is a polynomial with integer coefficients and then the proof is relatively straightforward. Let us suppose to the contrary that $P(n)$ is a prime for every n. Let α be an arbitrary positive integer, put $\beta = P(\alpha)$, and consider the sequence of values

$$P(\alpha + \beta), P(\alpha + 2\beta), P(\alpha + 3\beta), \ldots.$$

By virtue of the binomial theorem, for each $n = 1, 2, 3, \ldots,$

$$P(\alpha + n\beta) - P(\alpha) = \text{ an integral multiple of } \beta$$

(remember that P has integer coefficients). Therefore $P(\alpha + n\beta)$ is also an integral multiple of β, and, since it is a prime, it must be β. Thus $P(\alpha + n\beta) - \beta = 0$ for every positive integer n, contradicting the fact that a nonconstant polynomial can have only finitely many roots.

When the coefficients of P are not all integers the proof fails and a more subtle argument is needed. We first begin with a definition.

A polynomial P is said to be *integer-valued* if the numbers $P(0), P(1), P(2), \ldots$ are integers. If P has integer coefficients then it is certainly integer-valued, but the example

$$P(x) = \frac{x(x+1)}{2}$$

shows that this condition is by no means necessary.

A particularly useful set of integer-valued polynomials is the set of *binomial functions*

$$\binom{x}{0}, \binom{x}{1}, \binom{x}{2}, \ldots.$$

Here the symbol ( ) has its usual meaning, so that

$$\binom{x}{0} = 1$$

$$\binom{x}{1} = x$$

$$\binom{x}{2} = \frac{x(x-1)}{2!}$$

$$\binom{x}{3} = \frac{x(x-1)(x-2)}{3!}$$

$$\cdots$$

and so forth. It is clear that $\binom{x}{n}$ does not have integer coefficients when $n > 1$, but when viewed combinatorially it is certainly integer-valued. Notice also that $\binom{x}{n}$ also takes on integer values whenever x is a negative integer, since

$$\binom{-x}{n} = (-1)^n \binom{x+n-1}{n}.$$

The importance of the binomial functions is illustrated by the next lemma. It asserts, in the language of linear algebra, that the set of binomial functions forms a basis for the vector space of all polynomials.

Lemma. *For any polynomial P of degree n there are unique real numbers b_0, $b_1, \ldots, b_n$ such that*

$$P(x) = b_0 \binom{x}{0} + b_1 \binom{x}{1} + \cdots + b_n \binom{x}{n}. \tag{2}$$

Furthermore, if P is integer-valued then the coefficients $b_0, b_1, \ldots, b_n$ are all integers.

Proof. To show that P can be expressed *in at least one way* as a linear combination of the polynomials

$$\binom{x}{0}, \binom{x}{1}, \binom{x}{2}, \ldots, \binom{x}{n}$$

it is clearly enough to show that each of the functions

$$1, x, x^2, \ldots, x^n$$

can be so expressed. This is trivial for the first two functions 1 and x, and we proceed by induction. Having found $1, x, \ldots, x^k$ in terms of $\binom{x}{0}, \binom{x}{1}, \ldots, \binom{x}{k}$, we expand the polynomial

$$\binom{x}{k+1} = \frac{x(x-1)(x-2)\cdots(x-k)}{(k+1)!}$$

and then solve for x^{k+1} in terms of the lower powers of x. In this way the desired representation for $P(x)$ can be obtained.

To show that this representation is unique, we substitute successively in equation (2) the values $x = 0, 1, \ldots, n$. The resulting system of equations is then

$$P(0) = b_0$$

$$P(1) = b_0 + \binom{1}{1}b_1$$

$$P(2) = b_0 + \binom{2}{1}b_1 + \binom{2}{2}b_2$$

$$P(3) = b_0 + \binom{3}{1}b_1 + \binom{3}{2}b_2 + \binom{3}{3}b_3$$

$$\cdots$$

$$P(n) = b_0 + \binom{n}{1}b_1 + \binom{n}{2}b_2 + \cdots + \binom{n}{n}b_n.$$

The coefficients $b_0, b_1, \ldots, b_n$ may now be determined successively from the known values $P(0), P(1), \ldots, P(n)$. If these values are integers, then so are the coefficients. This completes the proof of the lemma.

It is interesting to observe that the second part of the proof uses only the fact that the values $P(0), P(1), \ldots, P(n)$ are integers and not that P is integer-valued. Therefore, we have established the following striking corollary.

Corollary. *If a polynomial of degree n assumes integer values for $n + 1$ consecutive integer values of the variable, then it assumes integer values for all integer values of the variable.*

It is now a simple matter to prove Goldbach's theorem.

Proof of the theorem. As before, we argue by contradiction. Suppose then that for some nonconstant polynomial P, the values $P(1), P(2), \ldots$ are all primes. Then the leading coefficient of P must be positive, so that ultimately (say for $x \geq \alpha$) P increases to infinity. (*Reason*: When x is sufficiently large, $P'(x) > 0$.) We may certainly take α to be a positive integer, so large that

$$\beta = P(\alpha) > \text{degree of } P = n.$$

It follows that

$$P(\alpha + \beta) > P(\alpha) = \beta. \tag{3}$$

But the polynomial $P(x + \alpha)$ is integer-valued and therefore, by the lemma, it may be represented in the form

$$P(x + \alpha) = b_0 \binom{x}{0} + b_1 \binom{x}{1} + \cdots + b_n \binom{x}{n}$$

where $b_0, b_1, \ldots, b_n$ are integers. Observe now that when $x = \beta$, every term on the right-hand side is divisible by β. Indeed, for the first term we see that

$$b_0 = P(\alpha) = \beta,$$

while for every other term the binomial coefficients

$$\binom{\beta}{k}, \qquad k = 1, 2, \ldots, n,$$

are all divisible by β, since β is a prime greater than n. Thus, $P(\alpha + \beta)$ is divisible by β. But, by condition (3), β is a *proper divisor* of $P(\alpha + \beta)$. This shows, contrary to assumption, that the values $P(1), P(2), \ldots$ cannot all be primes. The contradiction proves the theorem.

It is also known that no rational function can be prime for all integral values of the variable (see, for example, [345], volume 2, p. 130).

Suppose then that we moderate our demands even more and ask only for a function that assumes infinitely many prime values. In this case the problem is trivial for, by Euclid's theorem, $P(x) = x$ is such a function. Euclid's proof can be carried a bit further, and the same reasoning shows that certain simple linear functions, such as $4x - 1$ and $6x - 1$, also give rise to an infinity of primes (see problem 2). But such methods cannot cope with the general linear function. In an important memoir of 1837, Dirichlet proved that any linear function $ax + b$ has the required property, provided only that a and b are relatively prime. In other words, *every arithmetical progression*

$$b, b + a, b + 2a, b + 3a, \ldots$$

(subject to the aforementioned requirement) *contains infinitely many primes.* Dirichlet's theorem is easy to state but difficult to prove. The original proof made use of the most advanced methods of analysis (functions of a continuous variable, limits, and infinite series), and it was the first really important application of such methods to the theory of numbers.

Very little is known about other forms which represent infinitely many primes. No one has ever proved that an expression as simple as

$$x^2 + 1$$

(or any other polynomial of degree 2 or more) can take on an infinity of primes. The same is true for the exponentials

$$2^n - 1 \quad \text{and} \quad 2^n + 1$$

and all such problems appear to be exceedingly difficult.

All this suggests that the distribution of the primes is hopelessly chaotic, and that in all likelihood we shall never get to know all of them.[3]

---

[3] The mathematical grail may well be a formula for the primes, or a formula that produces prime numbers only. In 1947, Mills [302] showed that there is a real number A such that $[A^{3^n}]$ is prime for every natural number n. While the result seems astounding, the proof shows that the only way to find A is to construct it, and for that it is necessary to know the values of arbitrarily large primes. (A more transparent example is based on the real number

$$B = .20030000500000070\ldots = \sum p_n/10^{n^2}.$$

"Knowing" B we can easily and systematically retrieve all the primes.)

Wilson's theorem (see §3) provides another class of formulas. Among them is the function

$$f(n) = \sin^2 \pi n + \sin^2 \pi \left(\frac{1 + (n-1)!}{n} \right) \quad n > 1$$

which is zero if and only if n is a prime.

In a different direction, the work on Hilbert's tenth problem has made it possible to prove that there are polynomials whose positive values consist of all the primes, when the variables range over all nonnegative integers. A specific example of such a polynomial (of degree 25 in the 26 variables $a, b, c, \ldots, z$) is given in [239].

For a delightful discussion of these and other prime producing formulas, see Underwood Dudley, "Formulas for Primes" [138]. As Dudley points out, "Formulas should be useful. If not, they should be astounding, elegant, enlightening, simple, or have some other redeeming value." Although it is not a formula, the following result of Mann and Shanks [291] succeeds on all counts. Write Pascal's triangle with row n starting in column $2n$:

	0	1	2	3	4	5	6	7	8	9	10	11	12	13	14
0	1														
1			1	1											
2					1	2	1								
3							1	3	3	1					
4									1	4	6	4	1		
5											1	5	10	10	5
6													1	6	15
7															1....

Then *a column number is a prime if and only if each number in it is divisible by the corresponding row number.*

PROBLEMS

1. Show that the numbers

$$2 + 1 = 3$$

$$2 \cdot 3 + 1 = 7$$

$$2 \cdot 3 \cdot 5 + 1 = 31$$

$$2 \cdot 3 \cdot 5 \cdot 7 + 1 = 211$$

$$2 \cdot 3 \cdot 5 \cdot 7 \cdot 11 + 1 = 2311$$

appearing in Euclid's proof of the infinitude of the primes are all relatively prime in pairs. Are they all primes?

2. By modifying Euclid's proof, show that:
   a.   There are infinitely many primes of the form $4n - 1$.
   b.   There are infinitely many primes of the form $6n - 1$.
   Hint. For the proof of the first part, consider numbers of the form

$$N = 4(3 \cdot 5 \cdot 7 \cdots P) - 1.$$

The proof of the second part is similar.

3. It is not known whether the terms of the sequence

$$11, 111, 1111, \ldots$$

(called **repunits**) include infinitely many prime numbers. Show, however, that if a repunit is a prime, then the number of its digits must be a prime. (To see that the converse is false, note, for example, that 111 is divisible by 3.) The only known prime repunits are those with $2, 19, 23, 317$, and 1031 decimal digits (see Williams and Dubner [455]).

4. Prove that every number of the form $8^n + 1$ is composite.
   Remark. It is not known whether there are infinitely many prime numbers of the form $10^n + 1$, or whether every number of the form $12^n + 1$ $(n > 1)$ is composite.

5. Prove that every number of the form $2^{2^n} + 5$ is composite.

6. Show that if $2^n + 1$ is a prime then n is a power of 2.

7. Prove that if one of the numbers $2^n - 1$ and $2^n + 1$ is a prime, where $n > 2$, then the other number must be composite.

8. (A Putnam competition problem that no contestant solved!) If n^c is an integer for $n = 1, 2, 3, \ldots$, then c is an integer. (The result remains true if we assume merely that $2^c, 3^c, 5^c$ are integers. See, e.g., *American Mathematical Monthly*, 83 (1976) 473.)

9. (Unsolved) Do there exist infinitely many composite Fermat numbers? (For a list of the 84 that are known, see [385], p. 372. The largest of them is F_{23471}. It has more than 10^{7064} digits, so we cannot even write it down.)

10. (Unsolved) There are only three known primes of the form $n^n + 1$. They are $1^1 + 1 = 2$, $2^2 + 1 = 5$, $4^4 + 1 = 257$. Are there any others? Show that if there are, they must have more than 300,000 digits.

11. (Unsolved) Are there infinitely many primes of the form $n! + 1$?

12. (Unsolved) Does the Fibonacci sequence

$$1, 1, \mathbf{2}, \mathbf{3}, \mathbf{5}, 8, \mathbf{13}, 21, 34, 55, \mathbf{89}, 144, \ldots$$

where $f_1 = f_2 = 1$, $f_{n+1} = f_n + f_{n-1}$ ($n \geq 2$) contain infinitely many primes?

13. Use Bertrand's postulate (see chapter 1, §3, problem 3) to prove that there is a constant $b \approx 1.25$ such that the numbers

$$[2^b], \left[2^{2^b}\right], \left[2^{2^{2^b}}\right], \ldots$$

are all prime.

14. Prove that Euler's trinomial $P(x) = x^2 + x + 41$ has the following interesting property: For all integer values of x the number $P(x)$ is not divisible by any prime less than 41.

15. Prove that there is no polynomial that takes on prime values for all prime values of the variable (except, of course, the trivial polynomial $P(x) = x$).

16. The functions

$$1$$

$$x$$

$$x(x-1)$$

$$x(x-1)(x-2)$$

$$x(x-1)(x-2)(x-3)$$

$$\cdots$$

are called **factorial powers**; we denote them by $x^{\underline{0}}, x^{\underline{1}}, x^{\underline{2}}, x^{\underline{3}}, x^{\underline{4}}, \ldots.$

a. Prove the following analogue of the binomial theorem:

$$(x+y)^{\underline{n}} = \sum_{k=0}^{n} \binom{n}{k} x^{\underline{k}} y^{\underline{n-k}}.$$

b. In the calculus of finite differences the role of the derivative is played by the *difference operator* Δ, defined by

$$\Delta f(x) = f(x+1) - f(x).$$

Prove that

$$\Delta x^{\underline{n}} = n x^{\underline{n-1}} \qquad n = 1, 2, \ldots.$$

c. Find a simple formula for the sum

$$\sum_{0 < k < n} k^{\underline{p}}$$

where p is a positive integer. Compare the result with

$$\int_0^n x^p \, dx.$$

17. (Continuation) By virtue of the lemma, for each natural number n, there exist unique integers

$$\left\{ {n \atop 0} \right\}, \left\{ {n \atop 1} \right\}, \left\{ {n \atop 2} \right\}, \ldots, \left\{ {n \atop n} \right\}$$

such that

$$x^n = \sum_{k=0}^{n} \left\{ {n \atop k} \right\} x^{\underline{k}}.$$

The numbers $\left\{ {n \atop k} \right\}$ are called *Stirling numbers of the second kind*. (Compare chapter 3, §3, problem 12.)

    a. Show that $\left\{ {n \atop 0} \right\} = 0$ and $\left\{ {n \atop 1} \right\} = 1$ for all $n = 1, 2, \ldots$.

    b. Show that the Stirling numbers satisfy the recursion

$$\left\{ {n \atop k} \right\} = \left\{ {n-1 \atop k-1} \right\} + k\left\{ {n-1 \atop k} \right\}.$$

    c. Give a combinatorial argument to show that $\left\{ {n \atop k} \right\}$ is equal to the number of ways to partition a set of n objects into k nonempty subsets. For example, $\left\{ {3 \atop 2} \right\} = 3$ because there are three ways to partition $\{1, 2, 3\}$ into two subsets:

$$\{1, 2\} \cup \{3\}, \quad \{1, 3\} \cup \{2\}, \quad \{2, 3\} \cup \{1\}.$$

18. (Continuation) The higher-order differences $\Delta^2, \Delta^3, \Delta^4, \ldots$ are defined recursively by setting $\Delta^{n+1} f = \Delta(\Delta^n f)$. For example, we find readily that

$$\Delta^2 f(x) = f(x+2) - 2f(x+1) + f(x)$$
$$\Delta^3 f(x) = f(x+3) - 3f(x+2) + 3f(x+1) - f(x)$$
$$\Delta^4 f(x) = f(x+4) - 4f(x+3) + 6f(x+2) - 4f(x+1) + f(x).$$

    a. Show generally that

$$\Delta^n f(x) = \sum_{k=0}^{n} (-1)^{n-k} \binom{n}{k} f(x+k).$$

    b. Show that every polynomial $f(x)$ of degree n can be represented by the **Newton forward difference formula**

$$f(x) = f(0)\binom{x}{0} + \Delta f(0)\binom{x}{1} + \Delta^2 f(0)\binom{x}{2} + \cdots + \Delta^n f(0)\binom{x}{n}.$$

    c. Find a simple formula for the sum

$$\sum_{k=0}^{n} (-1)^k \binom{n}{k} \left(1 - \frac{k}{n}\right)^n.$$

2. Summing the Powers of the Integers

a. The Figurate Numbers

The classification of natural numbers according to the simple geometric shapes that may be used to represent them dates back to Pythagoras. Believing that whole numbers were the essence of the universe, the Pythagoreans elevated them above all else and they searched for patterns, such as those formed by dots or pebbles in the sand. Thus the numbers $1, 3 = 1 + 2, 6 = 1 + 2 + 3, 10 = 1+2+3+4, \ldots$ were called *triangular numbers* because the corresponding dots could be arranged as triangles (Figure 1) and the numbers $1, 4, 9, 16, \ldots$ were called *square numbers* because as dots they could be arranged as squares (Figure 2).

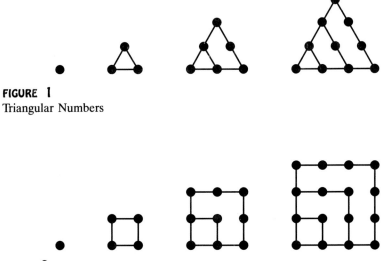

FIGURE 1
Triangular Numbers

FIGURE 2
Square Numbers

By means of such figures the Pythagoreans came to realize that, starting with any arithmetical progression of the form

$$1, 1 + d, 1 + 2d, 1 + 3d, \ldots$$

and then adding consecutive terms, one arrives at the *polygonal numbers* with

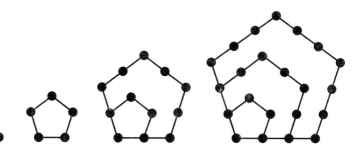

FIGURE 3
Pentagonal Numbers

$d + 2$ sides. For example, when $d = 3$, the progression is $1, 4, 7, 10, \ldots$ and the resulting sums $1, 5, 12, 22, \ldots$ are the *pentagonal numbers*.

From these simple origins a wealth of profound mathematics would emerge. In connection with Bachet's empirical theorem that every integer is the sum of four squares, Fermat made the famous comment:

> I was the first to discover the very beautiful and entirely general theorem that every number is either triangular or the sum of 2 or 3 triangular numbers; every number is either a square or the sum of 2, 3, or 4 squares; either pentagonal or the sum of 2, 3, 4, or 5 pentagonal numbers; and so on ad infinitum, whether it is a question of hexagonal, heptagonal or any polygonal numbers. I can not give the proof here, which depends upon numerous and abstruse mysteries of numbers; for I intend to devote an entire book to this subject and to effect in this part of arithmetic astonishing advances over the previously known limits.[4]

But such a book was never published and Fermat's proof, if he ever had one, was never found. Nearly two centuries later, Cauchy (in 1813–1815) supplied the first proof.

In the 18th century, in the hands of the great Euler, square numbers and pentagonal numbers would emerge in two recondite mathematical truths whose proofs would enrich and reshape both number theory and analysis. The celebrated pentagonal number theorem, which reveals an astonishing relation between pentagonal numbers and the function $\sigma(n)$, the sum of the divisors of n, as well as Euler's remarkable formula for the sum of the reciprocals of the

---

[4] *Oeuvres de Fermat,* volume I, p. 305; French translation, volume III, p. 252. The English translation given here is from Dickson, *History of the Theory of Numbers* [135], volume 2, p. 6.

squares

$$1 + \frac{1}{4} + \frac{1}{9} + \frac{1}{16} + \cdots = \frac{\pi^2}{6}$$

—both will be taken up in the final chapter.

In a different direction, triangular numbers lead naturally to the figurate numbers in three and higher dimensions. The *pyramidal* or *tetrahedral numbers* described by Theon of Smyrna and Nicomachus (each about A.D. 100) are formed by summing the first n triangular numbers, just as these numbers had been formed by summing the first n natural numbers. Thus the sequence of tetrahedral numbers begins 1, $4 = 1 + 3$, $10 = 1 + 3 + 6$, $20 = 1 + 3 + 6 + 10, \ldots$. Higher-dimensional figurate numbers are defined recursively in the same way, first by summing the first n tetrahedral numbers to arrive at the sequence $1, 5, 15, 35, 70, \ldots$, then by summing those, and so on, ad infinitum.

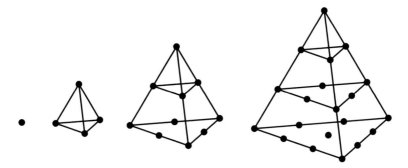

FIGURE 4
Tetrahedral Numbers

Fermat found inspiration in these numbers through Diophantus of Alexandria's *Book on Polygonal Numbers* (circa A.D. 250). There, in one of the margins, next to the formula for polygonal numbers, Fermat recorded without proof "a very beautiful and very remarkable property that I have discovered." He described it rhetorically as follows:

> Taking the natural numbers, the product of one of them with its successor gives twice the triangular number corresponding to the initial number; taking one of the natural numbers and multiplying it by the triangular number corresponding to its successor gives three times the pyramidal number corresponding to the initial number; indeed, taking one of the natural numbers and multiplying it by the pyrami-

dal number corresponding to its successor, one obtains four times the triangulo—triangulaire of the original number. *And so on indefinitely by this general rule.*[5]

In modern notation, Fermat's theorem on the higher-dimensional figurate numbers may be written as follows:

$$\sum_{1}^{n} i = \frac{n(n+1)}{1 \cdot 2}$$

$$\sum_{1}^{n} \frac{i(i+1)}{1 \cdot 2} = \frac{n(n+1)(n+2)}{1 \cdot 2 \cdot 3}$$

$$\sum_{1}^{n} \frac{i(i+1)(i+2)}{1 \cdot 2 \cdot 3} = \frac{n(n+1)(n+2)(n+3)}{1 \cdot 2 \cdot 3 \cdot 4}$$

and so on. Fermat gave no proof of his theorem, stating only that the margin of his copy of *Diophantus* was too small to contain it, just as he had done for his notorious statement about the equation $x^n + y^n = z^n$. Mathematical induction provides the simple proof.

If we arrange the figurate numbers in tabular form, writing those of dimension k in the kth column (as in Figure 5), then we arrive at the celebrated **arithmetical triangle** of Pascal. (For an excellent account of the history of this most famous of all number patterns, which did not originate with Pascal, see [144].)

1								
1	1							
1	2	1						
1	3	3	1					
1	4	6	4	1				
1	5	10	10	5	1			
1	6	15	20	15	6	1		
1	7	21	35	35	21	7	1	
1	8	28	56	70	56	28	8	1

FIGURE 5

---

[5] *Oeuvres de Fermat,* volume I, p. 341; French translation, volume III, p. 273. The English translation given here is from Edwards, *Pascal's Arithmetical Triangle* [144], p. 14.

"Pascal's triangle provides a laboratory for students to conduct mathematical experiments."[6] The patterns that lie within it, some transparent, others more or less concealed, seem almost boundless, and one hears continually of another new relation that has suddenly come to light. As Pascal himself wrote, "It is extraorinary how fertile in properties this is. Everyone can try his hand."[7]

With the appearance of Pascal's "Traité du Triangle Arithmétique" (1665), the figurate numbers gradually "lost their identity and became merged with the binomial coefficients."[8] They were resurrected during the next century by James Bernoulli (1654–1705). Bernoulli was unfamiliar with Pascal's treatise and he rediscovered the arithmetical triangle through his work on combinatorics, immediately recognizing the figurate numbers. Waxing eloquent about the arithmetical triangle, Bernoulli writes: "This Table has truly exceptional and admirable properties. The mysteries of combinations are concealed within it, but those who are more intimately acquainted with Geometry know also that capital secrets of all mathematics are hidden in it."[9]

Fermat, Pascal, and Bernoulli all saw the importance of the figurate numbers in summing the powers of the integers

$$1^k + 2^k + \cdots + n^k,$$

and it is to this problem that we shall now turn our attention.

b. The First Three Powers

The formulas

$$1 + 2 + \cdots + n = \frac{n(n+1)}{2}$$

$$1^2 + 2^2 + \cdots + n^2 = \frac{n(n+1)(2n+1)}{6}$$

$$1^3 + 2^3 + \cdots + n^3 = \left[\frac{n(n+1)}{2}\right]^2$$

for the sum of the first n natural numbers, their squares, and their cubes were all known in antiquity. They can be discovered and proved in a variety of ways. Proof by mathematical induction is simple but sterile—the results first must be known before they can be verified. It is therefore worth pointing out how

---

[6] P. Hilton and J. Pedersen [220], p. 307.

[7] "Traité du Triangle Arithmétique." See *Oeuvres de Blaise Pascal* [331], volume 3, p. 465.

[8] A. W. F. Edwards, *Pascal's Arithmetical Triangle* [144], p. 16.

[9] *Ars conjectandi* [37], p. 159.

we might be led to discover (and not merely verify) the patterns that underlie these three simple identities.

The first formula follows the moment we chance upon the idea of writing the sum in two different ways, first with the numbers $1, 2, 3, \ldots, n$ in their natural (ascending) order, and then with these same numbers in reverse (descending) order:

$$
\begin{array}{cc}
1 & n \\
2 & n-1 \\
  & n-2 \\
\vdots & \vdots \\
n-2 & 3 \\
n-1 & 2 \\
n & 1
\end{array}
$$

Since the sum of the entries in each row is $n + 1$, it follows that if S is the common sum of the columns then

$$
2S = n(n+1) \qquad \text{or} \qquad S = \frac{n(n+1)}{2}.
$$

This ingenious argument has been attributed to the young Gauss ([341], volume I, p. 60).

But the most elegant proof is the Pythagoreans' own—a "proof without words":

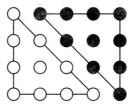

FIGURE 6

To discover a formula for the sum of the first n squares is not nearly so simple. If we try to imitate the first proof given above, beginning with the two columns of numbers

$$
\begin{array}{cc}
1 & n^2 \\
4 & (n-1)^2 \\
9 & (n-2)^2 \\
\vdots & \vdots \\
n^2 & 1
\end{array}
$$

then we see at once that the proof fails: the sums of the entries in every row are not all the same. We may, however, hit upon the idea of comparing the values of the sums $1^2+2^2+\cdots+n^2$ with those of the corresponding sums $1+2+\cdots+n$, whose values we now know. If, for example, we happen to compare their ratio

$$
\frac{1^2 + 2^2 + \cdots + n^2}{1 + 2 + \cdots + n}
$$

then an unmistakable pattern emerges. Choosing the first few values of n, we find these ratios to be

$$
\frac{3}{3}, \frac{5}{3}, \frac{7}{3}, \frac{9}{3}, \frac{11}{3}, \frac{13}{3}, \ldots
$$

and we cannot help but conjecture that, for all values of n,

$$
\frac{1^2 + 2^2 + \cdots + n^2}{1 + 2 + \cdots + n} = \frac{2n + 1}{3}.
$$

Therefore

$$
1^2 + 2^2 + \cdots + n^2 = \frac{n(n + 1)(2n + 1)}{6}.
$$

The pattern that underlies the sum of the first n cubes is far more transparent. Consider the evidence:

$$
\begin{array}{rcl}
1^3 & = & 1 \\
1^3 + 2^3 & = & 9 \\
1^3 + 2^3 + 3^3 & = & 36 \\
1^3 + 2^3 + 3^3 + 4^3 & = & 100 \\
1^3 + 2^3 + 3^3 + 4^3 + 5^3 & = & 225
\end{array}
$$

The resulting sums are not only perfect squares, but what is more, they are the squares of the first five triangular numbers 1, 3, 6, 10, 15. The evidence is again compelling and we are led to the beautiful relation

$$
1^3 + 2^3 + \cdots + n^3 = (1 + 2 + \cdots + n)^2 = \left[\frac{n(n + 1)}{2}\right]^2 \tag{1}
$$

valid for every natural number n.

Finally, let us mention a striking geometric proof of formula (1) which was known to Arab mathematicians almost a thousand years ago ([417], p. 52). It is based on a single figure, a square of side $1 + 2 + \cdots + n$ which has been partitioned into n *gnomons* (or angle irons)—see Figure 7.

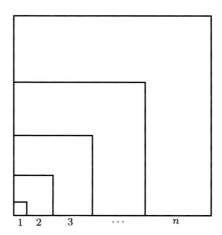

FIGURE 7

The area of the square can be calculated in two different ways,

$$A = (1 + 2 + \cdots + n)^2$$

$$= G_1 + G_2 + \cdots + G_n$$

where $G_1, \ldots, G_n$ denote the areas of the gnomons. But each gnomon is the difference of two squares: the area of the first is $1^2 - 0^2$, the area of the second $(1 + 2)^2 - 1^2$, that of the third $(1 + 2 + 3)^2 - (1 + 2)^2$, and so on up to the last,

$$G_n = (1 + 2 + \cdots + n)^2 - (1 + 2 + \cdots + n - 1)^2.$$

Therefore, by the formula for the nth triangular number,

$$G_n = \left[\frac{n(n+1)}{2} \right]^2 - \left[\frac{n(n-1)}{2} \right]^2 = n^3.$$

(*The difference of the squares of two consecutive triangular numbers is always a perfect cube*). Thus A is the sum of the first n cubes.

These formulas would play an important role in the continuing development of the calculus.

c. Higher Powers

The Methods of Fermat and Pascal. The procedures described in the last section for evaluating the sums of the integers, their squares, and their cubes are all too specialized to be of value for powers of higher degree. It was not until the seventeenth century that completely general and efficient methods for dealing with these sums were discovered.

In 1636, Fermat derived a simple recursive procedure for evaluating the sums

$$1^k + 2^k + \cdots + n^k$$

based on his formula for the figurate numbers. For example, starting with the formula for the sum of the first n tetrahedral numbers

$$\sum_1^n \frac{i(i+1)(i+2)}{1 \cdot 2 \cdot 3} = \frac{n(n+1)(n+2)(n+3)}{1 \cdot 2 \cdot 3 \cdot 4},$$

he expanded the left-hand side as $\frac{1}{6}\sum_1^n i^3 + \frac{1}{2}\sum_1^n i^2 + \frac{1}{3}\sum_1^n i$ and thereby obtained a formula for $\sum_1^n i^3$ in terms of the sums of the lower powers. The sum of the first n fourth powers can be obtained in similar fashion, first by starting with the formula

$$\sum_1^n \frac{i(i+1)(i+2)(i+3)}{1 \cdot 2 \cdot 3 \cdot 4} = \frac{n(n+1)(n+2)(n+3)(n+4)}{1 \cdot 2 \cdot 3 \cdot 4 \cdot 5},$$

then by expanding the left side as before, and finally by solving for the desired sum $\sum_1^n i^4$. And so on indefinitely. While the method is completely general, the algebraic computations soon become burdensome.

In 1654, Pascal devised a more explicit procedure which would become the standard technique used in algebra textbooks.[10] To illustrate the method in the particular case $k = 4$ we start from a special case of the binomial formula:

$$(i+1)^5 - i^5 = \binom{5}{1}i^4 + \binom{5}{2}i^3 + \binom{5}{3}i^2 + \binom{5}{4}i + 1.$$

This equation is valid for all values of i. Choosing successively $i = 1, 2, \ldots, n$ and then adding the results, we find

$$\sum_1^n [(i+1)^5 - i^5] = 5\sum_1^n i^4 + 10\sum_1^n i^3 + 10\sum_1^n i^2 + 5\sum_1^n i + \sum_1^n 1.$$

---

[10] *Oeuvres de Blaise Pascal* [331], volume III, pp. 360–363.

Therefore, since the sum on the left telescopes,

$$(n + 1)^5 - (n + 1) = 5 \sum_1^n i^4 + 10 \sum_1^n i^3 + 10 \sum_1^n i^2 + 5 \sum_1^n i.$$

Since every sum on the right other than the first is already known (as a function of n), it follows that a formula for the sum of the first n fourth powers can be found. By straightforward algebra we arrive at the result

$$\sum_1^n i^4 = \frac{n^5}{5} + \frac{n^4}{2} + \frac{n^3}{3} - \frac{n}{30}.$$

The general case follows at once by analogy.

On the basis of either of these two recursive procedures one can deduce (by a simple induction argument) that the sum of the kth powers of the first n positive integers,

$$1^k + 2^k + \cdots + n^k,$$

is expressible as a polynomial in n of degree $k + 1$, and that moreover the leading coefficient is always $1/(k + 1)$. However, neither Fermat nor Pascal would discover the general formula that links all of these simple sums.

The Quadrature of the Higher Parabolas. Both Fermat and Pascal applied their methods to give more or less rigorous proofs of what may be called the first general theorem of the integral calculus[11]

$$\int_0^a x^k \, dx = \frac{a^{k+1}}{k+1}, \tag{2}$$

or, in their language, the quadrature of the "higher parabolas" $y = x^k$ (k a positive integer). To prove (2) arithmetically requires the limit

$$\lim_{n \to \infty} \frac{1^k + 2^k + \cdots + n^k}{n^{k+1}} = \frac{1}{k+1} \tag{3}$$

---

[11] The first theorem of this sort is due to Archimedes, who in his treatise *On Spirals* used the formulas for the sum of the first n positive integers and their squares to establish quadrature results equivalent to the integrals

$$\int_0^a x \, dx = \frac{a^2}{2} \quad \text{and} \quad \int_0^a x^2 \, dx = \frac{a^3}{3}.$$

More than a thousand years later, the Arabs extended his work and their computations required formulas for the sum of the first n cubes and fourth powers. The general result (2) was first formulated by Cavalieri (1598–1647). (See, for example, [145], p. 106.)

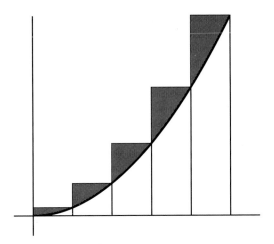

FIGURE 8

as Figure 8 makes clear.

To calculate the area of the region Ω bounded by the curve $y = x^k$ and the horizontal axis between $x = 0$ and $x = a$, we proceed in the usual fashion by subdividing the interval $[0, a]$ into n equal parts and then constructing two sets of approximating rectangles; one is a set of inscribed rectangles whose combined areas approximate the area of Ω from below, and the other is a set of circumscribed rectangles which approximate the area of Ω from above. Every circumscribed rectangle has a base equal to $\frac{a}{n}$ and their heights are $(\frac{a}{n})^k, (\frac{2a}{n})^k, \ldots, (\frac{na}{n})^k$. Therefore their combined area is

$$T_n = \frac{a}{n}\left(\frac{a}{n}\right)^k + \frac{a}{n}\left(\frac{2a}{n}\right)^k + \cdots + \frac{a}{n}\left(\frac{na}{n}\right)^k = a^{k+1}\left[\frac{1^k + 2^k + \cdots + n^k}{n^{k+1}}\right].$$

Similarly, the combined area of the inscribed rectangles is equal to

$$S_n = a^{k+1}\left[\frac{1^k + 2^k + \cdots + (n-1)^k}{n^{k+1}}\right].$$

Observe that for every n the inequalities

$$S_n < \text{Area}\,(\Omega) < T_n$$

hold. Observe also that the difference $T_n - S_n$, with all but one term canceling out, is at once seen to be equal to the area a^{k+1}/n of the largest rectangle. Thus as n increases the difference in area between the inscribed and circumscribed rectangles decreases to zero, and therefore

$$\text{Area } (\Omega) = \lim_{n \to \infty} T_n = a^{k+1} \cdot \lim_{n \to \infty} \frac{1^k + \cdots + n^k}{n^{k+1}}.$$

But we have already observed that the sum $1^k + \cdots + n^k$ can be expressed as a polynomial in n of degree $k + 1$, with leading coefficient $1/(k+1)$. Therefore,

$$1^k + 2^k + \cdots + n^k = \frac{n^{k+1}}{k+1} + \text{ lower powers of } n \qquad (4)$$

and this is precisely what is needed to establish the limit (3). Thus

$$\text{Area } (\Omega) = \frac{a^{k+1}}{k+1}.$$

Of course neither Fermat nor Pascal used limits. Fermat adhered to the strict and logically sound ancient *method of exhaustion*. [12] Generalizing the techniques introduced by Archimedes in his treatise *On Spirals,* Fermat based his proof on the inequalities

$$1^k + \cdots + (n-1)^k < \frac{n^{k+1}}{k+1} < 1^k + \cdots + n^k$$

from which (3) follows at once.

Pascal, on the other hand, made daring use of "infinitesimals." He argued roughly as follows. Imagine that the area under the curve $y = x^k$ between $x = 0$ and $x = a$ is made up of an infinite number n of "rectangular strips," each having an *infinitesimal* width $w = a/n$. The area under the curve is then

$$[w^k + (2w)^k + \cdots + (nw)^k]w = w^{k+1} \sum_{i=1}^{n} i^k$$

$$\approx \frac{(nw)^{k+1}}{k+1} = \frac{a^{k+1}}{k+1}.$$

To reach the second line Pascal has assumed that, when n is infinite, the lower powers of n appearing in the sum (4) are negligible in comparison with the first term $n^{k+1}/(k+1)$, and may therefore be omitted! "In answering the objections of those of his contemporaries who held that the omission of infinitely small quantities constituted a violation of common sense, Pascal had recourse to a favorite theme—that the heart intervenes to make this work clear"[13]

---

[12] For a discussion of the method, see [62], pp. 90ff.

[13] Carl Boyer, *The History of the Calculus and its Conceptual Development* [61], p. 150.

Ultimately the calculus was founded on the modern theory of *limits* toward which the geometrical method of exhaustion and the method of infinitesimals had both pointed.

What Bernoulli Saw.   The final step in the search for an explicit formula for all the sums

$$1^k + 2^k + \cdots + n^k$$

was taken by James Bernoulli, some fifty years after Fermat and Pascal had put forth their general recursive schemes. In his famous treatise *Ars conjectandi*, published posthumously in 1713, Bernoulli proceeded just as Fermat had done, showing how to compute the sums up to $1^3 + \cdots + n^3$. Observing that higher powers could be treated in the same way, he then gave a table *Summae Potestatum* (*Sums of Powers*) which lists all the formulas for the first ten exponents $k = 1, 2, \ldots, 10$.

Sums of Powers[14]

$$\int n \;= \tfrac{1}{2}nn \;+ \tfrac{1}{2}n,$$
$$\int nn = \tfrac{1}{3}n^3 \;+ \tfrac{1}{2}nn + \tfrac{1}{6}n,$$
$$\int n^3 \;= \tfrac{1}{4}n^4 \;+ \tfrac{1}{2}n^3 + \tfrac{1}{4}nn,$$
$$\int n^4 \;= \tfrac{1}{5}n^5 \;+ \tfrac{1}{2}n^4 + \tfrac{1}{3}n^3 \;*- \tfrac{1}{30}n,$$
$$\int n^5 \;= \tfrac{1}{6}n^6 \;+ \tfrac{1}{2}n^5 + \tfrac{5}{12}n^4 \;*- \tfrac{1}{12}nn,$$
$$\int n^6 \;= \tfrac{1}{7}n^7 \;+ \tfrac{1}{2}n^6 + \tfrac{1}{2}n^5 \;*- \tfrac{1}{6}n^3 \quad *+ \tfrac{1}{42}n,$$
$$\int n^7 \;= \tfrac{1}{8}n^8 \;+ \tfrac{1}{2}n^7 + \tfrac{7}{12}n^6 \;*- \tfrac{7}{24}n^4 \quad *+ \tfrac{1}{12}nn,$$
$$\int n^8 \;= \tfrac{1}{9}n^9 \;+ \tfrac{1}{2}n^8 + \tfrac{2}{3}n^7 \;*- \tfrac{7}{15}n^5 \quad *+ \tfrac{2}{9}n^3 \quad *- \tfrac{1}{30}n,$$
$$\int n^9 \;= \tfrac{1}{10}n^{10} + \tfrac{1}{2}n^9 + \tfrac{3}{4}n^8 \;*- \tfrac{7}{10}n^6 \quad *+ \tfrac{1}{2}n^4 \quad *- \tfrac{3}{20}nn$$
$$\int n^{10} = \tfrac{1}{11}n^{11} + \tfrac{1}{2}n^{10} + \tfrac{5}{6}n^9 \;*- 1n^7 \quad *+ 1n^5 \quad *- \tfrac{1}{2}n^3 \;*+ \tfrac{5}{66}n.$$

"Whoever will examine the series as to their regularity," wrote Bernoulli, "may be able to continue the table" ([408], p. 319). Then, offering neither proof nor explanation, he recorded the correct formula.

---

[14] Bernoulli used the symbol $\int$ for summation (for example, $\int n^3$ designates the sum of the first n cubes), and he used the symbol $*$ to indicate that there is a term with coefficient zero. His original table contains an error: the coefficient of nn in $\int n^9$ is listed as $-1/12$ instead of $-3/20$ (see, for example, [408], p. 319).

Can the reader discern the underlying rule? While the coefficients in each row appear to be unrelated, yet when they are viewed column by column an unmistakable pattern emerges. Notice to begin with that every column after the second alternates in sign, and that every formula beginning with the second terminates in either a multiple of n or n^2. The pattern that underlies the first two columns is obvious. Examining the numbers in the third column, we find

$$\frac{1}{6}, \frac{1}{4}, \frac{1}{3}, \frac{5}{12}, \frac{1}{2}, \frac{7}{12}, \frac{2}{3}, \frac{3}{4}, \frac{5}{6}.$$

Here the rule is also obvious and it is almost impossible to miss it if the foregoing ratios are expressed in terms of their least common denominator, 12. The corresponding numerators are then simply the natural numbers

$$2, 3, 4, 5, 6, 7, 8, 9, 10.$$

When the ratios in the next column are placed over their least common denominator, 120, the corresponding numerators are

$$4, 10, 20, 35, 56, 84, 120,$$

which anyone steeped in the figurate numbers will immediately recognize as the tetrahedral numbers (see Figure 5). Suspecting that Pascal's triangle may be involved throughout, we proceed to the fifth column and discover, true to form, that the coefficients are the sixth-order figurate numbers

$$6, 21, 56, 126, 252$$

divided by 252. The remaining columns maintain the pattern.

To continue the table indefinitely, all that remains is to identify the initial numbers of the columns beginning with the third. These numbers $\frac{1}{6}$, $-\frac{1}{30}$, $\frac{1}{42}$, $-\frac{1}{30}$, ... came to be known as Bernoulli numbers and they have found startling applications in both number theory and anlysis. They are not expressible by any simple formula, but they can be found recursively by the simple observation that *the sum of the coefficients in each row must be unity* (simply take $n = 1$ in each formula).

So far all this has been mere speculation and we must now provide the proof.

d. Bernoulli Polynomials

The polynomials

$$1, x, \frac{x^2}{2!}, \frac{x^3}{3!}, \frac{x^4}{4!}, \dots$$

which figure so prominently in Taylor series expansions, have the property that each is an antiderivative of the preceding one. Thus

$$\mathbf{D}(x) = 1$$
$$\mathbf{D}(x^2/2!) = x$$
$$\mathbf{D}(x^3/3!) = x^2/2!$$
$$\mathbf{D}(x^4/4!) = x^3/3!, \ldots$$

where $\mathbf{D}$ denotes the differentiation operator. In general, if P_n is the nth polynomial in the sequence, then

$$P_0(x) = 1 \quad \text{and} \quad P_n'(x) = P_{n-1}(x).$$

This important recursive formula does not determine P_n uniquely. Once $P_0, P_1, \ldots, P_{n-1}$ are known, P_n can be found only within an arbitrary constant of integration. In the simplest case, all of the constants are zero. A far more intriguing set of polynomials satisfying the same recursion is obtained by stipulating that

$$\int_0^1 P_n(x)dx = 0 \quad n = 1, 2, 3, \ldots.$$

The resulting polynomials are called **Bernoulli polynomials** and are denoted by $\mathbf{B}_n$ in honor of their discoverer, James Bernoulli.

It is at once evident that $\mathbf{B}_n$ is a polynomial of degree n with rational coefficients. The first few Bernoulli polynomials are found to be

$$\mathbf{B}_0(x) = 1$$
$$\mathbf{B}_1(x) = x - \frac{1}{2}$$
$$\mathbf{B}_2(x) = \frac{1}{2}x^2 - \frac{1}{2}x + \frac{1}{12} \tag{5}$$
$$\mathbf{B}_3(x) = \frac{1}{6}x^3 - \frac{1}{4}x^2 + \frac{1}{12}x$$
$$\mathbf{B}_4(x) = \frac{1}{24}x^4 - \frac{1}{12}x^3 + \frac{1}{24}x^2 - \frac{1}{720}.$$

To display the Bernoulli polynomials in a form in which their coefficients become particularly elegant, we shall write their constant terms as

$$\mathbf{B}_n(0) = \frac{B_n}{n!} \quad n = 0, 1, 2, \ldots.$$

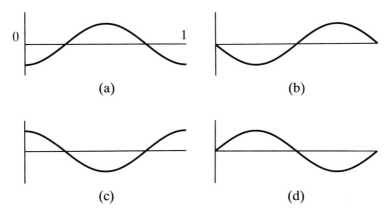

FIGURE 9
Bernoulli polynomials. The general shape of $\mathbf{B}_n$ is (a), (b), (c), or (d) depending on whether $n \equiv 0, 1, 2,$ or $3 \pmod 4$.

We then have

$$\mathbf{B}_0(x) = B_0$$

$$\mathbf{B}_1(x) = \frac{x}{1!} + \frac{B_1}{1!}$$

$$\mathbf{B}_2(x) = \frac{x^2}{2!} + \frac{B_1}{1!}\frac{x}{1!} + \frac{B_2}{2!}$$

$$\mathbf{B}_3(x) = \frac{x^3}{3!} + \frac{B_1}{1!}\frac{x^2}{2!} + \frac{B_2}{2!}\frac{x}{1!} + \frac{B_3}{3!}$$

$$\mathbf{B}_4(x) = \frac{x^4}{4!} + \frac{B_1}{1!}\frac{x^3}{3!} + \frac{B_2}{2!}\frac{x^2}{2!} + \frac{B_3}{3!}\frac{x}{1!} + \frac{B_4}{4!}$$

$$\cdots$$

and in general

$$\mathbf{B}_n(x) = \frac{1}{n!}\sum_{k=0}^{n}\binom{n}{k}B_k x^{n-k} \tag{6}$$

which is easily verified by induction.

The numbers $B_0, B_1, B_2, \ldots$ are called **Bernoulli numbers**. From (5), we see that the first few of their values are

$$B_0 = 1,\ B_1 = -\frac{1}{2},\ B_2 = \frac{1}{6},\ B_3 = 0,\ B_4 = -\frac{1}{30}.$$

They are a curious sequence of rational numbers, with a somewhat hidden law of formation. Their computation is facilitated by the observation that, for $n = 1, 2, 3, \ldots$,

$$0 = \int_0^1 \mathbf{B}_n(x)\, dx = \int_0^1 \mathbf{B}'_{n+1}(x)\, dx = \mathbf{B}_{n+1}(1) - \mathbf{B}_{n+1}(0),$$

and therefore

$$\mathbf{B}_{n+1}(0) = \mathbf{B}_{n+1}(1).$$

Therefore, by (6), it follows that

$$B_{n+1} = \sum_{k=0}^{n+1} \binom{n+1}{k} B_k \tag{7}$$

and so

$$0 = \sum_{k=0}^{n} \binom{n+1}{k} B_k.$$

This provides us with a simple recursive procedure for computing B_n once $B_0, B_1, \ldots, B_{n-1}$ are known. For example, knowing B_0, B_1, B_2, B_3, B_4, we find

$$B_5 = -\frac{\binom{6}{0}B_0 + \binom{6}{1}B_1 + \binom{6}{2}B_2 + \binom{6}{3}B_3 + \binom{6}{4}B_4}{\binom{6}{5}} = 0.$$

Table 2 gives the value of B_n for all even values of n up to 60.

It is not difficult to show that B_n is zero whenever n is odd and greater than 1 (problem 19), and that the B_n alternate in sign whenever n is even (problem 20).

We are now in a position to derive Bernoulli's elegant formula for the sum of consecutive powers. We first prove the following theorem.

Theorem. *For every real number x,*

$$\mathbf{B}_{n+1}(x+1) - \mathbf{B}_{n+1}(x) = \frac{x^n}{n!} \qquad n = 0, 1, 2, \ldots. \tag{8}$$

Proof. When $n = 0$, the right side is equal to unity and the stated relation is obvious. Proceeding by induction, we assume its truth for n and we show that

$$\mathbf{B}_{n+2}(x+1) - \mathbf{B}_{n+2}(x) = \frac{x^{n+1}}{(n+1)!}$$

for all x.

$$B_1 = -1/2$$
$$B_2 = 1/6$$
$$B_4 = -1/30$$
$$B_6 = 1/42$$
$$B_8 = -1/30$$
$$B_{10} = 5/66$$
$$B_{12} = -691/2730$$
$$B_{14} = 7/6$$
$$B_{16} = -3617/510$$
$$B_{18} = 43867/798$$
$$B_{20} = -174611/330$$
$$B_{22} = 854513/138$$
$$B_{24} = -236364091/2730$$
$$B_{26} = 8553103/6$$
$$B_{28} = -23749461029/870$$
$$B_{30} = 8615841276005/14322$$
$$B_{32} = -7709321041217/510$$
$$B_{34} = 2577687858367/6$$
$$B_{36} = -26315271553053477373/1919190$$
$$B_{38} = 2929993913841559/6$$
$$B_{40} = -261082718496449122051/13530$$
$$B_{42} = 1520097643918070802691/1806$$
$$B_{44} = -27833269579301024235023/690$$
$$B_{46} = 596451111593912163277961/282$$
$$B_{48} = -5609403368997817686249127547/46410$$
$$B_{50} = 495057205241079648212477525/66$$
$$B_{52} = -801165718135489957347924991853/1590$$
$$B_{54} = 29149963634884862421418123812691/798$$
$$B_{56} = -2479392929313226753685415739663229/870$$
$$B_{58} = 84483613334888004186204677599 4036021/354$$
$$B_{60} = -1215233140483755572040304994079820246041491/56786730$$

TABLE 2. BERNOULLI NUMBERS

We have already shown that $\mathbf{B}_n(0) = \mathbf{B}_n(1)$ whenever $n \geq 2$, and therefore both sides agree when $x = 0$. To see that they agree everywhere it suffices to show that they have the same derivative. But this is obvious, since, by the induction hypothesis,

$$\mathbf{D}\left\{\mathbf{B}_{n+2}(x+1) - \mathbf{B}_{n+2}(x)\right\} = \mathbf{B}_{n+1}(x+1) - \mathbf{B}_{n+1}(x)$$

$$= \mathbf{D}\left\{\frac{x^{n+1}}{(n+1)!}\right\}.$$

Corollary. *The sum of the nth powers of the first $N - 1$ positive integers is given by*

$$1^n + 2^n + \cdots + (N-1)^n = n!\left[\mathbf{B}_{n+1}(N) - \mathbf{B}_{n+1}(0)\right]$$

$$= n!\int_0^N \mathbf{B}_n(x)\,dx.$$

Proof. The first formula is an almost immediate consequence of (8): simply take $x = 0, 1, \ldots, N - 1$ and then add the resulting equations. We then have

$$1^n + 2^n + \cdots + (N-1)^n = n!\sum_{x=0}^{N-1}\left[\mathbf{B}_{n+1}(x+1) - \mathbf{B}_{n+1}(x)\right]$$

$$= n!\left[\mathbf{B}_{n+1}(N) - \mathbf{B}_{n+1}(0)\right].$$

The second formula follows at once from the first (by the fundamental theorem of calculus) since $\mathbf{B}'_{n+1} = \mathbf{B}_n$.

Bernoulli's formula may also be written symbolically as

$$1^n + 2^n + \cdots + (N-1)^n = \frac{1}{n+1}\left\{(N+B)^{n+1} - B^{n+1}\right\}. \qquad (9)$$

Here, the understanding is that the term $(N+B)^{n+1}$ is to be expanded formally by means of the binomial theorem, and each of the powers B^i is to be replaced by the corresponding Bernoulli number B_i. For example,

$$1^3 + 2^3 + \cdots + (N-1)^3 = \frac{1}{4}(N^4 + 4N^3B_1 + 6N^2B_2 + 4NB_3)$$

$$= \frac{1}{4}(N^4 - 2N^3 + N^2)$$

$$= \frac{1}{4}N^2(N-1)^2.$$

Notice also that by taking $N = 1$ in formula (9), we obtain the symbolic relation

$$\frac{1}{n+1}\left\{(1+B)^{n+1} - B^{n+1}\right\} = 0$$

or

$$(1+B)^{n+1} = B^{n+1} \qquad \text{for all } n = 1, 2, \ldots.$$

This is precisely the recursive formula given by (7).

But the story of the Bernoulli numbers has just begun. Tucked away within the second half of the corollary, almost as if it were an afterthought, is the statement

$$1^n + 2^n + \cdots + (N-1)^n = n! \int_0^N \mathbf{B}_n(x)\, dx. \tag{10}$$

And herein lies an idea of fundamental importance: *the representation of a sum in the form of an integral.* In the hands of the great Euler, formula (10) would lead to one of the most powerful and general known methods for approximating sums.[15]

Euler's work also led him to a beautiful relation that exists between Bernoulli numbers and the trigonometric funtions, and this in turn led him to the remarkable formula for the sum of the reciprocals of all the even powers of the integers:[16]

$$\sum_{n=1}^{\infty} \frac{1}{n^{2k}} = \frac{(-1)^{k-1} B_{2k} (2\pi)^{2k}}{2(2k)!}$$

for every positive integer k. In particular, for $k = 1, 2, 3$ we obtain the formulas

$$\sum_{1}^{\infty} \frac{1}{n^2} = \frac{\pi^2}{6}, \qquad \sum_{1}^{\infty} \frac{1}{n^4} = \frac{\pi^4}{90}, \qquad \sum_{1}^{\infty} \frac{1}{n^6} = \frac{\pi^6}{945}.$$

Despite 250 years of intense efforts, no one has ever found the exact value of any of the series

$$\sum_{1}^{\infty} \frac{1}{n^3}, \quad \sum_{1}^{\infty} \frac{1}{n^5}, \quad \sum_{1}^{\infty} \frac{1}{n^7}, \ldots$$

in terms of the familiar constants of mathematics.

---

[15] See chapter 6, §2, problem 21.

[16] See chapter 6, §2, problem 18.

In number theory, Bernoulli numbers made a startling and unexpected appearance when in 1850, Kummer showed that *Fermat's last theorem is true for every exponent that is a regular prime.* By definition, a prime $p > 2$ is regular if it does not divide any of the numerators of the Bernoulli numbers

$$B_0, B_2, B_4, B_6, \ldots, B_{p-3}$$

when these numbers are written in reduced form. With this criterion it becomes a matter of routine computation to prove, for example, that Fermat's last theorem is true for all prime exponents less than 100 except possibly for 37, 59, and 67. Unfortunately, the number of irregular primes is infinite (see, for example, [85]).

There are no solved problems; there are only problems that are more or less solved.

—Henri Poincaré

PROBLEMS

1. Consider the following table of polygonal numbers:

Number	1st	2nd	3rd	4th	5th	6th	7th	8th	9th	10th
Triangular	1	3	6	10	15	21	28	36	45	55
Square	1	4	9	16	25	36	49	64	81	100
Pentagonal	1	5	12	22	35	51	70	92	117	145
Hexagonal	1	6	15	28	45	66	91	120	153	190
Heptagonal	1	7	18	34	55	81	112	148	189	235
Octagonal	1	8	21	40	65	96	133	176	225	280

a. Show that the nth polygonal number with $d + 2$ sides is given by the formula

$$n + d\frac{n(n - 1)}{2}. \tag{11}$$

b. Show that each polygonal number equals the sum of the polygonal number immediately above it in the table and the triangular number in the preceding column. For example, the octagonal number 96 is the sum of the heptagonal number 81 and the triangular number 15.

c. Show that the numbers in each column form an arithmetical progression whose common difference is the triangular number in the preceding column.

d. Illustrate formula (11) geometrically when $d = 6$: Every octagonal number is equal to its rank plus the triangular number of the preceding rank multiplied by 6.

2. (Plutarch) If we multiply a triangular number by 8 and add 1, we obtain a square.

3. Prove geometrically that the sum of two consecutive triangular numbers is a square.

4. Prove that the product of four consecutive positive integers cannot be a square or a cube.

5. (A prize problem from the *Ladies' Diary* for 1792) Find n $(n > 1)$ such that

$$1^2 + 2^2 + \cdots + n^2 = \text{square}.$$

(There is only one solution.)

6. There are exactly 3 tetrahedral numbers that are squares. Find them. Can you use this information to solve the previous problem?

7. Given

$$b_n = \sum_{k=0}^{n} \binom{n}{k}^{-1}$$

for $n = 1, 2, 3, \ldots$, prove that

$$\lim_{n \to \infty} b_n = 2.$$

Hint. Find a simple recursive formula for b_n.

8. Let us return once again to the remarkable ladder of whole numbers

a	b
1	1
2	3
5	7
12	17
29	41
⋮	⋮

Show that by forming the product of the entries on each successive rung, 1×1, $2 \times 3, 5 \times 7, \ldots$, we obtain all the natural numbers whose squares are triangular. For example, in the first three cases,

$$T_1 = 1^2, \qquad T_8 = 6^2, \qquad T_{49} = 35^2.$$

Indication of the Proof. Begin by assuming that the alternate rungs of the ladder, $(2, 3), (12, 17), (70, 99), \ldots$, provide all the solutions of the Diophantine equation

$$b^2 - 2a^2 = 1 \tag{12}$$

(compare problem 7, chapter 1, § 1).

    a. Let a and b be natural numbers that satisfy equation (12). Show that by writing $a = 2y$ and $b = 2x + 1$, we obtain natural numbers x and y that satisfy the equation

$$\frac{x(x + 1)}{2} = y^2. \tag{13}$$

Show conversely that every solution of the Diophantine equation (13) arises in this way.

    b. Complete the proof by showing that

$$a_{2n} = 2a_n b_n \qquad n = 1, 2, 3, \ldots.$$

Remark. Euler discovered the explicit formula

$$a_n b_n = \frac{(3 + 2\sqrt{2})^n - (3 - 2\sqrt{2})^n}{4\sqrt{2}}$$

for the nth number whose square is triangular. (See, for example, the editorial note to problem E 1473 from *The American Mathematical Monthly*, 69 (1962), p. 169.)

9.   Prove that for each natural number n,

$$\binom{n+2}{3} - \binom{n}{3} = n^2.$$

Use this result to find the sum of the squares of the first n natural numbers.

10.   Try to discover a formula for the sum of the fourth powers of the first n natural numbers by computing the ratio

$$\frac{1^4 + 2^4 + \cdots + n^4}{1^2 + 2^2 + \cdots + n^2}$$

for several small values of n.

11.   Let S_k denote the sum of the kth powers of the first n natural numbers

$$S_k = 1^k + 2^k + 3^k + \cdots + n^k.$$

Show that

$$n = S_0$$

$$n^2 = 2S_1 - S_0$$

$$n^3 = 3S_2 - 3S_1 + S_0$$

$$n^4 = 4S_3 - 6S_2 + 4S_1 - S_0.$$

Guess the general law suggested by these examples, express it in suitable mathematical notation, and prove it by induction.

12.   **a.**   Show that

$$S_1 = S_1$$

$$2S_1^2 = 2S_3$$

$$4S_1^3 = 3S_5 + S_3$$

$$8S_1^4 = 4S_7 + 4S_5$$

$$16S_1^5 = 5S_9 + 10S_7 + S_5.$$

Guess the general law suggested by these examples, express it in suitable mathematical notation, and prove it by induction.

  b.   Show generally that S_{2k-1} is a polynomial in $S_1 = n(n+1)/2$, of degree k, divisible by S_1^2 provided that $2k - 1 \geq 3$. (See Pólya [341], volume I, p. 80 for the structure of S_{2k}.)

13. Observe that

$$\binom{1}{0} = \binom{2}{0}$$

$$\binom{1}{0} + \binom{2}{1} = \binom{3}{1}$$

$$\binom{1}{0} + \binom{2}{1} + \binom{3}{2} = \binom{4}{2}$$

$$\binom{1}{0} + \binom{2}{1} + \binom{3}{2} + \binom{4}{3} = \binom{5}{3}.$$

Guess the general law suggested by these examples, express it in suitable mathematical notation, and prove it by induction.

14. Find a simple formula for the sum

$$\binom{0}{k} + \binom{1}{k} + \binom{2}{k} + \cdots + \binom{n}{k}$$

for $0 \leq k \leq n$, $n = 1, 2, 3, \ldots$. In proving it by mathematical induction, is it simpler to proceed from n to $n+1$ or from k to $k+1$?

15. Establish the following identities:

  a. $\displaystyle \binom{n}{0} + \binom{n}{1} + \binom{n}{2} + \cdots + \binom{n}{n} = 2^n.$

  b. $\displaystyle \binom{n}{0}^2 + \binom{n}{1}^2 + \binom{n}{2}^2 + \cdots + \binom{n}{n}^2 = \binom{2n}{n}.$

16. Find a simple formula for the sum

$$\binom{n}{1} + 2\binom{n}{2} + 3\binom{n}{3} + \cdots + n\binom{n}{n}.$$

17. There is no closed form for the partial sum of a row of Pascal's triangle. Show, however, that there is a simple expression for the alternating sum

$$\binom{n}{0} - \binom{n}{1} + \binom{n}{2} - \binom{n}{3} + \cdots + (-1)^k \binom{n}{k}$$

for $0 < k \leq n$, $n = 1, 2, 3, \ldots$.

18. Establish the following remarkable property of the central elements of Pascal's triangle:

$$\sum_{k=0}^{n} \binom{2k}{k} \binom{2n - 2k}{n - k} = 4^n.$$

For example, $\binom{0}{0}\binom{6}{3} + \binom{2}{1}\binom{4}{2} + \binom{4}{2}\binom{2}{1} + \binom{6}{3}\binom{0}{0} = 4^3$.

19. Prove that $\mathbf{B}_n(1-x) = (-1)^n \mathbf{B}_n(x)$. Conclude that $B_{2n+1} = 0$ for $n > 0$.

20. Investigate the pattern of signs of $\mathbf{B}_n(x)$ for $0 \leq x \leq 1$. In particular, prove that the signs of B_{2n} and B_{2n+2} are opposite for $n > 0$.

21. A famous theorem due to Christian von Staudt (1798–1867) asserts that every Bernoulli number B_n (n even) can be expressed in the form

$$B_n = c_n - \sum \frac{1}{k + 1},$$

where c_n is an integer and the summation extends over all $k > 0$ such that k is a divisor of n and $k + 1$ is a prime number. For example, when $n = 6$, the divisors of n are 1, 2, 3, and 6. Increasing each of them by unity, we obtain the numbers 2, 3, 4, and 7. Of these, 2, 3, and 7 are primes, and therefore we find

$$B_6 = 1 - \frac{1}{2} - \frac{1}{3} - \frac{1}{7} = \frac{1}{42}.$$

Verify von Staudt's theorem for $n = 2, 4, \ldots, 20$.

3. Two Theorems of Fermat, the "Little" and the "Great"

a. The Little Theorem and the Search for Perfect Numbers

> It is difficult if not impossible to state why some theorems in arithmetic are considered "important" while others, equally difficult to prove, are dubbed trivial. One criterion, although not necessarily conclusive, is that the theorem shall be of use in other fields of mathematics. Another is that it shall suggest researches in arithmetic or in mathematics generally, and a third that it shall be in some respects universal. Fermat's theorem ... satisfies all of these somewhat arbitrary demands ... it is universal in the sense that it states a property of all prime numbers—such general statements are extremely difficult to find and very few are known.
>
> —E. T. Bell, Men of Mathematics

In order to put Fermat's "little" theorem in its proper historical context, and not simply have it be an isolated gem from the theory of numbers, we must first go back to classical antiquity and the numerology of the ancients.

The ancient Greeks attached special significance—mystical significance—to a class of numbers they identified with perfection. A natural number is said to be **perfect** if it is equal to the sum of all its divisors, other than itself. Thus the first perfect number is 6, for $6 = 1 + 2 + 3$.

The number mysticism of the ancients was echoed by St. Augustine who said: "Six is a number perfect in itself, and not because God created all things in six days; rather the inverse is true; God created all things in six days because this number is perfect. And it would remain perfect even if the work of the six days did not exist."

The next perfect number after 6 is $28 = 1 + 2 + 4 + 7 + 14$.

"Perfect numbers have engaged the attention of arithmeticians of every century of the Christian era."[17] While they no longer occupy a prominent role in modern mathematics, they have engendered ideas of fundamental importance. It was while investigating them that Fermat discovered the theorem that bears his name and upon which a large part of number theory now rests. A century later, Euler took up and extended the study of perfect numbers, and the

---

[17] L. E. Dickson, *History of the Theory of Numbers* [135], volume I, p. III.

methods he devised form the basis for the modern theory of the distribution of the primes.

There is only one known criterion for producing perfect numbers and it appears in Euclid's *Elements*: *The number*

$$N = 2^{n-1}(2^n - 1)$$

is perfect whenever the second factor $2^n - 1$ *is a prime*. The proof is simple. Setting $p = 2^n - 1$, we partition the divisors of N (including N itself) into two groups, according to whether or not they contain p as a factor:

$$1, 2, 2^2, 2^3, \ldots, 2^{n-1}$$

and

$$p, 2p, 2^2 p, 2^3 p, \ldots, 2^{n-1} p.$$

The sum of the elements in the first group is equal to $1+2+2^2+2^3+\cdots+2^{n-1} = 2^n - 1 = p$, and therefore the sum of those in the second group is just p^2. Adding the two results, we find that the sum of all the divisors of N is equal to

$$p + p^2 = (1 + p)p = 2^n p = 2N.$$

Thus N is perfect.

The first four values of n for which $2^n - 1$ is a prime are 2, 3, 5, and 7, and they give rise to the perfect numbers

$$6 = 2(2^2 - 1)$$

$$28 = 2^2(2^3 - 1)$$

$$496 = 2^4(2^5 - 1)$$

$$8128 = 2^6(2^7 - 1).$$

These were known to the ancients. Two thousand years later Euler showed that *every even perfect number is necessarily of Euclid's form* (see [211], p. 240). No one has ever found an odd perfect number, and it seems unlikely that one exists, but this has not been proved. The question remains one of the oldest unsolved problems in mathematics.[18]

Prior to the 17th century, little about perfect numbers was known. Patient and prodigious calculators labored with little more than trial division in their

---

[18] Based on the work of Brent, Cohen, and te Riele, it is now known that there are no odd perfect numbers less than 10^{300} [66].

efforts to factor numbers of the form $2^n - 1$, and the results they obtained were seldom free from error. Any theory of factorization was still lacking. To the four perfect numbers known since antiquity the Renaissance added just three more.

The 17th century brought forth Fermat and with him number theory as a systematic science. It was while investigating perfect numbers that Fermat was led to one of the remarkable properties of the integers.

Theorem (Fermat's Little Theorem).   *If p is any prime and a is any integer, then*

$$a^p - a$$

is divisible by p.

In the language of congruences, Fermat's theorem takes the form

$$a^p \equiv a \pmod{p}.$$

If a is a multiple of p, the result is trivial; otherwise, we may cancel the factor a and obtain the equivalent statement

$$a^{p-1} \equiv 1 \pmod{p}.$$

This means that the $(p-1)$st power of a leaves the remainder 1 upon division by p.

Before proceeding with the proof, we will find it instructive to indicate how one might discover the result "experimentally," for instance in the case $a = 2$, which is the relevant one for perfect numbers.

The most natural way to factor $2^n - 1$ (or any number for that matter) is to successively divide it by all the primes $2, 3, 5, 7, 11, \ldots$, up to the square root of the given number. However, even when n is small, the process of trial division may prove tedious at best or intractable at worst. Even to establish the primality of

$$2^{31} - 1 = 2147483647$$

by this method requires knowledge of nearly five thousand primes. The reasonable solution is not always the most effective. Fermat's disarmingly simple technique was to change the point of view—rather than fix the number, he fixed the divisor. The question then becomes: *For a given prime $p > 2$, which numbers of the form $2^n - 1$ are divisible by p?* Or, what is the same thing, which powers of 2 leave the remainder 1 when divided by p?

Accordingly, we begin by dividing each term of the geometric progression

$$2, 2^2, 2^3, 2^4, \ldots$$

by p. Then the remainders must repeat themselves, so that for some m and n, we have $2^{m+n} \equiv 2^m \pmod{p}$, or

$$2^n \equiv 1 \pmod{p}.$$

Thus for each prime $p > 2$ there is a value of n for which $2^n - 1$ is divisible by p. The smallest positive n for which this is true is called the *order* of 2 modulo p; we shall denote it by d.

For example, if $p = 11$, then the powers of 2 modulo p are

$$2, 4, 8, 5, 10, 9, 7, 3, 6, 1, 2, 4, \ldots.$$

Each one is twice the previous one, with 11 subtracted if necessary to make the result less than 11. The first power of 2 that is congruent to 1 is 2^{10}, so in this case $d = 10$.

It is clear that no matter what the value of p, the remainders always show a cyclic pattern; when we have reached the first number n for which $2^n \equiv 1$, then $2^{n+1} \equiv 2$ and the previous cycle is repeated. Thus $2^n \equiv 1 \pmod{p}$ *if and only if n is a multiple of d.*

The following table lists the values of d for various values of p:

p:	3	5	7	11	13	17	19	23	29	31	37	41	43	47
d:	2	4	3	10	12	8	18	11	28	5	36	20	14	23

For example, the last column shows that $2^n - 1$ is divisible by 47 if and only if n is a multiple of 23.

What the evidence in the table suggests, and what Fermat discovered, is that $p - 1$ is a multiple of d. Or, what is the same thing,

$$2^{p-1} \equiv 1 \pmod{p}.$$

This is Fermat's theorem for $a = 2$. But the number 2 has played no special role in our argument and the same reasoning leads to the general result

$$a^{p-1} \equiv 1 \pmod{p},$$

provided of course that a is not a multiple of p. "Tentando comprobari potest", Euler wrote in his very first paper on number theory. "Its truth can be verified by experiment."

Fermat communicated this result in a letter to Frénicle de Bessy dated 18 October 1640, stating that he had found a proof. Unfortunately, as with most of Fermat's discoveries, the proof was never published or preserved. The following simple proof, based only on the binomial theorem and mathematical induction was discovered by Euler in 1735. Write

$$(a+1)^p = a^p + \binom{p}{1}a^{p-1} + \binom{p}{2}a^{p-2} + \cdots + \binom{p}{p-1}a + 1.$$

Since p is a prime, all the binomial coefficients on the right-hand side are multiples of p, and therefore

$$(a+1)^p \equiv a^p + 1 \pmod{p}.$$

This shows that if $a^p \equiv a \pmod{p}$, then also $(a+1)^p \equiv a+1 \pmod{p}$. But $1^p \equiv 1$ is trivial, and hence by the principle of mathematical induction, $a^p \equiv a$ $\pmod{p}$ for every positive integer a. (The case in which a is negative is left to the reader.)

There is another more natural method of proof, due to Ivory (1806)—and later rediscovered by Dirichlet (1828)—which is based on the multiplicative properties of the integers modulo p, and which is applicable in many similar settings.

If p is a prime and a is not a multiple of p, then it is impossible that $ax \equiv ay$ $\pmod{p}$ unless $x \equiv y \pmod{p}$. Therefore, the integers

$$a, 2a, 3a, \ldots, (p-1)a$$

must be congruent (in some order) to the numbers

$$1, 2, 3, \ldots, p-1.$$

It follows that

$$a \cdot 2a \cdot 3a \cdots (p-1)a \equiv 1 \cdot 2 \cdot 3 \cdots (p-1) \pmod{p}.$$

Cancelling the factors $2, 3, \ldots, p-1$, as is permissible, we obtain

$$a^{p-1} \equiv 1 \pmod{p},$$

which is Fermat's theorem.

Numbers of the form $2^n - 1$ are called **Mersenne numbers** and are denoted by M_n in honor of Father Marin Mersenne (1588–1648), a frequent correspondent of Fermat and one of the many mathematicians who studied them. No complete rule for generating Mersenne primes has ever been found. For small values of n one finds relatively many such primes, but for larger n they seem to become more and more scarce. While it is easy to see that $2^n - 1$ can never be a prime when n is composite, it seems unreasonable that so many prime exponents should produce so many composite results. At present only 32 Mersenne primes are known.

In 1876, Lucas discovered a remarkable method for testing whether M_p is a prime and used it to establish the primality of M_{127}, a number of 39 digits:

$$M_{127} = 2^{127} - 1 = 170,141,183,460,469,231,731,687,303,715,884,105,727.$$

This was the largest prime discovered before the advent of computers. In 1930, D. H. Lehmer published an improved version of Lucas's algorithm, and since then the Lucas–Lehmer test has consistently produced the largest known primes.

Lucas-Lehmer Test. *If $p > 2$ is prime, then $M_p = 2^p - 1$ is prime if and only if it is a divisor of the $(p-1)$st term of the sequence $r_1, r_2, r_3, \ldots$, where*

$$r_1 = 4 \qquad \text{and} \qquad r_{n+1} = r_n^2 - 2.$$

For a proof using only elementary principles of number theory, see [368].

It is not known whether there are infinitely many Mersenne primes, nor is it known whether there are infinitely many Mersenne composites. It had been widely conjectured that if p is a Mersenne prime, then so is M_p, and this would have settled half the issue. Unfortunately, the conjecture proved false. In 1953, D. J. Wheeler showed that while $M_{13} = 8191$ is a prime, M_{8191} is not. The calculation involved was done by the electronic computer Illiac-I (utilizing the Lucas–Lehmer test) and required 100 hours. It is evidence of the extraordinary growth of computer power in the past 40 years that the same calculation performed today on the Cray-1 would take no more than 10 seconds!

Another famous conjecture, still undecided, asserts that every number in the sequence $u_1, u_2, u_3, \ldots$ which is defined recursively by

$$u_1 = 2 \qquad \text{and} \qquad u_{n+1} = 2^{u_n} - 1$$

is a prime number. This has been verified for the first five terms

$$u_1 = 2, \quad u_2 = 3, \quad u_3 = 7, \quad u_4 = 127,$$

$$u_5 = 2^{127} - 1,$$

but beyond this nothing is known. The next term, u_6, has more than 10^{37} digits and any analysis of its character seems hopelessly out of reach. The search continues.

The 32 Known Mersenne Primes

p	$2^p - 1$ (prime)	Date of Discovery	Discoverer	Machine Used
2	3			
3	7	Antiquity	Mentioned in Euclid's Elements	
5	31			
7	127			
13	8191	1461	Mentioned in Codex Lat. Monac. 14908	
17	131071	1588	P. Cataldi	
19	524287	1588	P. Cataldi	
31	2147483647	1750	L. Euler	
61	19 digits	1883	I. Pervouchine	
89	27 digits	1911	R. Powers	
107	33 digits	1914	E. Fauquembergue	
127	39 digits	1876	E. Lucas	
521	157 digits	1952	R. Robinson	SWAC
607	183 digits	1952	R. Robinson	SWAC
1279	386 digits	1952	R. Robinson	SWAC
2203	664 digits	1952	R. Robinson	SWAC
2281	687 digits	1952	R. Robinson	SWAC
3217	969 digits	1957	H. Riesel	BESK
4253	1281 digits	1961	A. Hurwitz	IBM 7090
4423	1332 digits	1961	A. Hurwitz	IBM 7090
9689	2917 digits	1963	D. Gillies	ILLIAC II
9941	2993 digits	1963	D. Gillies	ILLIAC II
11213	3376 digits	1963	D. Gillies	ILLIAC II
19937	6002 digits	1971	B. Tuckerman	IBM 360/91
21701	6533 digits	1978	L. Nickel, C. Noll	CDC Cyber 174
23209	6987 digits	1979	C. Noll	CDC Cyber 174
44497	13395 digits	1979	H. Nelson, D. Slowinski	Cray-1
86243	25962 digits	1982	D. Slowinski	Cray-1
110503	33265 digits	1988	W. Colquitt, L. Welsh, Jr.	NEC SX-2
132049	39751 digits	1983	D. Slowinski	Cray-1
216091	65050 digits	1985	D. Slowinski	Cray X-MP
756839	227832 digits	1992	D. Slowinski, P. Gage	Cray-2

b. Two Applications

Wilson's Theorem.   Among the many and varied applications of Fermat's little theorem we single out one that is itself of fundamental importance—Wilson's theorem. It was recorded without proof by the 18th century English mathematician Edward Waring in his book *Meditationes Algebraicae* (1770) and attributed to Sir John Wilson, a lawyer and amateur scientist who had studied mathematics at Cambridge. It is a most remarkable result—a simple property shared by all prime numbers and by prime numbers alone. Although it is now known that the same property was observed nearly a century earlier by Leibniz, it is still ascribed to Wilson. The first published proof of Wilson's theorem was given by Lagrange in 1771.

Wilson's Theorem.   *If p is a prime, then $(p-1)! + 1$ is always divisible by p, or, in congruence notation,*

$$(p-1)! \equiv -1 \pmod{p}. \tag{1}$$

Proof.   The following proof relies on an elementary property of algebraic congruences. *If P is a polynomial of degree n with integer coefficients and if the congruence*

$$P(x) \equiv 0 \pmod{p} \tag{2}$$

has n different solutions

$$\alpha_1, \alpha_2, \ldots, \alpha_n,$$

no two of which are congruent modulo p, then P can be represented as a product of n linear factors

$$P(x) \equiv A(x - \alpha_1)(x - \alpha_2) \cdots (x - \alpha_n) \pmod{p}.$$

The proof is the same as that of the corresponding result for equations. For if α satisfies the congruence (2), then $P(x) \equiv P(x) - P(\alpha) = (x - \alpha)Q(x)$, where Q is a polynomial of degree $n - 1$. Notice that

$$Q(x) = \frac{P(x) - P(\alpha)}{x - \alpha}$$

has integer coefficients since, by ordinary algebra, the polynomials

$$\frac{x^k - \alpha^k}{x - \alpha} = x^{k-1} + x^{k-2}\alpha + \cdots + x\alpha^{k-2} + \alpha^{k-1}$$

also have integer coefficients. Suppose now that β is any other solution of the congruence (2) such that $\beta \not\equiv \alpha$. Then $P(\beta) \equiv (\beta - \alpha)Q(\beta) \equiv 0$. But a prime cannot divide a product without dividing at least one of the factors. Since p does not divide $\beta - \alpha$ it must therefore divide $Q(\beta)$. Thus $Q(\beta) \equiv 0$ and β gives rise to a factor $x - \beta$ of Q; we then have two linear factors for the original polynomial. Continuing in this fashion we arrive, after n steps, at the desired representation.

By Fermat's little theorem we know that the congruence

$$x^{p-1} - 1 \equiv 0 \quad (\text{mod } p)$$

has the $p - 1$ solutions $x = 1, 2, 3, \ldots, p - 1$, and so, by the preceding remarks, we can write

$$x^{p-1} - 1 \equiv (x - 1)(x - 2)\cdots(x - (p - 1)) \quad (\text{mod } p)$$

where the relation is valid for every integer x. Choosing $x = 0$ we find

$$-1 \equiv (-1)(-2)(-3)\cdots(-(p - 1))$$
$$= (-1)^{p-1}(p - 1)! \quad (\text{mod } p),$$

and so, when p is an odd prime,

$$(p - 1)! \equiv -1 \quad (\text{mod } p).$$

This congruence is clearly satisfied when $p = 2$, and the proof is complete.

The converse of Wilson's theorem is almost trivial. For if p is not a prime, then it has a divisor d different from 1 and p. Since $(p - 1)!$ is divisible by d, it follows that $(p - 1)! + 1$ cannot be divisible by d, and so, *a fortiori,* neither can it be divisible by p. Expressed contrapositively: If $(p - 1)! + 1$ is divisible by p, then p is a prime.

Thus Wilson's congruence provides a complete and elegant characterization for primality. It is all the more remarkable that a congruence that cannot distinguish primes from composites should give rise to one that can. But the greater generality is achieved at the expense of utility. This happens not infrequently. Any method comprehensive enough to encompass all the primes is likely to be of little practical value, while the most useful techniques are invariably the most restrictive. Thus the efficacy of Wilson's theorem, which could at least in principle be universally applied is nil, while the Lucas–Lehmer test, applicable only to the Mersenne numbers, has for the past century uncovered the largest known primes. Great generality and great utility are usually incompatible.

On Primes in Certain Arithmetical Progressions.   Dirichlet's theorem on the existence of primes in certain arithmetical progressions (see §1) lies very deep, but there is one case of particular importance that is well within our reach. Let us suppose that the odd numbers have been partitioned into the two groups

$$1, 5, 9, 13, 17, 21, \ldots \quad \text{and} \quad 3, 7, 11, 15, 19, 23, \ldots.$$

The first group consists of all numbers of the form $4n + 1$, while the second group consists of all numbers of the form $4n - 1$. We shall show that each of these arithmetical progressions contains infinitely many primes. The second progression can be handled by means of a simple modification of Euclid's famous argument. Let P be any prime of the form $4n-1$ and consider the number

$$N = 4(3 \cdot 7 \cdot 11 \cdots P) - 1$$

which is four times the product of all the primes in the second progression up to P decreased by one. Then N is certainly not divisible by any of the primes $3, 7, 11, \ldots, P$. Accordingly, N is either a prime (also of the form $4n - 1$ and much larger than P) or else N has a prime factor different from $3, 7, 11, \ldots, P$. But not every prime factor of N can be of the form $4n + 1$ because the product of two such numbers is again of the same form: namely,

$$(4m + 1)(4n + 1) = 4(4mn + m + n) + 1.$$

Therefore N must have some prime factor of the form $4n - 1$ and this factor is larger than P. Thus no matter how large P may be, there is always a larger prime of the same form. This proves that the number of primes in the second progression is infinite.

   To prove that there are infinitely many primes in the first progression is not quite so simple. If we try to apply the same argument we fail, for a number of the form $4n + 1$ need not have a prime factor of the same form. The remedy is to look for another class of numbers which are rich in prime factors of the desired type.

Lemma.   *Any number of the form*

$$a^2 + 1 \qquad a > 1$$

has a prime factor of the form $4n + 1$.

Proof.   It is easy to see that $a^2 + 1$ cannot be a power of 2 and hence it must have at least one prime factor $p > 2$. Suppose that p were of the form $4n - 1$.

By assumption

$$a^2 \equiv -1 \pmod{p}$$

and therefore, upon raising each side to the power $(p-1)/2$, we see that

$$a^{p-1} \equiv (-1)^{(p-1)/2} \equiv -1 \pmod{p}$$

since $(p-1)/2$ is odd. But this contradicts Fermat's little theorem, and hence p cannot be of the form $4n-1$. This proves that any number of the form a^2+1 is composed entirely of primes of the form $4n+1$, together possibly with the number 2.

We can now prove that there are infinitely many primes of the form $4n+1$. Let P be any prime of this form and put

$$N = (5 \cdot 13 \cdot 17 \cdots P)^2 + 1,$$

where the numbers inside the parentheses are all the primes in the first progression up to P. By virtue of the lemma, N has at least one prime factor of the form $4n+1$. Clearly this factor cannot be any of the primes $5, 13, 17, \ldots, P$. Thus, no matter how large P may be there is always a larger prime of the same form. This proves Dirichlet's theorem for numbers of the form $4n+1$.

c. Fermat's Great Theorem

In the Theory of Numbers it happens rather frequently that, by some unexpected luck, the most elegant new truths spring up by induction.

—Carl Friedrich Gauss[19]

Let us return once again to the Pythagorean relation

$$a^2 + b^2 = c^2$$

whose solutions in integers we have already found (see chapter 1, §4). If we omit the requirement that the hypotenuse c be an integer, then

$$n = c^2 = a^2 + b^2$$

is the sum of two squares, and we are led to ask what numbers n possess this property.

---

[19] *Werke,* volume 2, p. 3. The English translation given here is from [338], p. 59.

The question is a very old one—it originates in the *Arithmetica* of Diophantus—but the answer was not found until the seventeenth century. It was given by the Dutch mathematician Albert Girard in 1625, but it is Fermat who first claims possession of the proof.

Since the primes are the multiplicative building blocks for all the integers, it is only natural to first ask: *When is a prime representable as a sum of two squares?* Trivially, $2 = 1^2 + 1^2$. Examining all the odd primes up to 100, we obtain the following empirical evidence (the dash indicates that such a representation is impossible).

$$
\begin{array}{ll}
3 = \; - & 43 = \; - \\
5 = 1 + 4 & 47 = \; - \\
7 = \; - & 53 = 4 + 49 \\
11 = \; - & 59 = \; - \\
13 = 4 + 9 & 61 = 25 + 36 \\
17 = 1 + 16 & 67 = \; - \\
19 = \; - & 71 = \; - \\
23 = \; - & 73 = 9 + 64 \\
29 = 4 + 25 & 79 = \; - \\
31 = \; - & 83 = \; - \\
37 = 1 + 36 & 89 = 25 + 64 \\
41 = 16 + 25 & 97 = 16 + 81
\end{array}
$$

What does the evidence suggest? What distinguishes the two sets of primes

$$5, 13, 17, 29, \ldots \quad \text{and} \quad 3, 7, 11, 19, \ldots?$$

We may notice that the primes of the first set, beginning with 5, are all of the form $4n + 1$ with integral n. The primes of the second set, beginning with 3, are all of the form $4n - 1$. Is this the relevant distinction we are looking for? On the basis of the evidence, it seems not unreasonable that we should formulate the following conjecture: *A prime of the form $4n + 1$ is representable as the sum of two squares; a prime of the form $4n - 1$ is never representable in this way.*

The second part of the conjecture is plainly obvious, since any square leaves the remainder 0 or 1 when divided by 4. The first part is one of the most famous theorems in the theory of numbers. Fermat called it the "fundamental theorem on right-angled triangles."

Theorem (Fermat's Great Theorem).   *Every prime of the form $4n + 1$ is a sum of two squares.*

Proof.   It is convenient to present the proof in two stages.

First stage.   The first stage consists in showing that there is a number x such that

$$x^2 \equiv -1 \quad (\text{mod } p) \tag{3}$$

whenever p is a prime of the form $p = 4n + 1$.

By virtue of Wilson's theorem,

$$1 \cdot 2 \cdot 3 \cdots 4n \equiv -1 \quad (\text{mod } p).$$

But $4n \equiv -1$, $4n - 1 \equiv -2$, and so on down to $2n + 1 \equiv -2n$. When these values are substituted above, the result is

$$(1 \cdot 2 \cdot 3 \cdots 2n)^2 \equiv -1 \quad (\text{mod } p)$$

since the number of negative signs introduced is $2n$, which is even. Thus, we have provided an explicit solution for the congruence (3).

We now know that *some multiple of p is equal to the sum of two squares* (one of which is the number 1). One way to proceed is to show that if

$$mp = x^2 + y^2$$

for some natural number $m > 1$, then there is always a smaller (natural) multiple of p which is also the sum of two squares. Applying this result repeatedly, we ultimately arrive at p itself, which must then be the sum of two squares. We shall not pursue this method here, but the details may be found in Davenport [120], pp. 117ff.

Paul Erdős once remarked that every theorem has at least one "right" proof.[20] In this case, the right proof of Fermat's great theorem—the most natural proof—resides in the domain of the *Gaussian integers*. These are the complex numbers of the form

$$a + bi$$

where a and b are ordinary integers and $i = \sqrt{-1}$ is the imaginary unit.

Gaussian integers behave algebraically in much the same way as do ordinary integers. They can be added and multiplied according to the usual rules of

---

[20] Erdős, an agnostic, likes to speak of God's having a transfinite book of all mathematical theorems and their best possible proofs. "And if he is well intentioned, he gives us the book for a moment." ([2], p. 87) Gauss's proof of Fermat's great theorem is certainly "straight from the book." (Compare problem 24.)

complex arithmetic, and the results obtained are again of the same form. Division can also be defined within this class, as can the notion of a *prime*. There are four numbers in the new system, ± 1 and $\pm i$, which play a role comparable to that of 1 and -1 in ordinary arithmetic. They are the Gaussian *units*. If the *norm* of $\alpha = a + bi$ is defined by

$$N(\alpha) = a^2 + b^2$$

—so that geometrically $N(\alpha)$ represents the square of the distance from α to the origin—then the units are precisely those Gaussian integers whose norm is 1. They are considered not to be primes, in the same way that in ordinary arithmetic the number 1 is considered not to be a prime.

Now, it can be shown that the analogue of the fundamental theorem of arithmetic remains valid within the class of Gaussian integers. That is to say: *Every Gaussian integer that is not a unit can be represented as a product of Gaussian primes, and, apart from the order of the factors (and any complications that may arise from the units) in one way only.* (See, for example, [211], pp. 185ff.)

With these preliminary remarks, we are now ready to give Gauss's remarkable proof of Fermat's great theorem.

Second stage.   From the first part of the proof, we know that, if p is a prime of the form $4n + 1$, then there is a natural number x such that

$$1 + x^2 = (1 + ix)(1 - ix)$$

is divisible by p. Clearly p does not divide either $1 + ix$ or $1 - ix$. But *a number that divides a product without dividing any of its factors cannot be a prime.* Therefore, p cannot be a Gaussian prime, and hence we may write

$$p = \alpha\beta$$

where $N(\alpha) > 1$ and $N(\beta) > 1$. Now, it is a simple matter to show that the norm of a product of two Gaussian integers is equal to the product of their respective norms. Therefore

$$N(\alpha)N(\beta) = N(p) = p^2.$$

It follows from this that $N(\alpha)$ and $N(\beta)$ must each be equal to p. Writing

$$\alpha = a + bi,$$

we have

$$p = N(\alpha) = a^2 + b^2,$$

which is Fermat's theorem.

Thus Fermat's great theorem is equivalent to the fact that all primes of the form $4n + 1$ can be factored with complex integers. Primes of the form $4n - 1$ cannot be factored and remain prime within the larger system.

"Algebra is generous, she often gives more than is asked of her."[21]

Like many great artists, whose work is valued only after their time, so Fermat had to reconcile himself to the total lack of interest that his contemporaries showed toward his number theory. In one of his final letters to Huygens (August 1659), he concludes with the following melancholy words:

> Such is in brief the tale of my musings on numbers. I have put it down only because I fear that I shall never find the leisure to write out and expand properly all these proofs and methods. Anyway this will serve as a pointer to men of science for finding by themselves what I am not writing out, particularly if Monsieur de Carcavi and Frenicle communicate to them a few proofs by descent that I have sent them on the subject of some negative propositions. Maybe posterity will be grateful to me for having shown that the ancients did not know everything, and this account may come to be regarded by my successors as the "handing on of the torch", in the words of the great Chancellor of England [Bacon], following whose intention and motto I shall add: many will pass away, science will grow.[22]

PROBLEMS

1. a. Show that every even perfect number is triangular.

b. Show that every even perfect number greater than 6 is the sum of consecutive odd cubes.

2. With the single exception of the number 6, if one adds up the digits of any even perfect number, then adds up the digits of the resulting sum, and so on, the net result is always 1. For example: $4 + 9 + 6 = 19$, $1 + 9 = 10$, $1 + 0 = 1$.

---

[21] D'Alembert, quoted in *Bulletin of the American Mathematical Society,* 2 (1905), p. 285.

[22] *Oeuvres de Fermat,* volume II, p. 436. The English translation given here is from Weil, *Number Theory: An Approach through History* [444], pp. 118ff.

3. Show that no proper divisor of a perfect number is perfect.

Hint. Let $\sigma(n)$ denote the sum of the divisors of a natural number n. Show that if m is a proper divisor of n, then

$$\frac{\sigma(n)}{n} > \frac{\sigma(m)}{m}.$$

(The quotient $\sigma(n)/n$ is the sum of the reciprocals of all the divisors of n.)

4. Prove that a square integer is not a perfect number.

5. Unsolved Conjecture of Catalan. Let $f(n) = \sigma(n) - n$, where $\sigma(n)$ denotes the sum of the divisors of n. Is it true that the infinite sequence

$$n, f(n), f(f(n)), f(f(f(n))), \ldots$$

of consecutive iterates of f is ultimately periodic for all values of n? For example, when $n = 95$, we find $f(95) = 25$, $f(25) = 6$, $f(6) = 6$, and the sequence is periodic from the fourth term onwards, the period consisting of one term only.

6. Show that every prime number except 2 and 5 divides infinitely many of the integers

$$1, 11, 111, 1111, \ldots.$$

7. Fermat's little theorem is the cornerstone of almost all modern primality tests and compositeness tests that do not depend on factoring. If a and n are relatively prime and if

$$a^{n-1} \not\equiv 1 \pmod{n},$$

then n is certainly not a prime.

a. Prove in this way that the fifth Fermat number $F_5 = 2^{32} + 1$ is composite.

b. The solution to part (a) tells us that F_5 is composite without offering us the slightest clue as to what its factors are. Show, however, that if p is a prime divisor of F_5, then $p - 1$ must be a multiple of 64.

Hint. Observe that p divides

$$2^{64} - 1 = (2^{32} + 1)(2^{32} - 1).$$

8. Prove that $2^{37} - 1$ is not a prime.

Hint. If p is a prime divisor of $2^n - 1$ (n a prime), then $p - 1$ is a multiple of n.

9. The converse of Fermat's little theorem is false: The congruence

$$2^{n-1} \equiv 1 \pmod{n} \tag{4}$$

cannot guarantee that n is a prime. Odd composite numbers n for which (4) is true are called **pseudoprimes**, meaning "false primes." The smallest pseudoprime, 341, was discovered in 1819 by the French mathematician Pierre Sarrus (see footnote 1, chapter 1).

    a.   Find the next smallest pseudoprime.

    b.   Show that the number of pseudoprimes is infinite.

    Hint. If n is a pseudoprime, then $2^n - 1$ is also a pseudoprime.

    c.   Show that all Fermat numbers and all Mersenne numbers $2^p - 1$ (p a prime) satisfy the congruence (4).

    Remarks. Pomerance, Selfridge, and Wagstaff [348] have tabulated all the pseudoprimes below 25×10^9. They are very rare, much rarer than primes. For example, there are only 245 pseudoprimes below a million compared with 78498 primes. It has been shown that there are infinitely many even numbers n for which $2^n - 2$ is divisible by n [30], but it was not until 1950 that one of them was found (see [385], p. 230).

10. (Continuation) Prove that the number $n = 561$ has the remarkable property that

$$a^{n-1} \equiv 1 \pmod{n} \tag{5}$$

for every integer a that is relatively prime to n. (It is the smallest number with this property.) Odd composite numbers n for which (5) is true are called **Carmichael numbers**, after the American mathematician R. D. Carmichael, who was the first to notice their existence in 1909. They are extraordinarily rare. For example, there are only 2163 Carmichael numbers below 25×10^9 (compared with 21853 pseudoprimes). It is widely believed that there are infinitely many Carmichael numbers, but this has not been proved.

11. Show that if p is a prime, then

$$1^{p-1} + 2^{p-1} + \cdots + (p-1)^{p-1} \equiv -1 \pmod{p}.$$

It is not known whether this relation holds for any composite number p. (See Guy, *Unsolved Problems in Number Theory* [191], p. 22.)

12. (Euler) Prove that a natural number is composite if it has two different representations as the sum of two squares. For example, the fifth Fermat number $F_5 = 2^{32} + 1$ cannot be a prime since

$$(2^{16})^2 + 1 = 62264^2 + 20449^2.$$

13. Euler's generalization of Fermat's little theorem. For a positive integer n, the **Euler phi-function** $\phi(n)$ is defined to be the number of positive integers not exceeding n that are relatively prime to n (see problem 11, chapter 1, §3). The first twenty values of $\phi(n)$ are tabulated below. Notice that $\phi(n) = n - 1$ if and only if n is a prime.

n	$\phi(n)$	n	$\phi(n)$
1	1	11	10
2	1	12	4
3	2	13	12
4	2	14	6
5	4	15	8
6	2	16	8
7	6	17	16
8	4	18	6
9	6	19	18
10	4	20	8

By modifying Ivory's proof of Fermat's little theorem, show that

$$a^{\phi(n)} \equiv 1 \pmod{n}$$

whenever a is relatively prime to n.

14. Lucas's converse of Fermat's little theorem. The existence of Carmichael numbers (problem 10) puts to rest any hope that Fermat's little theorem can serve as an ironclad test for primality. Prove, however, the following partial converse due to Lucas (1876): *If there is a number a such that the congruence*

$$a^x \equiv 1 \pmod{n}$$

is satisfied when $x = n - 1$ but not when x is a proper divisor of $n - 1$, then n is a prime.

Remarks. An important extension of Lucas's theorem asserts that, if for each prime p dividing $n - 1$ there exists an a for which $a^{n-1} \equiv 1 \pmod{n}$, but $a^{(n-1)/p} \not\equiv 1 \pmod{n}$, then n is a prime (see, for example, [363], p. 32). The repeated use of this theorem is an efficient method for generating large random primes for use in cryptography (see problem 25).

15. Show that there are infinitely many natural numbers n for which $n! + 1$ is composite. (It is not known whether there are infinitely many prime numbers of the form $n! + 1$.)

16. (Leibniz) A natural number $n > 1$ is prime if and only if

$$(n - 2)! \equiv 1 \pmod{n}.$$

17. Show that there are infinitely many natural numbers n for which $n! - 1$ is composite. (It is not known whether there are infinitely many prime numbers of the form $n! - 1$.)

18. Prove: The product of the first n positive integers $(1 \cdot 2 \cdots n)$ is divisible by their sum $(1 + 2 + \cdots + n)$ if and only if $n + 1$ is not an odd prime. Formulate a conjecture about the remainder obtained in the quotient

$$\frac{1 \cdot 2 \cdots n}{1 + 2 + \cdots + n}.$$

19. Prove that every odd prime can be expressed uniquely as the difference of two integral squares.

20. Prove that a natural number n is the sum of two triangular numbers if and only if $4n + 1$ is the sum of two squares.

21. (Unsolved) Are there infinitely many primes, each of them being the sum of two consecutive squares? For example,

$$5 = 1^2 + 2^2$$
$$13 = 2^2 + 3^2$$
$$41 = 4^2 + 5^2$$
$$61 = 5^2 + 6^2.$$

22. In 1796, Gauss recorded in his diary (the eighteenth entry),

$$EYPHKA! \text{ num } = \Delta + \Delta + \Delta.$$

Every natural number is the sum of three triangular numbers. Deduce from this that every natural number is the sum of four squares.

Hint. Show that every number of the form $8n + 3$ is the sum of three squares.

23. Formulate a conjecture about when a natural number is the hypotenuse of a Pythagorean triangle.

24. Don Zagier [462] has given the following "one-sentence proof" that every prime number p of the form $4n + 1$ is a sum of two squares:

> The involution on the finite set $S = \{(x, y, z) \in \mathbf{N}^3 : x^2 + 4yz = p\}$ defined by
>
> $$(x, y, z) \to \begin{cases} (x + 2z, z, y - x - z) & \text{if } x < y - z; \\ (2y - x, y, x - y + z) & \text{if } y - z < x < 2y; \\ (x - 2y, x - y + z, y) & \text{if } x > 2y \end{cases}$$
>
> has exactly one fixed point, so $|S|$ is odd and the involution defined by $(x, y, z) \to (x, z, y)$ also has a fixed point.

Verify the following three implicitly made assertions:

1. S is finite,
2. the map is well-defined and involutory (i.e., equal to its own inverse),
3. the map has exactly one fixed point.

Remarks. The underlying combinatorial principle used is this: *The cardinalities of a finite set and its fixed-point set under any involution have the same parity.* For a discussion of various constructive proofs of Fermat's two-squares theorem, see [22], p. 239, [120], pp. 120ff., and [440]. An elegant geometric proof, based on Minkowski's theorem is given in [401]. (Minkowski's theorem asserts that *any convex region in n-dimensional space, symmetric about the origin and of volume greater than 2^n, contains a point with integral coordinates, not all zero.*)

25. Public Key Cryptography

The problem of distinguishing prime numbers from composite numbers and of resolving the latter into their prime factors is known to be one of the most important and useful in arithmetic. It has engaged the industry and wisdom of ancient and modern geometers to such an extent that it would be superfluous to discuss the problem at length.

> ... Further, the dignity of the science itself seems to require that every possible means be explored for the solution of a problem so elegant and so celebrated.
>
> —Gauss, *Disquisitiones Arithmeticae* (1801)

Number theory, once considered the purest branch of mathematics, has increasingly found practical applications in the secure transmission of information. The following elegant system, proposed in 1978 by Rivest, Shamir, and Adleman—and known as the RSA cryptosystem—represents one of the great advances in the field. It is based on the remarkable disparity in difficulty between multiplying and factoring.

The algorithm is quite simple. Anyone who wishes to receive secret information first selects a pair of numbers, r and s, called the *encryption key*. The number r, the *modulus,* is taken to be the product

$$r = pq$$

of two large prime numbers p and q; the number s, the *encryption exponent,* is chosen so that it is relatively prime to

$$\phi(r) = (p-1)(q-1),$$

where ϕ denotes the Euler phi-function (problem 13). The encryption key is made public, but the two prime factors of r are kept secret.

To encode a message M—assumed to be in the form of a positive integer less than r—simply raise M to the power s and then reduce the answer modulo r. The encrypted message is then

$$E = M^s \pmod{r}.$$

Now, while modular exponentiation is simple (see the Appendix), there is no reasonably fast algorithm known for extracting the corresponding root. To recover the original message, the recipient needs a *decryption exponent* t, determined by the congruence

$$st \equiv 1 \pmod{\phi(r)}. \tag{6}$$

Once t is known, decryption proceeds as before: simply raise E to the power t and then reduce the answer modulo r. Provided that the original message M was relatively prime to r, we then have

$$M = E^t \pmod{r}. \tag{7}$$

a.   Show that (7) is valid by applying Euler's generalization of Fermat's little theorem (problem 13).

b.   Find the probability that a message is not relatively prime to r.

c.   Show that if $\phi(r)$ is known, then (6) can readily be solved for t by applying the Euclidean algorithm.

d.   Show that the problem of finding $\phi(r)$—and thereby cracking the RSA code—is equivalent to the problem of finding the factorization of r. This is what makes the RSA system so effective. Using the fastest computers, primes of about 100 decimal digits can be found in less than a minute. Factoring, on the other hand, is an exceedingly difficult task: There is no known algorithm that runs in polynomial time. With present technology, using the best algorithms available, factoring a 200-digit number would take more than a billion years! At the same time, it has not yet been proved that factoring is intractable. For further information about the RSA system and its implementation, see [268] and [373]. Recent advances in factorization are discussed in [68], [136], [366], and [454].

CHAPTER 3

FIBONACCI NUMBERS: FUNCTION AND FORM

It may appear surprising that sensibility should be introduced in connexion with mathematical demonstrations, which, it would seem, can only interest the intellect. But not if we bear in mind the feeling of mathematical beauty, of the harmony of numbers and forms and of geometric elegance. It is a real aesthetic feeling that all true mathematicians recognize, and this is truly sensibility.

— Henri Poincaré[1]

---

[1] *Science and Method,* Dover, New York, 1952, p. 59.

The greatest mathematician in Europe during the Middle Ages, Leonardo of Pisa (circa 1180–1250), better known as *Fibonacci,* is remembered best not for the mathematics he produced but for a single problem that he posed. It appears in his great work *Liber abaci* (1202), in which Leonardo introduces the ten Hindu-Arabic numerals to the West, 9 8 7 6 5 4 3 2 1 and the sign 0. The problem stands alone, tucked away among other unrelated problems, and yet it is the one problem from that work that has most inspired future mathematicians. It happens to concern rabbits.

> How many pairs of rabbits can be produced from a single pair in a year if every month each pair begets a new pair which from the second month on becomes productive, and if death does not occur?

Fibonacci certainly did not pose the question as a serious problem in population biology; it was simply an exercise in addition. If the experiment begins with a newborn pair of rabbits in the first month, then in the second month there is still just one pair. In the third month there are 2 pairs, in the fourth month, 3 pairs, in the fifth, 5, and so on. Thus, the rabbit problem gives rise to the famous sequence

$$1, \ 1, \ 2, \ 3, \ 5, \ 8, \ 13, \ 21, \ 34, \ 55, \ldots$$

in which the first two terms are 1 and thereafter each succeeding term is the sum of the two immediately preceding ones. If we let f_n denote the nth term of the sequence, then the Fibonacci numbers satisfy the recursive formula

$$f_{n+2} = f_{n+1} + f_n,$$

where $f_1 = f_2 = 1$.

This celebrated sequence, with its simple law of formation, appears unexpectedly in an astonishing number of different settings. In art, in music, and in nature, no less than in mathematics, it has been made to play a role. It is present in the tiling patterns of ancient mosaics and in the spiral shapes of many seashells; Kepler wrote about it in connection with phyllotaxis (the study of the arrangement of leaves and flowers in plant life); and Bartók used it as the basis for durations in his music ([199], [267], [303], [307], [388], [422], [434]).

More recently, the Fibonacci numbers have played a crucial role in the solution of Hilbert's famous tenth problem: *There is no algorithm that can determine in advance whether an arbitrary prescribed polynomial Diophantine equation has integer solutions.*[2] The proof, which was discovered in 1970 by the 22-year-

---

[2] See footnote 25 of chapter 1.

old Russian mathematician Yuri Matijasevich, makes use of the rate at which the Fibonacci numbers grow [121].

The first practical application of the Fibonacci numbers came to light in 1844 when the French mathematician Gabriel Lamé (1795–1870) used them to study the behavior of the Euclidean algorithm in the "worst case." Lamé proved that *the number of divisions required to find the greatest common divisor of two positive integers is never greater than five times the number of digits in the smaller number* (see §1b). Since then the Fibonacci numbers have been intimately connected with algorithms and the study of algorithms. In the latter half of the nineteenth century, Lucas uncovered deep and beautiful number-theoretic properties of the Fibonacci numbers, which he used in particular to prove that the 39-digit Mersenne number $2^{127} - 1$ is a prime. (This was the first new Mersenne prime discovered in over a century and the largest ever to be found before the advent of electronic computing.) It was Lucas who coined the term "Fibonacci numbers," in honor of Leonardo, and so they are known today.

Fibonacci Numbers

f_1	1	f_{11}	89	f_{21}	10946	f_{31}	1346269
f_2	1	f_{12}	144	f_{22}	17711	f_{32}	2178309
f_3	2	f_{13}	233	f_{23}	28657	f_{33}	3524578
f_4	3	f_{14}	377	f_{24}	46368	f_{34}	5702887
f_5	5	f_{15}	610	f_{25}	75025	f_{35}	9227465
f_6	8	f_{16}	987	f_{26}	121393	f_{36}	14930352
f_7	13	f_{17}	1597	f_{27}	196418	f_{37}	24157817
f_8	21	f_{18}	2584	f_{28}	317811	f_{38}	39088169
f_9	34	f_{19}	4181	f_{29}	514229	f_{39}	63245986
f_{10}	55	f_{20}	6765	f_{30}	832040	f_{40}	102334155

1. Elementary Properties

a. Algebraic Identities

The Fibonacci numbers give rise to a host of interesting and unexpected patterns. Among the simplest, we mention three that are particularly easy to discover, and, once discovered, just as easy to prove by induction: a formula for the sum of the first n Fibonacci numbers, a formula for the sum of their squares, and an important relation that exists between any three consecutive Fibonacci numbers. The reader may find it profitable to attempt to uncover these patterns on his own before reading on.

For each pattern we shall leave it to the reader to supply the simple inductive proof, preferring instead to present three different and distinctive arguments.

1. The sum of the first n Fibonacci numbers.

$$f_1 + f_2 + \cdots + f_n = f_{n+2} - 1.$$

Proof. We start off with the defining relation for the Fibonacci numbers in the form $f_k = f_{k+2} - f_{k+1}$. Putting $k = 1, 2, \ldots, n$ and adding, we obtain

$$
\begin{aligned}
f_1 + f_2 + f_3 &+ \cdots + f_n \\
&= (f_3 - f_2) + (f_4 - f_3) + (f_5 - f_4) + \cdots + (f_{n+2} - f_{n+1}) \\
&= f_{n+2} - f_2 \\
&= f_{n+2} - 1.
\end{aligned}
$$

2. The sum of the squares of the first n Fibonacci numbers.

$$f_1^2 + f_2^2 + \cdots + f_n^2 = f_n f_{n+1}. \tag{1}$$

Proof. The most transparent proof is accomplished with a single picture. Figure 1 below shows clearly how, starting with a single square of side one (darkened in the figure), we may systematically adjoin squares of sides 1, 2, 3, 5, 8 to obtain a rectangle of dimensions 8 by 13. The total area of the rectangle is then

$$1^2 + 1^2 + 2^2 + 3^2 + 5^2 + 8^2 = 8 \times 13.$$

It is clear that this construction can be extended as far as desired, and therefore (1) is true generally.

This and many other formulas are derived geometrically in an article by Brother Alfred Brousseau (*The Fibonacci Quarterly*, 1972, pp. 303–318).

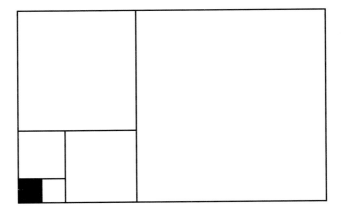

FIGURE 1

3. Cassini's Identity[3].   *The square of any term of the Fibonacci sequence differs from the product of the two adjacent terms by 1 or −1:*

$$f_{n-1}f_{n+1} - f_n^2 = (-1)^n.$$

Proof.   As before, we leave the straightforward inductive argument to the reader, this time favoring a somewhat esoteric method of proof based on the matrix $\left(\begin{smallmatrix}1&1\\1&0\end{smallmatrix}\right)$. The successive powers of this matrix are readily found (by induction!) to be

$$\begin{pmatrix} 1 & 1 \\ 1 & 0 \end{pmatrix}^n = \begin{pmatrix} f_{n+1} & f_n \\ f_n & f_{n-1} \end{pmatrix}.$$

We need now only take the determinant of both sides, remembering that, if two square matrices A and B have the same size, then $\det(AB) = \det(A) \cdot \det(B)$, and the result follows.

Further applications of the matrix $\left(\begin{smallmatrix}1&1\\1&0\end{smallmatrix}\right)$ and its relation to the Fibonacci numbers may be found in Part II of the excellent series of articles in the *Fibonacci Quarterly* entitled "A Primer on the Fibonacci Sequence" (1963, volume 1, number 2, pp. 61–68).

Cassini's identity forms the basis of an ingenious geometrical paradox, said to be one of Lewis Carroll's favorite puzzles ([98], [442]). Illustrated below for

---

[3] This relation, one of the oldest involving Fibonacci numbers, was discovered by the distinguished French astronomer Jean-Dominique Cassini in 1680 [88].

the case $n = 6$, it claims to prove by visual demonstration that an 8×8 square can be divided into four pieces and the pieces reassembled to form a rectangle of dimensions 5×13 (see Figure 2). The original area of 64 squares has thereby been transformed into 65 squares! Explain the deception. (See [434], p. 45.) A similar dissection transforms any square of dimensions $f_n \times f_n$ into a rectangle of dimensions $f_{n-1} \times f_{n+1}$; by Cassini's identity, one square has been lost or gained depending on whether n is odd or even.

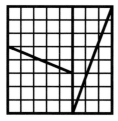 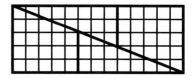

FIGURE 2

b. Number Theory: Relation to the Euclidean Algorithm

Fibonacci posed his famous rabbit problem in 1202, but it would take nearly six and a half centuries before it was finally put to practical use. This was done in 1844 by Gabriel Lamé, a French mathematician, who discovered a beautiful and surprising connection between the Fibonacci numbers and the Euclidean algorithm. This important algorithm finds the greatest common divisor of two positive integers through a simple series of divisions (see chapter 1). It is only natural to investigate the behavior of the algorithm in the "worst case," that is, to determine an upper bound on the required number of divisions, based on the size of the inputs.

The following theorem provides the answer.

Theorem.   *If a and b are positive integers, with a < b, and if the Euclidean algorithm requires exactly n division steps to find their greatest common divisor, then*

$$a \geq f_{n+1} \quad \text{and} \quad b \geq f_{n+2}.$$

Proof.   Let a and b satisfy the conditions of the theorem. To simplify notation, we set $a_n = a$ and $a_{n+1} = b$. The n steps of the algorithm may then be represented by the system of equations

$$a_{n+1} = q_1 \cdot a_n + a_{n-1}$$

$$a_n = q_2 \cdot a_{n-1} + a_{n-2}$$

$$\vdots$$

$$a_4 = q_{n-2} \cdot a_3 + a_2$$

$$a_3 = q_{n-1} \cdot a_2 + a_1$$

$$a_2 = q_n \cdot a_1.$$

Observe that all of the numbers involved are positive integers, and therefore each is at least 1. Observe also, however, that the final quotient, q_n, must be at least 2; otherwise the last equation would reduce to $a_2 = a_1$, thereby contradicting the implied condition $(0 < a_1 < a_2)$ of the previous equation.

Working backwards through the display, we find successively that

$$a_2 \geq 2 \cdot 1 \quad\;\; = 2$$

$$a_3 \geq 1 \cdot 2 + 1 = 3$$

$$a_4 \geq 1 \cdot 3 + 2 = 5$$

$$a_5 \geq 1 \cdot 5 + 3 = 8$$

and so on. The pattern is clear, and we see, in general, that a_k is at least as great as f_{k+1}. Thus $a \geq f_{n+1}$ and $b \geq f_{n+2}$.

The theorem may also be phrased according to its contrapositive: If the two inputs a and b do not exceed f_{n+2} then at most n division steps are needed. Now, it is a simple matter to verify that when $a = f_{n+1}$ and $b = f_{n+2}$, precisely n division steps are needed. Thus, the "worst case" behavior of the Euclidean algorithm occurs when the inputs are consecutive Fibonacci numbers.

We are now in a position to establish Lamé's upper bound on the number of divisions, based on the size of the smaller of the two inputs.

Corollary.   *The number of divisions needed to find the greatest common divisor of two positive integers using the Euclidean algorithm does not exceed five times the number of digits in the smaller number.*

Proof.   Suppose that the smaller number, call it a, has k decimal digits, so that $a < 10^k$. If the Euclidean algorithm requires n division steps, then we know from the theorem that $a \geq f_{n+1}$. Therefore

$$10^k > f_{n+1}.$$

Now a simple induction argument shows that

$$f_m > (8/5)^{m-2}$$

for every $m > 2$, and therefore

$$10^k > (8/5)^{n-1}.$$

Raising each side to the fifth power, and observing that $(8/5)^5 > 10$, we get

$$10^{5k} > 10^{n-1}.$$

Thus, $5k > n - 1$, and, since k is an integer, $n \leq 5k$, as asserted.

The result of the corollary is, in fact, the best possible, since the constant 5 cannot be replaced by any smaller number. Why?

c. Binet's Formula

This important "closed form" expression for the Fibonacci numbers was first discovered by De Moivre in 1718. It was rediscovered more than a century later by the French scholar Binet (1843), whose name it now bears.

Binet's Formula.   *The nth term of the Fibonacci sequence is given by the formula*

$$f_n = \frac{1}{\sqrt{5}} \left[\left(\frac{1 + \sqrt{5}}{2} \right)^n - \left(\frac{1 - \sqrt{5}}{2} \right)^n \right].$$

It seems surprising that a sequence of numbers that is so easy to define recursively should give rise to so complicated a formula. It is not even apparent that the expression on the right-hand side is an integer!

Before proceeding with the proof, we shall find it instructive to make a few simple observations.

The Fibonacci sequence is only one of arbitrarily many different number sequences that satisfy the recursive formula

$$u_{n+2} = u_{n+1} + u_n. \tag{2}$$

Any such sequence will be called a *solution* of equation (2).

It is clear that a solution is completely determined once we have specified its first two terms. For example, the choice $u_1 = 1$ and $u_2 = 3$ gives rise to the important *Lucas sequence*

$$1,\ 3,\ 4,\ 7,\ 11,\ 18, \ldots$$

which was used by Lucas to investigate the primality of the Mersenne numbers (see, for example, [374]).

The following simple lemma shows how solutions of equation (2) can be combined to produce new solutions. The proof, which is entirely obvious, is left to the reader.

Lemma. (a) *If* $A = (a_1, a_2, a_3, \ldots)$ *is a solution of equation* (2) *and* c *is an arbitrary real number, then the sequence*

$$cA = (ca_1, ca_2, ca_3, \ldots)$$

is also a solution of equation (2).

(b) *If the sequences* $A = (a_1, a_2, a_3, \ldots)$ *and* $B = (b_1, b_2, b_3, \ldots)$ *are solutions of equation* (2), *then their sum*

$$A + B = (a_1 + b_1, a_2 + b_2, a_3 + b_3, \ldots)$$

is also a solution of equation (2).

Proof of Binet's formula. Perhaps the simplest proof of Binet's formula is based on the observation that the geometric progression

$$\tau, \tau^2, \tau^3, \ldots$$

is a solution of equation (2) if τ is the positive root of the equation $x^2 = x + 1$, namely, $\tau = (1 + \sqrt{5})/2$. Reason: since

$$\tau^2 = \tau + 1,$$

multiplication by τ^n yields

$$\tau^{n+2} = \tau^{n+1} + \tau^n$$

for every $n = 1, 2, 3, \ldots$.

If $\mu = 1 - \tau$ denotes the other root of the quadratic equation $x^2 = x + 1$, then it follows in the same way that the geometric progression

$$\mu, \mu^2, \mu^3, \ldots$$

is also a solution of equation (2). Therefore, by virtue of the lemma, the sequence of differences

$$\tau^n - \mu^n \qquad n = 1, 2, 3, \ldots$$

also satisfies the recursion. But the first two differences are equal, since

$$\tau^2 - \mu^2 = (\tau - \mu)(\tau + \mu) = \tau - \mu,$$

and therefore the sequence

$$\frac{\tau^n - \mu^n}{\tau - \mu} \qquad n = 1, 2, 3, \ldots$$

must be the Fibonacci sequence. This proves Binet's formula.

Most of the important properties of the Fibonacci numbers can be derived by means of Binet's formula. For example, it follows at once from that formula that

$$f_n \text{ is approximately equal to } \tau^n / \sqrt{5}$$

with an absolute error that tends to zero as n tends to infinity. To prove this, we need only observe that

$$\left| f_n - \frac{\tau^n}{\sqrt{5}} \right| = \frac{|1 - \tau|^n}{\sqrt{5}} \to 0$$

(because $|1 - \tau| < 1$). Thus, the Fibonacci numbers grow exponentially, and, as n increases, they approach the terms of the geometric progression

$$\frac{\tau}{\sqrt{5}}, \frac{\tau^2}{\sqrt{5}}, \frac{\tau^3}{\sqrt{5}}, \ldots$$

Example. The Transmission of Information.   Imagine a communication system that can process only two different signals—say the binary digits 0 and 1. Messages are transmitted over some channel by first encoding them into sequences of these two digits. Let us suppose, for example, that it requires exactly 1 unit of time to transmit a zero and exactly twice as long to transmit a one. We wish to determine the total number N_t of possible message sequences of duration t. If a message of duration t begins with a zero, then that digit may be followed by any one of N_{t-1} different sequences of duration $t - 1$. Hence, the total number of messages of duration t that begin with a zero is just N_{t-1}. Similarly, the total number of messages of duration t that begin with a one is given by N_{t-2}. But a message of duration t must begin with either a zero or a one. Therefore,

$$N_t = N_{t-1} + N_{t-2} \qquad t \geq 3$$

which is the same recursive formula satisfied by the Fibonacci numbers. In view of the definition of N_t, we have $N_1 = 1$ and $N_2 = 2$, so that

$$N_t = f_{t+1} \qquad \text{for every } t.$$

The *capacity* C of the channel is defined by

$$C = \lim_{t \to \infty} \frac{\log_2 N_t}{t}.$$

It is a measure of the rate at which information can be transmitted over the channel.[4]

The true significance of the channel capacity becomes apparent only through the *fundamental theorem of information theory*. This asserts, loosely speaking, that, by a proper encoding of messages, it is possible to transmit information through a "noisy" channel at any rate less than the channel capacity with an arbitrarily small probability of error. Information theory is largely the study of this one theorem ([357], [375]).

While it is quite difficult to calculate the capacity of the channel in general, in the present case it is trivial. By virtue of Binet's formula,

$$C = \lim_{t \to \infty} \frac{\log_2(\alpha \tau^{t+1})}{t},$$

where $\alpha = 1/\sqrt{5}$. Since $\log_2 \alpha \tau^{t+1} = \log_2 \alpha + (t+1) \log_2 \tau$, it follows that

$$C = \lim_{t \to \infty} \left(\frac{\log_2 \alpha}{t} + \log_2 \tau \right)$$

$$= \log_2 \tau.$$

The capacity of the channel is found to be roughly 0.7.

---

[4] It may at first seem strange that information is defined in terms of the logarithm of the number of possible message sequences. But, in fact, it is quite natural. For example, doubling the time roughly squares the number of messages, and therefore doubles the logarithm.

PROBLEMS

1. In how many ways is it possible to climb a staircase with n steps, taking only one or two steps at a time?

2. How many permutations of the first n positive integers are there in which each integer either stays fixed or changes place with one of its neighbors?

3. In how many ways is it possible to express a positive integer n as a sum of positive integers greater than 1 if different orderings are regarded as distinct? (For example, $6 = 2 + 4 = 4 + 2 = 3 + 3 = 2 + 2 + 2$.)

4. **a.** A fair coin is tossed until two consecutive heads appear. What is the probability that the sequence terminates after n tosses?
  b. A fair coin is tossed until either three consecutive heads or three consecutive tails appear. What is the probability that the sequence terminates after n tosses?

5. Find the number of possible sequences of 0's and 1's of length n such that no two 0's are consecutive.

6. Show that the sum $f_n^2 + f_{n+1}^2$ is always a Fibonacci number.

7. Show that the difference $f_{n+1}^2 - f_{n-1}^2$ is always a Fibonacci number.

8. Establish the identity $f_1 f_2 + f_2 f_3 + \cdots + f_{2n-1} f_{2n} = f_{2n}^2$.

9. Establish simple formulas for the sums

$$f_1 + f_3 + f_5 + \cdots + f_{2n-1} \quad \text{and} \quad f_2 + f_4 + f_6 + \cdots + f_{2n}.$$

10. Let g_n denote the nth term of the Lucas sequence $\{1, 3, 4, 7, 11, 18, 29, \ldots\}$. Show that

$$f_{n-1} + f_{n+1} = g_n \quad \text{and} \quad g_{n-1} + g_{n+1} = 5f_n.$$

11. Show that $f_n g_n$ is always a Fibonacci number, and that $f_n + g_n$ is always twice a Fibonacci number.

12.   Show that

$$\sum_{i=1}^{n} g_i^2 = g_n g_{n+1} - 2.$$

13.   Lucas's original test for the primality of the Mersenne numbers was based on the sequence $\{a_n\}$ defined recursively by $a_1 = 3$ and

$$a_{n+1} = a_n^2 - 2 \qquad \text{for } n = 1, 2, 3, \ldots.$$

(See, for example, [211], p. 223.) Show that

$$a_n = f_{2^{n+1}}/f_{2^n} = g_{2^n}.$$

14.   Fibonacci Numbers and Binomial Coefficients.   The Fibonacci numbers appear unexpectedly in Pascal's triangle, when viewed from the right angle (see Figure 3). Prove that

$$f_{n+1} = \binom{n}{0} + \binom{n-1}{1} + \binom{n-2}{2} + \cdots.$$

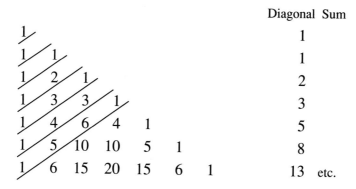

Diagonal Sum

1							1
1	1						1
1	2	1					2
1	3	3	1				3
1	4	6	4	1			5
1	5	10	10	5	1		8
1	6	15	20	15	6	1	13 etc.

FIGURE 3

15.   (E. Cesàro) Show that for every natural number n,

$$\sum_{k=1}^{n} \binom{n}{k} f_k = f_{2n}.$$

16.   Show that

$$\sum_{n=2}^{\infty} \frac{1}{f_{n-1}f_{n+1}} = 1.$$

17.   Use Binet's formula to show that the infinite sum

$$\sum_{n=1}^{\infty} \frac{f_n}{10^n}$$

converges to a rational number.

18.   Show that

$$\sum_{n=1}^{\infty} \tan^{-1} \frac{1}{f_{2n+1}} = \frac{\pi}{4}.$$

19.   A Problem of Steinhaus.   Find a sequence $a_0, a_1, a_2, \ldots$ whose elements are positive and such that $a_0 = 1$ and $a_{n+2} = a_n - a_{n+1}$ for $n = 0, 1, 2, \ldots$. Show that there is only one such sequence!

20.   *Prove*: For any four consecutive Fibonacci numbers, the product of the outer terms and twice the product of the inner terms form the sides of a Pythagorean triangle. For example: 3, 5, 8, 13 produces the sides 39 and 80 of the right-triangle 39, 80, 89. Notice that the hypotenuse, 89, is also a Fibonacci number! How is its subscript related to the subscripts of the four original numbers? How is the area of the triangle related to those four numbers?

21.   Show that any two consecutive Fibonacci numbers are relatively prime.

22.   a.   Observing that

$$f_{n+2} = f_{n+1} + f_n$$
$$f_{n+3} = 2f_{n+1} + f_n$$
$$f_{n+4} = 3f_{n+1} + 2f_n$$
$$f_{n+5} = 5f_{n+1} + 3f_n$$

discern the general pattern and verify it by induction.
   b.   Show that f_{kn} is a multiple of f_n for all k and n.

23.   a.   Prove the following theorem due to Lucas (1876): For all m and n,

$$\gcd(f_m, f_n) = f_{\gcd(m,n)}.$$

    b.   Use Lucas's theorem to give another proof of the infinitude of the primes.

24.   Prove that f_m is divisible by f_n if and only if m is divisible by n.

    Remark. In his famous solution of Hilbert's tenth problem, Matijasevich proved that f_m is divisible by f_n^2 ($n > 2$) if and only if m is divisible by $n f_n$. (For a proof, see [186], p. 280.)

25.   Show that every positive integer is a factor of some Fibonacci number. In particular, show that for any integer m there is at least one Fibonacci number among the first m^2 Fibonacci numbers that is divisible by m.

    Hint. For any positive integer k, let $\bar{k}$ be the remainder obtained by dividing k by m. Consider the following sequence of pairs of remainders:

$$(\bar{f}_1, \bar{f}_2), (\bar{f}_2, \bar{f}_3), (\bar{f}_3, \bar{f}_4), \dots .$$

Show that the first pair that appears more than once is $(1, 1)$.

26.   a.   Prove that the sequence of indices of those Fibonacci numbers that are divisible by some natural number m form an arithmetic progression.

    b.   A Fibonacci number is divisible by 2 if and only if its index is divisible by 3.

        A Fibonacci number is divisible by 3 if and only if its index is divisible by 4.

        A Fibonacci number is divisible by 5 if and only if its index is divisible by 5.

27.   The original Lucas test for the primality of the Mersenne numbers was based on the following theorem: *If p is an odd prime different from 5, then either f_{p-1} or f_{p+1} is divisible by p.* Prove the theorem.

Indication of the Proof.   Using the binomial theorem to expand the powers in Binet's formula, show that

$$f_n = \frac{1}{2^{n-1}} \left[\binom{n}{1} + 5 \binom{n}{3} + 5^2 \binom{n}{5} + \cdots \right],$$

a formula known to Catalan (1857). Now show that

$$\binom{p-1}{n} \equiv (-1)^n \pmod{p} \qquad \text{for } 1 \le n \le p - 1,$$

and

$$\binom{p+1}{n} \equiv 0 \pmod{p} \qquad \text{for } 2 \le n \le p-1.$$

28. If p is a prime number, is f_p necessarily a prime number?

29. Find all positive integers n such that either $f_n + 1$ or $f_n - 1$ is a prime number.

30. (Lagrange) The final digits of the Fibonacci numbers recur after a cycle of sixty:

$$1, 1, 2, 3, 5, 8, 3, 1, 4, \dots, 7, 2, 9, 1, 0.$$

31. Evaluate the $n \times n$ determinant

$$\begin{vmatrix} 1 & -1 & 0 & 0 & \cdots & 0 & 0 & 0 \\ 1 & 1 & -1 & 0 & \cdots & 0 & 0 & 0 \\ 0 & 1 & 1 & -1 & \cdots & 0 & 0 & 0 \\ & & & \cdots & & & & \\ 0 & 0 & 0 & 0 & \cdots & 1 & 1 & -1 \\ 0 & 0 & 0 & 0 & \cdots & 0 & 1 & 1 \end{vmatrix}.$$

32. Cassini's identity shows that the points $(x, y) = (f_n, f_{n+1})$ lie on the two hyperbolas

$$y^2 - xy - x^2 = \pm 1.$$

Show, conversely, that these are the only points with positive integral coordinates on the hyperbolas.

33. The Fibonacci Number System. Prove Zeckendorf's theorem [463]: *Every positive integer has a unique representation as the sum of nonconsecutive Fibonacci numbers.*

  Hint. Given n, find the largest Fibonacci number f not exceeding n; then find the largest not exceeding $n - f$, and so on.

34. The Fibonacci Search Algorithm. Let f be a function defined and continuous on the closed interval $[a, b]$. Any subinterval of $[a, b]$ on which f assumes its maximum or minimum value will be called an *interval of uncertainty*. Establish the following theorem due to Kiefer [250].

Theorem. *Let f be a convex function on the closed interval [a, b] and let n be a fixed positive integer. If ε is any number greater than*

$$\frac{b-a}{f_{n+1}}$$

then we can locate the minimum of f on an interval of uncertainty of length not exceeding ε by computing n functional values.

Indication of the proof. The following algorithm provides an optimal search procedure: the number $(b-a)/f_{n+1}$ cannot be replaced by any smaller number ([34], pp. 34ff.). It is best illustrated by an example. Let us suppose, to simplify notation, that $b - a$ is a Fibonacci number, say, $f_7 = 13$. Then we can locate the minimum of f on an interval of uncertainty of length arbitrarily close to 1 by computing $n = 6$ functional values.

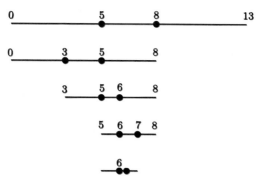

FIGURE 4
The Fibonacci search algorithm

a.   The algorithm begins by computing two functional values at the first stage. Let the given interval be [0, 13]. Express its length as the sum of the two Fibonacci numbers 5 and 8, and then compare $f(5)$ and $f(8)$. If $f(5) < f(8)$, then the minimum must occur on the subinterval [0, 8] (as pictured in Figure 4); otherwise it must occur on [5, 13]. In either case, the length of the new interval of uncertainty is 8.

b.   At each subsequent iteration, the algorithm computes only one new functional value. For the example given, the next comparison is between $f(3)$ and $f(5)$, since $3 + 5 = 8$. But $f(5)$ is already known, so only $f(3)$ must be determined. Suppose, for example, that $f(3) \geq f(5)$. Then the new interval of uncertainty is [3, 8]. Since its length is 5, and since 5 is the sum of the two

Fibonacci numbers 2 and 3, the new points of subdivision are $3 + 2$ and $3 + 3$. Continue.

(For an important application to Phase I clinical trials, the goal of which is to find the maximum dose of a drug that humans can tolerate, see M. A. Schneiderman, "Mouse to Man: Statistical Problems in Bringing a Drug to Clinical Trial," in *Proceedings of the Fifth Berkeley Symposium on Mathematical Statistics and Probability,* 1967.)

2. The Golden Ratio

> Geometry has two great treasures: one is the Theorem of Pythagoras; the other, the division of a line into extreme and mean ratio. The first we may compare to a measure of gold; the second we may name a precious jewel.
>
> —Johannes Kepler

Suppose that we form the ratios

$$r_n = f_{n+1}/f_n \qquad n = 1, 2, 3, \ldots$$

of consecutive Fibonacci numbers. The resulting sequence is then

$$\frac{1}{1}, \frac{2}{1}, \frac{3}{2}, \frac{5}{3}, \frac{8}{5}, \ldots$$

It is a most fascinating sequence of numbers.

To begin with, it appears that every other term, beginning with the first, gets larger

$$r_1 < r_3 < r_5 < \cdots$$

while every other term, beginning with the second, gets smaller

$$r_2 > r_4 > r_6 > \cdots.$$

In addition, the terms appear to alternate in size, so that $r_1 < r_2$ but $r_2 > r_3$, $r_3 < r_4$ but $r_4 > r_5$, and so on. These observations are readily confirmed by induction, and we may summarize our findings by saying that the sequence of closed intervals

$$[r_1, r_2], [r_3, r_4], [r_5, r_6], \ldots$$

is *nested,* that is, each one is entirely contained within the preceding one.

Furthermore, it is easy to see that the limiting lengths of these intervals is zero. For, by Cassini's identity,

$$r_n - r_{n-1} = \frac{(-1)^n}{f_n f_{n-1}}$$

which clearly tends to zero as n tends to infinity.

The stage is set, the nested interval principle can now make its appearance.

Nested Interval Principle. *If $I_1, I_2, I_3, \ldots$ is a nested sequence of closed and bounded intervals, and if the length of I_n tends to zero as n tends to infinity, then there is one and only one real number that belongs to every interval in the sequence.*

The nested interval principle asserts, in descriptive geometric terms, that the real number system is *complete,* that it is without gaps. In our present framework, it allows us to conclude that there is a unique real number L common to every closed interval

$$[r_{2n-1}, r_{2n}] \qquad n = 1, 2, 3, \ldots,$$

and hence that

$$L = \lim_{n \to \infty} \frac{f_{n+1}}{f_n}.$$

Knowing that the limit exists, we can now find it. Starting off with the recursive formula $f_{n+2} = f_{n+1} + f_n$, we divide throughout by f_{n+1} and obtain the relation

$$r_{n+1} = 1 + \frac{1}{r_n}.$$

Letting n tend to infinity, we then obtain

$$L = 1 + \frac{1}{L} \tag{1}$$

so that L is the positive root of the equation $L^2 = L + 1$, namely,

$$L = \frac{1 + \sqrt{5}}{2}.$$

This is the **golden ratio**—customarily denoted by τ—which we first encountered in Binet's formula for the Fibonacci numbers. To four decimal places, $\tau = 1.6180.\ldots$

The number τ has its origins in classical antiquity. Euclid called it the "mean and extreme ratio," an odd-sounding phrase signifying *the ratio obtained when a line segment is divided into two unequal parts such that the ratio of the whole to the larger is equal to the ratio of the larger to the smaller.* Thus, if the given segment is AB, then we are to find C such that

$$\frac{AB}{AC} = \frac{AC}{CB}.$$

If the common ratio is called x, then we find

$$x = \frac{AB}{AC} = \frac{AC + CB}{AC} = 1 + \frac{CB}{AC} = 1 + \frac{1}{x}.$$

Comparing this with the defining relation (1) for τ, we see that $x = \tau$ (after τομή, the *section,* later known as the *golden section,* or, in the still more fanciful language of the Renaissance, the *divine proportion*).

Just a few examples will serve to illustrate the wonder that has surrounded this simple algebraic number.[5]

a. Constructing a Regular Pentagon

The classical construction of a regular pentagon using only Euclidean tools depends on the division of a line segment in the ratio $\tau : 1$. Figure 5 shows why.

Observe to begin with that it is sufficient to inscribe a regular decagon in a given circle, for the pentagon may then be formed by connecting alternate vertices.

Following the analytic method of ancient Greece, we suppose at the outset that the decagon has already been constructed—so that the angle BOA is known to be $\pi/5$—and then try to clarify the underlying geometry.

Locate X on the radius OA so that BX bisects angle OBA. Since triangle OBA is isosceles, its base angles measure $2\pi/5$, and therefore $\beta = \pi/5$. Then $\alpha = \pi/5 + \beta = 2\pi/5$, so that triangles ABX and OBX are also isosceles. Hence,

$$OX = BX = AB.$$

Taking the radius of the circle to be 1 and the side of the decagon to be x, we see from the similarity of triangles ABO and ABX that

$$\frac{1}{x} = \frac{x}{1 - x}.$$

---

[5] See also [23], [164], [199], [218], [235], [294], [328], and [422].

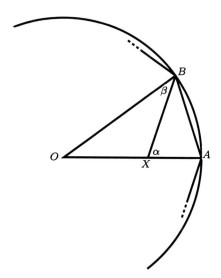

FIGURE 5

Thus, X divides the radius OA into extreme and mean ratio, and it follows that

$$x = \frac{1}{\tau} = \frac{\sqrt{5}-1}{2}.$$

Since a segment of length $\sqrt{5}$ can be constructed, so can the side of the decagon.

b. The Simplest Continued Fraction

The golden ratio gives rise to the simplest (and also the slowest to converge) infinite continued fraction:

$$\tau = 1 + \cfrac{1}{1 + \cfrac{1}{1 + \cfrac{1}{1 + \cdots}}}.$$

The value of the continued fraction is, by definition, the limiting value of the sequence of its "convergents." These are the partial quotients

$$1, \quad 1 + 1 = 2, \quad 1 + \frac{1}{1+1} = \frac{3}{2}, \quad 1 + \frac{1}{1 + \frac{1}{1+1}} = \frac{5}{3}, \cdots.$$

Notice that the nth convergent is equal to the ratio $r_n = f_{n+1}/f_n$ of consecutive Fibonacci numbers. (To verify this, we need only recall that

$$r_{n+1} = 1 + \frac{1}{r_n}$$

for every n.) Since we have already shown that r_n approaches τ, so must the convergents.

It can be shown that every positive real number has a representation as a continued fraction

$$a_0 + \cfrac{1}{a_1 + \cfrac{1}{a_2 + \cfrac{1}{a_3 + \cdots}}}$$

where $a_1, a_2, a_3, \ldots$ are positive integers, and a_0 is a nonnegative integer. The expansion terminates if and only if the given number is rational, a fact that follows readily from the Euclidean algorithm.

The theory of the expansion of irrational numbers as continued fractions constitutes a rich and extensive branch of number theory. We shall not penetrate more deeply into this area here, except to mention the truly extraordinary expansion

$$\frac{4}{\pi} = 1 + \cfrac{1^2}{2 + \cfrac{3^2}{2 + \cfrac{5^2}{2 + \cfrac{7^2}{2 + \cfrac{9^2}{2 + \cdots}}}}}.$$

This remarkable formula links π with the integers in a far more striking way than does the decimal expansion of π, which appears to exhibit no regularity in the succession of its digits. It was discovered by Lord Brouncker (circa 1620–1684), the first president of the Royal Society. How Brouncker arrived at this result remains a mystery, but a proof based on Euler's work during the next century can be found in *The Mathematics of Great Amateurs* [107], chapter XI.

c. Golden Rectangles and Aesthetics

Perhaps the greatest fascination with the golden ratio lies in its connection with aesthetics. In the art world, the **golden rectangle**—any rectangle whose sides are in the ratio of τ to 1—is said to be among the most pleasing of all geometric forms. From the rectangular facades of ancient Greek temples (most notably the Parthenon at Athens), to the modern architectural designs of Le Corbusier, who asserted that human life is "comforted" by mathematics, from the impressionist paintings of Seurat, to the modern abstractions of Mondrian (see Color Plates 1 and 2), golden rectangles abound.

Cautioning against the almost mystical aura that has surrounded the golden ratio in the past, N. N. Vorobyov writes: "Various idealistic philosophers in ancient times and in the Middle Ages elevated to the rank of an aesthetic or even a philosophical principle the beauty of the golden rectangle and other figures in which the golden section is found. They sought to explain the phenomena of nature and society by means of the golden section and several other number ratios, while the number τ itself and its convergents were used to perform various mystical 'operations.' Such 'theories,' of course, have nothing to do with science."[6]

While the mysticism behind the golden section has been dispelled, its links to beauty persist.

The relation

$$\tau = 1 + \frac{1}{\tau}$$

shows that when a square is removed from one end of a golden rectangle, the remaining rectangle is again golden. When this process is continued indefinitely the result is a sequence of squares, of ever-diminishing size, converging on a single point O. Circular arcs, inscribed in the successive squares, as in Figure 6, give rise to a composite spiral of quite pleasing proportions.

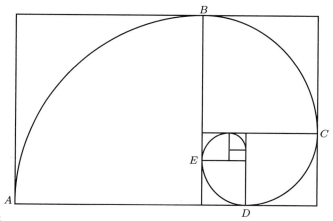

FIGURE 6

---

[6] *The Fibonacci Numbers* [434], p. 44.

COLOR PLATE 1

Parade de Cirque (1887–1888), Georges Seurat

Courtesy of The Metropolitan Museum of Art, Bequest of Stephen C. Clark, 1960. (61.101.17)

The composition is rigorously constructed with verticals and horizontals parallel to the picture frame. There is thus a curious mixture of encoded realism and formalism, which in no way prevents the participation of the imagination, and even seems to stimulate it. It is as if the painter's imagination had taken advantage of his preoccupation with calculating golden means, mobilizing the expressive resources of light, and adjusting contrasts of color and value, to overwhelm the picture and give it a fantasmagoric aspect. And surely also to express a deeper intention.

—Alain Madeleine-Perdrillat [287], p. 110.

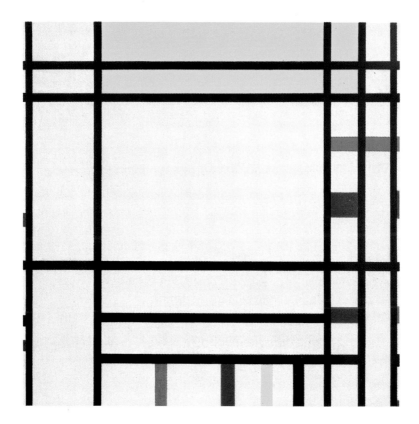

COLOR PLATE 2
Place de la Concorde (1938–1943), Piet Mondrian

Courtesy of the Dallas Museum of Art, Foundation for the Arts Collection, gift of the James H. and Lillian Clark Foundation.

The Dutch circle founded in 1917 a periodical called simply *De Stijl* (The Style), consecrated to the exaltation of mathematical simplicity in opposition to all "baroque" styles, among which they included Impressionism. Sternly, they reduced all formal elements to flat surfaces bounded by straight lines intersecting at right angles, and all colors to black, white, and gray and the three primary hues, red, yellow, and blue. ... The leading theorist and finest artist of the movement, indeed one of the most sensitive, imaginative, and influential painters of the twentieth century, was Piet Mondrian (1872–1944).
> —Frederick Hartt [212], p. 911.

It can be shown, furthermore, that O is the pole of a true logarithmic spiral that passes through the successive points of division $A, B, C, D, E, \ldots$ (see [116], p. 164). This is the "golden spiral"—the *spira mirabilis* of James Bernoulli, which appears so often and so unexpectedly in natural forms. In the next section, we shall explore some of its miraculous properties.[7]

d. Spira Mirabilus

In the growth of a shell, we can conceive no simpler law than this, namely, that it shall widen and lengthen in the same unvarying proportions: and this simplest of laws is that which Nature tends to follow. The shell, like the creature within it, grows in size *but does not change its shape*; and the existence of this constant relativity of growth, or constant similarity of form, is of the essence, and may be made the basis of a definition, of the equiangular spiral.

—D'Arcy Thompson, *On Growth and Form*

We have seen that the nth term of the Fibonacci sequence is approximately equal to

$$\frac{\tau^n}{\sqrt{5}} \tag{2}$$

and that the absolute error $|f_n - \tau^n/\sqrt{5}|$ approaches zero as $n \to \infty$. It is quite natural to replace n by a *continuous* variable in (2) and to introduce the polar equation

$$r = \alpha \beta^\theta.$$

This beautiful curve traces out a logarithmic spiral, and when $\alpha = 1/\sqrt{5}$ and $\beta = \tau^{2/\pi}$ every crossing point on the coordinate axes (when $\theta = n\pi/2$, $n = 1, 2, 3, \ldots$) produces a value of the radius which draws nearer and nearer to the terms of the Fibonacci sequence.

Unlike other important mathematical spirals which were known in ancient times (of which the *spiral of Archimedes* is the most important example), the logarithmic spiral was first recognized by Descartes, and discussed in the year 1638 in his letters to Mersenne. It later came to be known as the *equiangular spiral* (see Figure 7) because the growing curve intersects every radius vector

---

[7] The golden spiral is closely approximated by the "artificial" one, but, unlike the latter, it is not tangent to the sides of the squares ([118], p. 70).

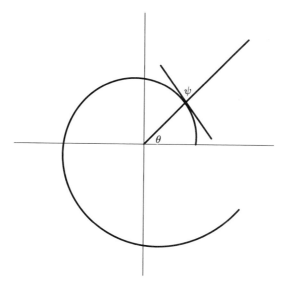

FIGURE 7
The equiangular spiral: for every value of θ, the angle ψ remains the same.

at a constant angle. Indeed, by elementary calculus, the angle of intersection, ψ, is given by the formula

$$\tan \psi = \frac{r}{dr/d\theta} = \frac{\alpha\beta^\theta}{\alpha\beta^\theta \log \beta} = \frac{1}{\log \beta}$$

which is independent of the polar angle θ.

But the most striking feature of the logarithmic spiral is its "self-similarity," by which we mean that *any two arcs of the curve that subtend equal angles at the origin are similar to one another.* In analytic terms, this self-similarity may be expressed by saying that for any pair of angles A and B and all increments λ, the ratio of the corresponding radii

$$\frac{r(A+\lambda)}{r(B+\lambda)}$$

does not depend on λ (see Figure 8). Thus, *as the curve grows in size its shape remains the same.*

It is this simple law of growth that characterizes the logarithmic spiral, and so gives rise to the beautiful spiral shapes that many organisms exhibit in their horns or shells. Perhaps its most perfect realization is to be found in the chambers of the *Nautilus* (see Figure 9).

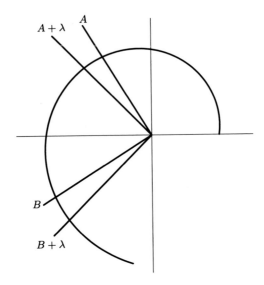

FIGURE 8

Theorem.   *Let f be a positive function, defined and continuous on the entire real line, and suppose that f does not reduce to a constant. If the polar curve*

$$r = f(\theta)$$

has the property that any two arcs of the curve that subtend equal angles at the origin are similar to each other, then the curve must be a logarithmic spiral.

Proof.   The asserted self-similarity of the curve can be formulated analytically as

$$\frac{f(A + \lambda)}{f(B + \lambda)} = \frac{f(A)}{f(B)} \tag{3}$$

for all A, B, and λ. For simplicity of notation, let us normalize f by defining

$$g(x) = \frac{1}{\alpha} f(x) \qquad \text{where } \alpha = f(0).$$

Then g is continuous, $g(0) = 1$, and equation (3) remains valid when f is replaced by g. Taking B to be zero in (3), we find that g must satisfy the important functional equation

$$g(A + \lambda) = g(A)g(\lambda) \tag{4}$$

for all values of A and λ.

FIGURE 9
Radiograph of the shell of the chambered nautilus (*Nautilus pompilius*).
Reprinted by permission of the publisher from H. E. Huntley, *The Divine Proportion,* p. iv. Copyright ©1970 by Dover.

The most obvious solutions of this equation are the exponential functions $g(x) = b^x$, where b is any nonnegative number. We shall show that there can be no other solutions than these.

Put $\beta = g(1)$. Then relation (4) shows in turn that

$$g(2) = g(1 + 1) = g(1)g(1) = \beta^2$$
$$g(3) = g(1 + 2) = g(1)g(2) = \beta^3$$
$$g(4) = g(1 + 3) = g(1)g(3) = \beta^4.$$

Continuing in this fashion (that is to say, inductively), we find that

$$g(n) = \beta^n$$

for every natural number n.

Similarly,

$$\beta = g(1) = \left[g\left(\frac{1}{2}\right) \right]^2 = \left[g\left(\frac{1}{3}\right) \right]^3 = \left[g\left(\frac{1}{4}\right) \right]^4 = \cdots$$

so that

$$g\left(\frac{1}{m}\right) = \beta^{1/m} \qquad m = 1, 2, 3, \ldots,$$

and hence also

$$g\left(\frac{n}{m}\right) = g\left(\frac{1}{m} + \frac{1}{m} + \frac{1}{m} + \cdots + \frac{1}{m}\right) = \left[g\left(\frac{1}{m}\right) \right]^n = \beta^{n/m}.$$

Thus, the functions $g(x)$ and β^x have the same values whenever x is a positive rational number. To see that they also agree whenever x is negative and rational, we need only observe that

$$1 = g(0) = g(x - x) = g(x)g(-x)$$

for any choice of x. But it is fundamental that *if two continuous functions agree on all the rationals, then they must necessarily agree everywhere.* (Hint: The rationals are dense in the reals.) Therefore,

$$g(x) = \beta^x$$

and we have found the most general solution of the functional equation (4).

The proof is over: since

$$r = f(\theta) = \alpha \beta^\theta$$

and since f does not reduce to a constant (so that β can be neither 0 nor 1), it follows that the polar curve must describe a logarithmic spiral.

Speaking of the many structures which display the logarithmic spiral—be they the shell of nautilus or snail, the elephant's tusk, the beaver's tooth, the cat's claw or the canary's beak—D'Arcy Thompson writes: "For it is peculiarly characteristic of the spiral shell, for instance, that it does not alter as it grows; each increment is similar to its predecessor, and the whole, after every spurt of growth, is just like what it was before. We feel no surprise when the animal which secretes the shell, or any animal whatsoever, grows by such symmetrical expansion as to preserve its form unchanged . . . ." [8] And so too, by virtue of the theorem, we feel no surprise when the resulting shape is that of a logarithmic spiral.

So enamored was James Bernoulli of the self-similarity of the logarithmic spiral—his *spira mirabilis*—that he requested that it be engraved on his tombstone, together with the inscription "*Eadem numero mutata resurgo*" ("Though changed, I arise again exactly the same").[9]

It has been held that, among all logarithmic spirals, those whose constant angle ψ lies in the neighborhood of 75° produce the most pleasing spiral shapes, which occur most frequently in art and nature ([118], p. 71; [414], volume II). The spiral

$$r = \alpha\beta^\theta$$

with $\alpha = 1/\sqrt{5}$ and $\beta = \tau^{2/\pi}$, which was introduced at the beginning of this section, and for which the values of the radius draw nearer and nearer to the terms of the Fibonacci sequence (as the curve crosses each successive coordinate axis), has an angle ψ which is given by the formula

$$\tan\psi = \frac{1}{\log\beta} = \frac{\pi}{2\log\tau}.$$

Solving for ψ we find that

$$\psi = 73° \text{ (approximately)}.$$

Is all this happenstance? Is the order there, waiting to be uncovered, or do we put it there to fulfill a need?

> We have found that where science has progressed the farthest, the mind has but regained from nature that which the mind has put into nature. We have found a strange footprint on the shores of the unknown. We have devised profound theories, one after another, to account for its origin. At last, we have succeeded in reconstructing the creature that made the footprint. And Lo! It is our own.
>
> —Sir Arthur Eddington

---

[8] *On Growth and Form* [414], volume II, p. 757.

[9] Quoted in [308], p. 145.

PROBLEMS

1. Show that for every positive integer n,

$$\tau^n = \tau f_n + f_{n-1}$$

(define $f_0 = 0$). Use this relation to give another proof of Binet's formula.

2. Establish the relation

$$\frac{1}{1 \times 2} - \frac{1}{2 \times 3} + \frac{1}{3 \times 5} - \frac{1}{5 \times 8} + \cdots = \tau^{-2}.$$

3. Establish the relation

$$\frac{1}{f_1} + \frac{1}{f_2} + \frac{1}{f_4} + \frac{1}{f_8} + \cdots = 4 - \tau.$$

    Hint. Find a simple formula for the nth partial sum of the series.

4. Show that for every positive integer n,

$$\lim_{k \to \infty} \frac{f_{k+n}}{f_k} = \tau^n.$$

5.   The diagonals of a regular pentagon form a figure called a *pentagram* (see Figure 10). It is the ancient symbol of the Pythagorean Brotherhood, a figure rich in golden ratios.

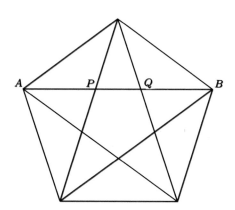

FIGURE 10

a. Show that if the side of the pentagon is 1, then the length of each diagonal is τ and the length of a side of the inner pentagon is τ^{-2}.

b. Calculate the ratios AB/AQ and AQ/AP.

c. When the five isosceles triangles of the pentagram are folded up to form a pyramid (see Figure 11), what is the ratio of the height to the radius of the circumcircle of the base?

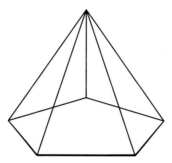

FIGURE 11

6. Five disks of unit radius are placed symmetrically as shown in Figure 12 so that their centers form the vertices of a regular pentagon and their circumferences all pass through the center of the pentagon. What is the radius of the largest circle covered by the five disks?

FIGURE 12

7. Prove that a regular icosahedron can be inscribed in a regular octahedron in such a way that each vertex of the former divides an edge of the latter in the golden ratio.

8. Wythoff's Game. Two players A and B alternately remove counters from two piles according to the following rules: Each player must take at least one counter at his turn. He may choose any number of counters from either pile, but, having decided to draw from both piles, he must take the same number of counters from each. The player who takes the last counter wins.

    a.   Show that in order for player A to win he should, after his move, leave one of the following "safe" combinations:

$$(1,2), (3,5), (4,7), (6,10), (8,13), (9,15), (11,18), (12,20), \ldots.$$

Show that no matter what B does on his next move, A can always convert this back into a safe combination, and thereby win the game.

    b.   Show that the nth pair of numbers forming a safe combination is given by

$$([n\tau], [n\tau^2]) \qquad n = 1, 2, 3, \ldots,$$

where $[x]$ denotes the greatest integer not exceeding x.

    Hint. A natural approach to the solution is provided by the following theorem of Beatty: *If x and y are positive irrational numbers such that $\frac{1}{x} + \frac{1}{y} = 1$, then the sequences*

$$[x], [2x], [3x], \ldots \qquad and \qquad [y], [2y], [3y], \ldots$$

together include every positive integer just once. (See, for example, [171], p. 57. For more details about Wythoff's game and related subjects, see [114] and [115].)

9. (Bombelli, 1572) Find the value of the continued fraction

$$a + \cfrac{b}{2a + \cfrac{b}{2a + \cfrac{b}{2a + \cdots}}}$$

for positive integers a and b.

10. Prove that it is impossible to list all the real numbers in the form of an infinite sequence $\mathbf{R} = \{x_1, x_2, x_3, \ldots\}$.

Indication of the proof.   Suppose that such an enumeration were possible. Choose a closed interval I_1 of length less than 1 that does not contain x_1. Within I_1, choose a closed interval I_2 of length less than $1/2$ that does not contain x_2. Continue. Now invoke the nested interval principle.

11.   Mazur's Game.   A game between two players A and B is defined as follows: A selects an arbitrary closed interval I_1 of length less than 1, B then selects an arbitrary closed interval I_2 contained in I_1, of length less than $1/2$; then A in his turn selects an arbitrary closed interval I_3 contained in I_2, of length less than $1/3$, and so on. By the nested interval principle, the intersection of $I_1, I_2, I_3, \ldots$ contains a single real number r. If r is irrational, then A wins the game; otherwise he loses. Show that A can always win the game, no matter what strategy B may adopt.

   Hint. It is possible to list all the rational numbers in the form of an infinite sequence $r_1, r_2, r_3, \ldots$.

12.   Show that every continuous function has the intermediate value property: *If f is continuous on $[a, b]$, $f(a) < 0$ and $f(b) > 0$, then $f(x) = 0$ for some x between a and b.*

   Hint. Bisect $[a, b]$ and evaluate f at the midpoint.

13.   Show that every derivative has the intermediate value property: *If f is differentiable on $[a, b]$, $f'(a) < 0$ and $f'(b) > 0$, then $f'(x) = 0$ for some x between a and b.*

14.   Prove that every continuous function on a closed interval is bounded.

   Hint. If f were continuous on $[a, b]$, yet unbounded, then it would be unbounded on either $[a, (a + b)/2]$ or $[(a + b)/2, b]$.

15.   a.   Prove that there does not exist a continuous function f defined on **R** that takes on every value exactly twice.

   Hint. If $f(a) = f(b)$ for $a < b$, then either $f(x) > f(a)$ for all x in (a, b) or $f(x) < f(a)$ for all x in (a, b).

   b.   Find a continuous function that takes on every value exactly 3 times.

   c.   More generally, show that there is a continuous function that takes on every value exactly n times if and only if n is odd.

16.   a. **Universal Chord Theorem.**   Let f be a continuous function defined on the interval $[0, 1]$. We say that f has a *horizontal chord* of length a if there is a point x such that x and $x + a$ are both in the domain of f and $f(x) = f(x + a)$.

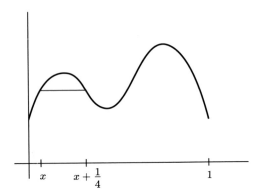

FIGURE 13

Show that if $f(0) = f(1)$, so that f has a horizontal chord of length 1, then it also has horizontal chords of lengths $1/2, 1/3, 1/4, \ldots$.

Hint. Let n be a positive integer and consider the telescoping sum

$$\sum_{k=0}^{n-1} \left[f\left(\frac{k+1}{n}\right) - f\left(\frac{k}{n}\right) \right].$$

b. Suppose that $0 < a < 1$, but that a is not equal to $1/n$ for any positive integer n. Find a function f continuous on $[0, 1]$, with $f(0) = f(1)$, but having no horizontal chord of length a.

(For related examples and references, see [46].)

17. A set A of real numbers is said to be **dense** if every open interval contains a point of A. For example, the set of rational numbers and the set of irrational numbers are both dense. Prove that if two continuous functions agree on a dense set, then they agree everywhere.

18. Prove that if f is continuous and $f(x + y) = f(x) + f(y)$ for all x and y, then there is a number c such that $f(x) = cx$ for all x.

19. Find all functions that (simultaneously) satisfy the equations

$$f(x + y) = f(x) + f(y) \qquad \text{and} \qquad f(xy) = f(x)f(y)$$

for all x and y.

Hint. Show that, unless f reduces to the zero function, $f(x) < f(y)$ whenever $x < y$.

3. Generating Functions

If we wish to obtain information about a sequence of numbers

$$a_1, a_2, a_3, \ldots$$

we can often do so indirectly by forming the power series

$$F(x) = \sum_{n=1}^{\infty} a_n x^n. \tag{1}$$

The resulting function $F(x)$ is a single quantity which represents the entire sequence $\{a_n\}$, and knowledge about the function frequently sheds light on the sequence that was used to define it. If, for example, $F(x)$ turns out to be a known function, then its coefficients can usually be determined by elementary calculus.

If we are to justify the use of the calculus, then we must be sure that the power series (1) is not just a "formal power series" but a well-defined function that converges for all x in some neighborhood of the origin. In other words, we must be certain that *the power series has a positive radius of convergence*.

The function $F(x)$ is called the **generating function** for the sequence a_1, $a_2, a_3, \ldots$.

Example 1. Binet's Formula Revisited. As an illustration of what is possible, let us take $\{a_n\}$ to be the sequence of Fibonacci numbers. The resulting generating function is then

$$F(x) = x + x^2 + 2x^3 + 3x^4 + 5x^5 + \cdots, \tag{2}$$

and its radius of convergence is given by the formula

$$R = \lim_{n \to \infty} \frac{f_n}{f_{n+1}} = \frac{1}{\tau}$$

where τ is the *golden ratio* (see §2). Since R is positive we may proceed.

Multiplying $F(x)$ successively by x and x^2, we find

$$xF(x) = x^2 + x^3 + 2x^4 + 3x^5 + 5x^6 + \cdots$$

and

$$x^2 F(x) = x^3 + x^4 + 2x^5 + 3x^6 + \cdots.$$

Adding these equations, we then find that

$$xF(x) + x^2 F(x) = F(x) - x,$$

and therefore

$$F(x) = \frac{x}{1 - x - x^2}.$$

This then is the elementary function whose Maclaurin series generates the Fibonacci numbers.

To obtain further information about the Fibonacci numbers, we shall derive another power series expansion for $F(x)$. Using the partial fractions decomposition for $x/(1 - x - x^2)$, we find readily that

$$F(x) = \frac{1}{\sqrt{5}} \left(\frac{1}{1 - \tau x} - \frac{1}{1 - \mu x} \right)$$

where $\mu = 1 - \tau$. Since each of the two terms in parentheses on the right is the sum of a convergent geometric series, it follows that

$$F(x) = \frac{1}{\sqrt{5}} \left[(\tau - \mu)x + (\tau^2 - \mu^2)x^2 + (\tau^3 - \mu^3)x^3 + \cdots \right]. \qquad (3)$$

But the Maclaurin expansion of a function is unique, and therefore the two representations (2) and (3) for $F(x)$ must be the same. Equating coefficients of like powers of x, we find that

$$f_n = \frac{\tau^n - \mu^n}{\sqrt{5}}$$

for every value of n. We have rediscovered Binet's formula.

This elegant and general method for investigating sequences was first introduced by Laplace in his "Théorie analytique des probabilités," 2nd ed. (1814). It was further pioneered during the eighteenth century by Nicolas Bernoulli, Euler, and Stirling. The method continues to play a prominent role in many areas of mathematics, especially in probability and combinatorics.

We have emphasized the fact that a generating function is not just a formal power series and that the convergence of the series in a neighborhood of the origin is crucial. We must now admit that most (if not all) of the operations usually performed with generating functions can be rigorously justified without regard to the convergence of the series (see, for example, [319]). Nevertheless, once the solution to a problem has been discovered, no matter how flawed the method, it may then be possible to justify the solution by other more rigorous means.

Example 2.   A Property of the Bernoulli Numbers.   Many of the striking properties of the Bernoulli numbers can be gleaned from the simple function that generates them. We begin by showing that

$$\frac{x}{e^x - 1} = \sum_{n=0}^{\infty} B_n x^n / n!. \tag{4}$$

(The presence of $n!$ in the denominator makes the resulting power series far simpler. It is called the *exponential generating function* of the sequence $\{B_n\}$.) Applying the *method of undetermined coefficients*, we suppose that there are numbers $b_0, b_1, b_2, \ldots$ such that the expansion $x/(e^x - 1) = b_0 + b_1 x + b_2 x^2/2! + \cdots$ is valid throughout some neighborhood of the origin (a proof that such a representation is possible can be found in [262], pp. 179ff.). Therefore,

$$x = (b_0 + b_1 x + b_2 x^2/2! + \cdots)(x + x^2/2! + x^3/3! + \cdots)$$

by virtue of the known power series for the exponential. By formally expanding the right-hand side and then equating coefficients of like powers of x, we find

$$b_0 = 1$$

$$b_0 \frac{1}{2!} + b_1 = 0$$

$$b_0 \frac{1}{3!} + b_1 \frac{1}{2!} + \frac{b_2}{2!} = 0$$

$$b_0 \frac{1}{4!} + b_1 \frac{1}{3!} + \frac{b_2}{2!} \frac{1}{2!} + \frac{b_3}{3!} = 0$$

$$\cdots$$

But every equation beyond the first can be put in the form

$$\sum_{i=0}^{n-1} b_i \binom{n}{i} = 0,$$

and this is precisely the recursive formula satisfied by the Bernoulli numbers (see chapter 2, §2). Since $b_0 = B_0 = 1$, it follows that $b_n = B_n$ for every n.

We can exploit formula (4) by writing it in the form

$$1 + \sum_{n=2}^{\infty} B_n x^n / n! = \frac{x}{e^x - 1} + \frac{x}{2}$$

(remember that $B_1 = -\frac{1}{2}$). Since the function on the right is just

$$\frac{x}{2} \frac{e^x + 1}{e^x - 1} = -\frac{x}{2} \frac{e^{-x} + 1}{e^{-x} - 1},$$

and hence an even function of x, it follows that the series on the left cannot contain a single odd power. Thus, *every Bernoulli number B_n of odd index $n > 1$ is equal to zero.*

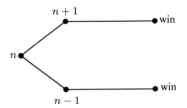

FIGURE 14

Example 3. Gambler's Ruin.   Consider the problem of a gambler who wins or loses a dollar in each play of a "fair" game. By fair we mean that the probability of a win or a loss is the same for both players and is equal to $1/2$ for each trial. Let the initial capital of the gambler be a and that of his opponent b. The game continues until one of the players has won all the money of his adversary, that is, until one of the players has been "ruined." We wish to determine the probability that the gambler will be the winner.

It is possible, of course, that neither player will win, that the game will continue indefinitely, but, as we shall see, the probability that this occurs is zero.

Let $P(n)$ denote the probability that the gambler is ultimately the winner, given that his current fortune is n dollars. There are two ways in which this may happen (illustrated by Figure 14). First, he may win the next game. The probability of this event is $1/2$. His fortune will then be $n + 1$ and his probability of winning $P(n+1)$. Secondly, he may lose the next game, also with a probability of $1/2$. His fortune will then be $n - 1$ and his probability of winning $P(n - 1)$. Therefore, by the law of total probability,

$$P(n) = \frac{1}{2}P(n + 1) + \frac{1}{2}P(n - 1) \tag{5}$$

provided $0 < n < a + b$.

The game terminates when either $n = 0$ or $n = a + b$. In the first case the gambler is ruined, and in the second case his opponent is ruined. Therefore, we must have

$$P(0) = 0 \qquad \text{and} \qquad P(a + b) = 1. \tag{6}$$

Systems of the form (5) are known as *difference equations,* and (6) represents the *boundary conditions* on $P(n)$. While it is a simple matter to derive an explicit expression for $P(n)$ by the method of particular solutions (see [156], volume I, pp. 344ff.), we shall prefer the method of generating functions instead.

For this purpose, let us regard $P(n)$ as an arbitrary function that satisfies conditions (5) and (6) for *all* values of the variable and not just for $n = 0, 1, \ldots, a + b$. Let

$$F(x) = P(1)x + P(2)x^2 + \cdots$$

be the corresponding generating function for $\{P(n)\}$. If we multiply $F(x)$ successively by x and x^2 then we obtain in the same way as before

$$F(x) - 2xF(x) + x^2F(x) = P(1)x.$$

Therefore,

$$F(x) = P(1)\frac{x}{(1-x)^2}.$$

To derive a second series representation for $F(x)$, we start with the geometric series

$$\frac{1}{1-x} = 1 + x + x^2 + x^3 + \cdots$$

and differentiate term-by-term. The result is $1/(1-x)^2 = 1 + 2x + 3x^2 + \cdots$, so that

$$\frac{x}{(1-x)^2} = x + 2x^2 + 3x^3 + \cdots.$$

Therefore

$$\sum_{n=1}^{\infty} P(n)x^n = \sum_{n=1}^{\infty} P(1)nx^n$$

and it follows that

$$P(n) = P(1)n$$

for $n = 0, 1, 2, \ldots$. Substituting $n = a + b$, we find that $P(1) = \frac{1}{a+b}$; therefore

$$P(n) = \frac{n}{a+b}.$$

Thus the probability that the gambler is ultimately the winner is given by $P(a) = \frac{a}{a+b}$, the ratio of his initial capital to the combined capital in the game. By symmetry, the probability that his opponent will be the winner is just $\frac{b}{a+b}$. But the two numbers $\frac{a}{a+b}$ and $\frac{b}{a+b}$ add up to 1, and therefore the probability of an unending game is zero, as previously asserted.

Using a similar argument, it is not difficult to show that *the expected duration of the game is equal to the product ab of each player's initial capital* (see

[156], volume I, pp. 348ff.). For example, if both players begin with 500 dollars, then the average duration of the game is 250,000 trials. If one player has 1000 dollars and his adversary only one dollar, the average duration is 1000 trials. These results are considerably longer than one might intuitively expect.

   For a detailed account of the classical ruin problem and its relation to "random walks" (especially when the game is not fair), the reader should consult volume I of William Feller's classic work *An Introduction to Probability Theory and Its Applications* [156].

PROBLEMS

1. A sequence of numbers $a_0, a_1, \ldots$ is defined recursively by setting $a_0 = 1$ and

$$a_{n+1} = 2a_n + n \qquad n = 0, 1, 2, \ldots.$$

Use generating functions to find an explicit formula for a_n.

2. Use generating functions to prove that

$$\binom{n}{0} + \binom{n}{1} + \binom{n}{2} + \cdots + \binom{n}{n} = 2^n.$$

3. Prove the binomial theorem

$$(x + y)^n = \sum_{k=0}^{n} \binom{n}{k} x^k y^{n-k}$$

by comparing the coefficient of $t^n/n!$ on both sides of the equation $e^{t(x+y)} = e^{tx}e^{ty}$.

4. The **convolution** of two sequences $\{a_n\}$ and $\{b_n\}$ is the sequence defined by

$$\left\{ \sum_k a_k b_{n-k} \right\}_{n=0}^{\infty}$$

Show that the generating function of the convolution is the product of the generating functions of the given sequences.

5. The nth **harmonic number** H_n is defined by

$$H_n = 1 + \frac{1}{2} + \frac{1}{3} + \cdots + \frac{1}{n} \qquad n = 1, 2, 3, \ldots.$$

Prove that the generating function for the sequence $\{H_n\}$ is given by

$$\sum_{n=1}^{\infty} H_n x^n = \frac{1}{1-x} \log\left(\frac{1}{1-x}\right).$$

6. Evaluate the sum $\sum_{0<k<n} 1/k(n-k)$ in two ways:
1. by expanding the summand in partial fractions, and
2. by treating the sum as a convolution and using generating functions.

7. Express the sum $\sum_{0<k<n} f_k f_{n-k}$ in closed form.

8. Use generating functions to show that

$$\sum_{k=0}^{n} \binom{r}{k}\binom{s}{n-k} = \binom{r+s}{n}.$$

This is *Vandermonde's convolution.*

9. Show that if $f(x)$ is the exponential generating function of $\{a_n\}$ and $g(x)$ is the exponential generating function of $\{b_n\}$, then $f(x)g(x)$ is the exponential generating function of the sequence

$$\left\{\sum_{k} \binom{n}{k} a_k b_{n-k}\right\}_{n=0}^{\infty}$$

(compare problem 4).

10. By viewing the quotient

$$\frac{x(e^{mx} - 1)}{e^x - 1}$$

in two ways, derive Bernoulli's formula for the sum of the nth powers of the first $m - 1$ positive integers (compare chapter 2, §2).

11. A permutation of the set of integers $\{1, 2, 3, \ldots, n\}$ that has no fixed points is called a **derangement**. (Compare problem 14, chapter 1, §3.) Let D_n denote the number of derangements of $\{1, 2, 3, \ldots, n\}$ and let

$$D(x) = \sum_{n=0}^{\infty} D_n x^n / n!$$

be the corresponding exponential generating function (take $D_0 = 1$).
    a.   Using a combinatorial argument, show that

$$n! = \sum_k \binom{n}{k} D_{n-k} \qquad n \geq 0. \qquad\qquad (7)$$

    b.   By computing the exponential generating function of both sides of (7), show that

$$\frac{1}{1-x} = e^x D(x).$$

    c.   Conclude that

$$D_n / n! = 1 - \frac{1}{1!} + \frac{1}{2!} - \frac{1}{3!} + \cdots + (-1)^n \frac{1}{n!}.$$

12.   Stirling Numbers of the Second Kind.   By a **partition** of a set S we mean a collection of nonempty, pairwise disjoint subsets of S whose union is the entire set. The symbol $\left\{ {n \atop k} \right\}$ represents the number of partitions of an n-element set into k subsets. These are the Stirling numbers of the second kind. For example, $\left\{ {4 \atop 2} \right\} = 7$ because there are seven ways to partition the four-element set $\{1, 2, 3, 4\}$ into two subsets:

$$\{1, 2, 3\} \cup \{4\}, \ \{1, 2, 4\} \cup \{3\}, \ \{1, 3, 4\} \cup \{2\}, \ \{2, 3, 4\} \cup \{1\},$$

$$\{1, 2\} \cup \{3, 4\}, \ \{1, 3\} \cup \{2, 4\}, \ \{1, 4\} \cup \{2, 3\}.$$

It is convenient to define $\left\{ {0 \atop 0} \right\} = 1$ and $\left\{ {n \atop 0} \right\} = 0$ for $n > 0$.
    a.   By using a combinatorial argument, show that $\left\{ {n \atop 2} \right\} = 2^{n-1} - 1$ for all $n > 0$.
    b.   Establish the recurrence relation

$$\left\{ {n \atop k} \right\} = \left\{ {n-1 \atop k-1} \right\} + k \left\{ {n-1 \atop k} \right\}$$

for all positive n and k.

Stirling's Triangle for Subsets

n	$\genfrac\{\}{0pt}{}{n}{0}$	$\genfrac\{\}{0pt}{}{n}{1}$	$\genfrac\{\}{0pt}{}{n}{2}$	$\genfrac\{\}{0pt}{}{n}{3}$	$\genfrac\{\}{0pt}{}{n}{4}$	$\genfrac\{\}{0pt}{}{n}{5}$	$\genfrac\{\}{0pt}{}{n}{6}$	$\genfrac\{\}{0pt}{}{n}{7}$	$\genfrac\{\}{0pt}{}{n}{8}$	$\genfrac\{\}{0pt}{}{n}{9}$
0	1									
1	0	1								
2	0	1	1							
3	0	1	3	1						
4	0	1	7	6	1					
5	0	1	15	25	10	1				
6	0	1	31	90	65	15	1			
7	0	1	63	301	350	140	21	1		
8	0	1	127	966	1701	1050	266	28	1	
9	0	1	255	3025	7770	6951	2646	462	36	1

c. Derive the expansion

$$\sum_{n=0}^{\infty} \genfrac\{\}{0pt}{}{n}{k} x^n = \frac{x^k}{(1-x)(1-2x)(1-3x)\cdots(1-kx)}$$

for all $k > 0$.

d. By using the partial fractions decomposition of

$$\frac{1}{(1-x)(1-2x)\cdots(1-kx)},$$

derive an explicit formula for $\genfrac\{\}{0pt}{}{n}{k}$.

13. The nth **Bell number** b_n is the number of ways of partitioning a set of n objects into subsets. For example, $b_3 = 5$ since the partitions of $\{1, 2, 3\}$ are:

$$\{1, 2, 3\}, \ \{1, 2\} \cup \{3\}, \ \{1, 3\} \cup \{2\}, \ \{2, 3\} \cup \{1\}, \ \{1\} \cup \{2\} \cup \{3\}.$$

It is convenient to define $b_0 = 1$. Thus the sequence of Bell numbers begins $1, 1, 2, 5, 15, 52, \ldots$.

a. Prove that $b_{n+1} = \sum_{k=0}^{n} \binom{n}{k} b_k$.

b. Use the recurrence in part (a) to show that the exponential generating function of the sequence of Bell numbers is $e^{e^x - 1}$.

c. Derive the following remarkable formula for the Bell numbers:

$$b_n = \frac{1}{e} \sum_{k=1}^{\infty} \frac{k^{n-1}}{(k-1)!}.$$

14. Problème de parties. This historically important problem was one of the first problems on probability discussed and solved by Fermat and Pascal in their correspondence. Two players, A and B, agree to play a series of games on the condition that A wins if he succeeds in winning a games before B wins b games. The probability of winning a single game is p for A and $q = 1 - p$ for B. What is the probability that A will win the series?

Hint. Let $u_{s,t}$ denote the probability that A wins the series when he has s games left to win and his adversary B has t games left to win.

a. Show that $u_{s,t}$ satisfies the recursion

$$u_{s,t} = p u_{s-1,t} + q u_{s,t-1}.$$

b. Use the generating function

$$F_s(x) = u_{s,0} + u_{s,1} x + u_{s,2} x^2 + \cdots$$

of the sequence $\{u_{s,t}\}_{t=0}^{\infty}$ to find an explicit formula for $u_{s,t}$.

15. Let c_n denote the number of ways to make n cents change using pennies, nickels, dimes, quarters, and half-dollars. Show that the generating function of $\{c_n\}$ is given by

$$\frac{1}{(1-x)(1-x^5)(1-x^{10})(1-x^{25})(1-x^{50})} = 1 + \sum_{n=1}^{\infty} c_n x^n.$$

How many ways are there to change a dollar?

Hint. Write the quotient as

$$(1 + x + x^2 + \cdots)$$
$$(1 + x^5 + x^{10} + \cdots)$$
$$(1 + x^{10} + x^{20} + \cdots)$$
$$(1 + x^{25} + x^{50} + \cdots)$$
$$(1 + x^{50} + x^{100} + \cdots)$$

and formally multiply the series together.

16. (Continuation) Find the generating function for the number of ways to make n cents change using pennies, nickels, dimes, quarters, and half-dollars, using each type of coin at least once.

17. Use generating functions to show that every positive integer can be expressed as a unique sum of distinct powers of 2.

Hint. Establish the identity

$$\frac{1}{1-x} = (1+x)(1+x^2)(1+x^4)(1+x^8)\cdots.$$

18. A **partition** of a number n is a representation of n as a sum of any number of positive integers. The order of the terms is irrelevant. For example, the seven partitions of the number 5 are

$$5 = 4+1 = 3+2 = 3+1+1 = 2+2+1$$

$$= 2+1+1+1 = 1+1+1+1+1.$$

If $p(n)$ denotes the number of partitions of n, establish Euler's formula for the generating function of $\{p(n)\}$:

$$\frac{1}{(1-x)(1-x^2)(1-x^3)\cdots} = 1 + \sum_{n=1}^{\infty} p(n)x^n.$$

Hint. Write the left side as

$$(1+x+x^2+\cdots)$$
$$(1+x^2+x^4+\cdots)$$
$$(1+x^3+x^6+\cdots)$$
$$\cdots$$

and formally multiply the series together.

19. (Continuation) Show that the generating function for the number of partitions of n into distinct integers is

$$(1+x)(1+x^2)(1+x^3)\cdots.$$

20. (Continuation) Show that the number of partitions of an integer into odd integers is equal to the number of partitions into distinct integers.

21. Example 2 shows that

$$\sum_{n=0}^{\infty} B_{2n} \frac{x^{2n}}{(2n)!} = \frac{x}{2} \frac{e^x+1}{e^x-1}$$

$$= \frac{x}{2} \frac{e^{x/2}+e^{-x/2}}{e^{x/2}-e^{-x/2}}.$$

Replace x by $2ix$ and thereby derive the expansion

$$x \cot x = \sum_{n=0}^{\infty} (-4)^n B_{2n} \frac{x^{2n}}{(2n)!}.$$

22. (Continuation) Derive the expansion

$$\tan x = \sum_{n=1}^{\infty} (-1)^{n-1} 4^n (4^n - 1) B_{2n} \frac{x^{2n-1}}{(2n)!}.$$

Remark. The formula can be used to show that all of the numbers

$$(-1)^{n-1} \frac{4^n (4^n - 1)}{2n} B_{2n}$$

are positive integers; they are called *tangent numbers* (see [186], p. 273).

4. Iterated Functions: From Order to Chaos

Let I be an interval of the real line and $f: I \to I$ a continuous function from I into itself. One of the simplest yet most profoundly useful operations one can perform on f is to apply it repeatedly. The process can be viewed in terms of a simple feedback loop as pictured below: a given input x results in an output $y = f(x)$, which is then taken as the new input.

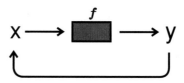

FIGURE 15

The feedback mechanism just described is an example of what is known generally as a **dynamical system**,[10] a process that occurs naturally throughout mathematics and the natural sciences. We may, for example, regard x as an element of a system that changes at discrete intervals of time—such as the size of a population or the position of a moving particle—the function f being the

---

[10] In its greatest generality, a dynamical system is any mapping of a set into itself. For purposes of application, however, it is customary to endow both the underlying set and the function that acts on it with some additional structure—algebraic, analytic, or geometric.

rule that specifies how the system is transformed from one time period to the next. After a single period, the new value of x is $f(x)$, after two periods it is $f(f(x)) = f^2(x)$, and so on ad infinitum. Iterative techniques arise in the study of the evolution of any system in any science.

The sequence of iterates

$$x, f(x), f^2(x), f^3(x), \ldots$$

is called the **orbit** of x, and it is natural to ask whether one can predict the fate of all the orbits. Will an orbit converge to some limit? Will it be periodic in the sense that $f^n(x) = x$ for some positive integer n? Will its values be dense in the interval I? It is usually quite difficult to tell by looking at a function what its infinite iterates will do.

As a simple example of a one-dimensional dynamical system, let I be the set of nonnegative real numbers and define

$$f(x) = \sqrt{1+x}$$

for every x in I. In this case, the behavior of the iterates

$$x_{n+1} = f(x_n) \qquad n = 1, 2, 3, \ldots$$

is quite easy to describe, but the reader will find it instructive to first take a hand-held calculator that has a square-root button, start with any initial input x, and perform the necessary calculations repeatedly.

Assertion: *Every orbit approaches a common limit, the golden ratio.*

For simplicity, let us begin by investigating the orbit of 1:

$$1, \sqrt{2}, \sqrt{1 + \sqrt{2}}, \sqrt{1 + \sqrt{1 + \sqrt{2}}}, \ldots.$$

It is at once clear that this sequence is increasing, since every term after the second is obtained from the previous one by replacing the final 2 with $1 + \sqrt{2}$. Boundedness follows readily by induction. Indeed, the first term is less than 2, and if it has already been established that $x_n < 2$, then $x_{n+1} = \sqrt{1 + x_n} < \sqrt{3} < 2$. Thus, the orbit, being both bounded and monotonic, is convergent. If we denote its limit by L, then the recursive formula

$$x_{n+1} = \sqrt{1 + x_n}$$

shows that, as $n \to \infty$,

$$L = \sqrt{1 + L}.$$

This tells us that L is a **fixed point** of f, namely, a solution of the equation $x = f(x)$. Routine algebra is all that is now needed to show that the unique fixed point is $L = (1 + \sqrt{5})/2$, the golden ratio.

The argument just given applies generally, but there is an even simpler graphical solution, of great utility, which enables us to predict the qualitative behavior of all the orbits. To compute the orbit of x, simply draw the diagonal $y = x$ together with the graph of f and then apply the following procedure repeatedly: *Draw a vertical line from (x, x) to the graph of f, followed by a horizontal line back to the diagonal, thereby reaching the point $(f(x), f(x))$.* When this procedure is repeated, we arrive next at the point $(f^2(x), f^2(x))$, thereafter at $(f^3(x), f^3(x))$, and so on ad infinitum. The orbit of x is now visibly displayed along the diagonal. As Figure 16 suggests, each polygonal path converges to the intersection point (L, L) of the curve and the diagonal—thus every orbit approaches the golden ratio. When the initial point x is smaller than L, the orbit increases, and when x is larger than L it decreases. As before, these assertions are readily verified by induction.

The study of continuous functions under iteration goes back to the English mathematician Arthur Cayley (1821–1895), who toward the end of the nineteenth century investigated Newton's method in the complex plane. Early in the twentieth century, Gaston Julia (1893–1978) and Pierre Fatou (1878–1929) in France studied iterations of polynomials and rational functions. However, it is only with the advent of the computer and the possibility of high-resolution graphics that the subject has recently come back to life.

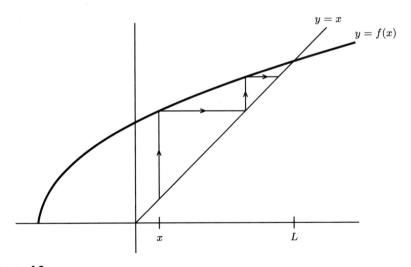

FIGURE 16

One of the startling discoveries of the past few decades is the realization that even the simplest of functions $f\colon \mathbf{R} \to \mathbf{R}$ may exhibit, under iteration, the most complex and unexpected behavior. The quadratic function $x \to x^2 + c$ is typical, and the reader will find it profitable to experiment with various values of the parameter c and the initial input x. While some orbits are easy to describe, others appear random and unpredictable.

However, it is only when we pass to the complex plane that the true beauty and mystery of the subject becomes visible. The most famous example—and in some sense a universal one, since it arises in so many different dynamic processes—is the **Mandelbrot set** M. The set is named for Benoit Mandelbrot, the first to produce pictures of it on a computer and to investigate its structure. By definition, M consists of all complex numbers c with the property that the orbit of c under the complex quadratic mapping $z \to z^2 + c$ remains bounded. In other words, c belongs to M if and only if the sequence

$$c, c^2 + c, (c^2 + c)^2 + c, \ldots$$

does not tend to infinity.

Figure 17 reveals a first approximation of what has been called the most complicated yet most beautiful object in mathematics.

Here is a description given by Adrien Douady, one of the pioneers in the field:

> When you look at the Mandelbrot set, the first thing you see is a region limited by a cardioïd, with a cusp at the point .25 and its round top at the point $-.75$. Then there is a disk centered at the point -1 with radius .25, tangent to the cardioïd. Then you see an infinity of smaller disk-like components, tangent to the cardioïd, most of which are very small. Attached to each of those components, there is again an infinity of smaller disk-like components, and on each of these there is attached an infinity of smaller disk-like components, and so on. But that is not all! If you start from the big cardioïd, go to the disk on the left, again to the component on the left, and keep going, you will tend to a point called the Myrberg–Feigenbaum point, situated at $-1.401\ldots$. Now, the segment from this point to the point -2 is contained in M. And on this segment, there is a small cardioïd-like component, with its cusp at -1.75 (its center is at $-1.754877666\ldots$), accompanied by its family of disk-like satellites just like the big cardioïd. Actually there are infinitely many such cardioïd-like components. There are also cardioïd-like components off of the real axis. B. Mandelbrot discovered one centered at $-0.1565201668 + 1.032247109\, i$,

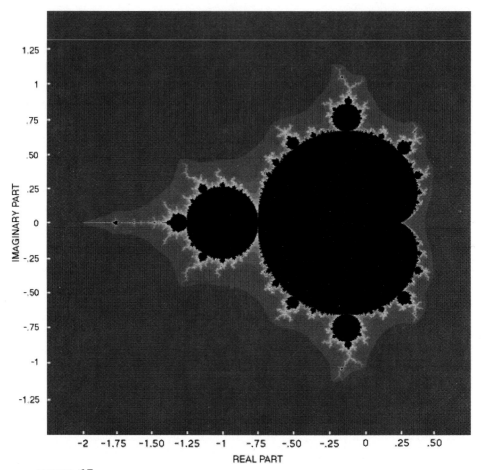

FIGURE 17
The Mandelbrot set

and many others. In fact he showed that there are an infinite number
of them. They are so tiny that it is hard to distinguish them from stains
on the computer picture (except for the fact that they arise symmetrically). However, if you make an enlarged picture, you will discover for
each of them the cardioïd shape and its company of disk-like components. And this is not all. . . . All of these cardioïd-like components are
linked to the main cardioïd by filaments, charged with small cardioïd-
like components, each of which is accompanied by its family of satel-
lites. These filaments are branched according to a very sophisticated
pattern, of which detailed combinatorial studies have been made. Be-

cause of these filaments, the set M is itself connected. The proof of this fact is by no means obvious, and there are still lots of open questions. For instance, one cannot prove up to now that the "filaments" I mentioned are actually arcs of curves that one can parametrize continuously.[11]

By successively enlarging portions of the Mandelbrot set, one can begin to glimpse its strange, concealed beauty and incredible intricacy of detail (see Color Plates 3–6).[12] Miniature versions of the Mandelbrot set become visible within it, and within these smaller copies still, no two exactly alike, each with its own distinctive patterns. The boundary of M is an example of what Mandelbrot has termed a **fractal**, a shape having the property of being "rough but self-similar."[13] Mandelbrot has shown that such shapes abound in nature, and in the final chapter we shall say more about these strange and wondrous forms.

---

[11] Adrien Douady, "Julia Sets and the Mandelbrot Set," in *The Beauty of Fractals* [333], pp. 161–173.

[12] Color Plates 3–5 display successive enlargements of the "shepherd's crook" lying in the region $-0.75104 < \text{Re}\, c < -0.74080, 0.10511 < \text{Im}\, c < 0.11536$. Color Plate 6 shows a "compound eye" peering out from a region in Color Plate 3. These images were produced with the kind assistance of Mok Oh.

In an expository account of the Mandelbrot set, A. K. Dewdney (*Scientific American,* August 1985) describes a simple program, called MANDELZOOM, for generating pictures such as these on a microcomputer. It is important to remember, however, that points of the Mandelbrot set appear in black. The enhanced beauty of the images results from the halo of colors that are assigned to the complement. This assignment is based on the number of iterations necessary for an orbit to escape from the circle $|z| = 2$. (After leaving the circle the orbit diverges rapidly to infinity and c does not belong to the Mandelbrot set.) Values of c for which escape occurs after only a few iterations are colored red. As the number of iterations increases the colors move toward the violet, at the opposite end of the spectrum. It was John Hubbard's inspiration to color the complement.

[13] Benoit B. Mandelbrot,"Fractals and the Rebirth of Iteration Theory," in *The Beauty of Fractals* [333], pp. 151–160. According to Mandelbrot, "The word similar does not always have the pedantic sense of 'linearly expanded or reduced', but it always conforms to the convenient loose sense of 'alike'." Mandelbrot's ideas are set forth in his classic work *The Fractal Geometry of Nature* [290]. For introductory accounts of this new and growing field, see [132] and [172]. More advanced treatments are given in [24], [131], [133], [333], and [334].

COLOR PLATE 3

COLOR PLATE 4

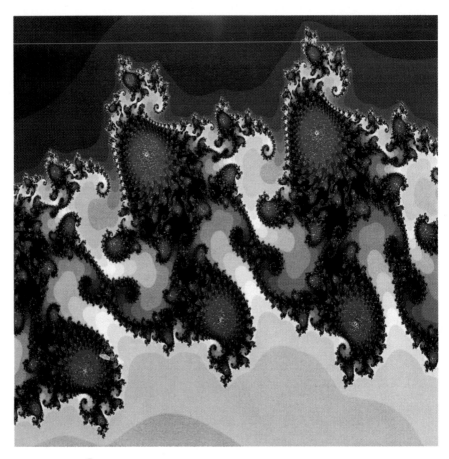

COLOR PLATE 5

COLOR PLATE 6

PROBLEMS

1. Division on a Computer. The following algorithm provides an approximate solution of the equation $ax = b$ $(a, b > 0)$ using only the operations of addition, subtraction, and multiplication. Since a binary computer can multiply and divide by a power of 2, we can assume at the outset that $1/2 \leq a < 1$. Rewrite the equation $ax = b$ in the form

$$x = (1 - a)x + b.$$

Show—with and without graphical analysis—that the orbit of 0 under the mapping $x \to (1 - a)x + b$ converges to b/a. (For an excellent account of efficient algorithms for doing arithmetic, see *The Art of Computer Programming* [263], volume 2.)

2. Prove that every continuous function $f \colon [a, b] \to [a, b]$ has a fixed point, i.e., a point x_0 such that $f(x_0) = x_0$.

3. A function $f \colon [a, b] \to [a, b]$ is called a **contraction mapping** if there is a number λ, where $0 < \lambda < 1$, such that for any two points x, y belonging to the interval $[a, b]$ the inequality

$$|f(x) - f(y)| \leq \lambda|x - y| \tag{1}$$

is satisfied.

    a.   Show that every contraction mapping is continuous. Give a geometric interpretation of condition (1).

    b.   Show that if f is differentiable and

$$|f'(x)| \leq \lambda < 1$$

for all x on $[a, b]$, then condition (1) is satisfied.

    Hint. "... the real nature of the mean value theorem is exhibited by writing it as an *inequality,* and not as an equality." (Dieudonné, *Foundations of Modern Analysis,* Academic Press, New York, 1969, p. 148.)

    c.   Show that every contraction mapping has a unique fixed point.

4. Fixed-Point Iteration. The method of proof used in problem 1 is known generally as fixed-point iteration. Suppose that we wish to find a root of the equation $g(x) = 0$. We begin by rewriting the equation in the form

$$x = f(x). \tag{2}$$

(Such a representation is, of course, never unique.) Next, starting with an initial approximation x_0, we substitute x_0 into the right side of (2). The value $x_1 = f(x_0)$ so obtained is taken as the next approximation for the solution. In general, if the approximation x_n has been found the next approximation is obtained from the formula

$$x_{n+1} = f(x_n).$$

a. Show that if $f: [a, b] \to [a, b]$ is a contraction mapping (see problem 3), then for any point x_0 belonging to the interval $[a, b]$ the sequence of iterates $x_0, x_1, x_2, \ldots$ converges to the unique root of the equation $x = f(x)$.

b. Show that this method can be applied to find the positive root of the equation $x^2 - 2 = 0$.

Hint. Write the equation in the form

$$x = \frac{1}{2}\left(x + \frac{2}{x}\right).$$

This is the Babylonian method for obtaining better and better approximations to $\sqrt{2}$. (Compare problem 10, chapter 1, §2.)

5. Prove: If $f: [a, b] \to [a, b]$ is continuous and increasing, then for any x in $[a, b]$, the sequence

$$x, f(x), f(f(x)), \ldots$$

has a limit (which is necessarily a fixed point of f).

6. Suppose that p is a fixed point of f (assumed to be continuously differentiable). We say that p is an *attracting fixed point* if there is an open interval containing p in which all points have orbits that tend to p. We say that p is a *repelling fixed point* if there is an open interval containing p in which all points (except p) have orbits that ultimately leave the interval. Prove that p is an attracting fixed point if $|f'(p)| < 1$ and a repelling fixed point if $|f'(p)| > 1$. (See Figure 18.)

7. The quadratic function $f(x) = cx(1 - x)$ is known generally in dynamics as the **logistic function**. It plays an important role in many growth processes (see, for example, [131] and [172]). Depending on the value of the parameter

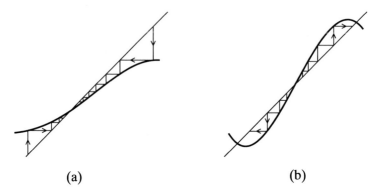

(a)   (b)

FIGURE 18
(a) attracting fixed point; (b) repelling fixed point.

c and the initial input x, the behavior of the system

$$x_{n+1} = cx_n(1 - x_n)$$

can be quite unpredictable. In fact, the dynamics of this system are still not completely understood. (By contrast, elementary calculus shows that the continuous model

$$\frac{dx}{dt} = cx(1 - x)$$

exhibits no unexpected behavior. See, for example, [387], pp. 230ff.)

The table below lists the first 32 values of the orbit of $x = 1/2$ for various values of c.

a. Show that, when $1 < c < 3$, f has an attracting fixed point at $(c-1)/c$ and a repelling fixed point at 0 (see problem 6). Show furthermore that all orbits with initial value $0 < x < 1$ converge to the attracting fixed point.

b. Investigate the case $c = 3$.

c. When c increases beyond 3, the dynamics of f become increasingly more complicated. The entries in the table indicate that the orbit of $1/2$ is attracted to a cycle of period 2 when $c = 3.2$, and to a cycle of period 4 when $c = 3.5$. When $c = 4$, most orbits appear to be "chaotic." Show that the function $f(x) = 4x(1 - x)$ has a periodic orbit of period 3. (A remarkable theorem due to Sarkovskii states that if a continuous function $f: \mathbf{R} \to \mathbf{R}$ has an orbit of period 3, then it has orbits of all other periods. For an elementary proof, based on an ingenious application of the intermediate value theorem, see [131], pp. 60–62.)

Iterate	$c = 1.5$	$c = 3.2$	$c = 3.5$
1	0.375	0.8	0.875
2	0.3515625	0.512	0.3828125
3	0.341949462	0.7995392	0.826934814
4	0.337530041	0.512884056	0.500897694
5	0.335405268	0.799468803	0.874997179
6	0.334362861	0.513018994	0.382819903
7	0.333846507	0.799457618	0.826940887
8	0.333589525	0.513040431	0.500883795
9	0.33346133	0.79945583	0.874997266
10	0.333397307	0.513043857	0.382819676
11	0.333365314	0.799455544	0.826940701
12	0.333349322	0.513044405	0.500884222
13	0.333341327	0.799455499	0.874997263
14	0.33333733	0.513044492	0.382819683
15	0.333335331	0.799455491	0.826940706
16	0.333334332	0.513044506	0.500884209
17	0.333333832	0.79945549	0.874997263
18	0.333333583	0.513044509	0.382819683
19	0.333333458	0.79945549	0.826940706
20	0.333333395	0.513044509	0.50088421
21	0.333333364	0.79945549	0.874997263
22	0.333333348	0.513044509	0.382819683
23	0.333333341	0.79945549	0.826940706
24	0.333333337	0.513044509	0.50088421
25	0.333333335	0.79945549	0.874997263
26	0.333333334	0.513044509	0.382819683
27	0.333333333	0.79945549	0.826940706
28	0.333333333	0.513044509	0.50088421
29	0.333333333	0.79945549	0.874997263
30	0.333333333	0.513044509	0.382819683
31	0.333333333	0.79945549	0.826940706
32	0.333333333	0.513044509	0.50088421

TABLE OF ITERATES

8. (Continuation) Take $c = 1$ as the parameter of the logistic function. Prove that for any initial value x, with $0 < x < 1$,

$$\lim_{n \to \infty} n x_n = 1.$$

9. (Continuation) Take $c = 4$ as the parameter of the logistic function. Determine all values of x $(0 < x < 1)$ for which the orbit of x is convergent, and find the limit of the orbit.

10. For which real values of x does the sequence

$$x, x^2 - 2, (x^2 - 2)^2 - 2, \left[(x^2 - 2)^2 - 2\right]^2 - 2, \ldots$$

converge?

11. Show that when the function $\cos x$ is iterated infinitely often, every orbit approaches the same limit.

12. Show that the series

$$x + \sin x + \sin(\sin x) + \sin[\sin(\sin x)] + \cdots$$

diverges for $0 < x < \pi$.

13. Show that the sequence

$$\sqrt{7}, \sqrt{7 - \sqrt{7}}, \sqrt{7 - \sqrt{7 + \sqrt{7}}}, \sqrt{7 - \sqrt{7 + \sqrt{7 - \sqrt{7}}}}, \ldots$$

converges, and evaluate the limit.

CHAPTER 4

ON THE AVERAGE

It is difficult to understand why statisticians commonly limit their inquiries to Averages, and do not revel in more comprehensive views. Their souls seem as dull to the charm of variety as that of the native of one of our flat English counties, whose retrospect of Switzerland was that, if its mountains could be thrown into its lakes, two nuisances would be got rid of at once.

—Sir Francis Galton[1]

---

[1] *Natural Inheritance,* MacMillan, London, 1889.

Average. Few words have had their origins so thoroughly investigated. While the precise origin of the word remains uncertain, current usage seems to date from the early sixteenth century when it appeared in connection with the maritime trade of the Mediterranean Sea. [2]

> In former times, when the hazards of sea voyages were much more serious than they are today, when ships buffeted by storms threw a portion of their cargo overboard, it was recognized that those whose goods were sacrificed had a claim in equity to indemnification at the expense of those whose goods were safely delivered. The value of the lost goods was paid for by agreement between all those whose merchandise had been in the same ship. The sea damage to cargo in transit was known as 'havaria' and the word came naturally to be applied to the compensation money which each individual was called upon to pay. From this Latin word derives our modern word *average*.[3]

In this way the notion of an average came to be identified with fairness. The ordinary average, or more formally the *arithmetic mean,* of n numbers $x_1, x_2, \ldots, x_n$ is

$$\frac{x_1 + x_2 + \cdots + x_n}{n};$$

their *geometric mean* is

$$(x_1 x_2 \cdots x_n)^{1/n}.$$

Connecting these two averages is an inequality of fundamental importance in analysis, and it is to this that we first turn our attention.

1. The Theorem of the Means

> If the Line AB is divided into any Number of Parts, AC, CD, DE, EB, the Product of all those Parts multiplied into one another will be a *Maximum* when the Parts are equal amongst themselves.
> —Colin Maclaurin (1698–1746)

The arithmetic and geometric means of two positive magnitudes had already appeared in antiquity in connection with a fundamental problem in Greek ge-

---

[2] *Oxford English Dictionary,* 2nd edition, vol. I, p. 817.

[3] M. J. Moroney, *Facts from Figures*, Penguin Books, 1951, p. 34.

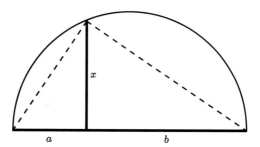

FIGURE 1

ometry: *To construct a square equal in area to a given rectangle.* If the dimensions of the rectangle are a and b, then the problem is to determine x such that

$$x^2 = ab.$$

A simple and elegant solution is contained within the semi-circle pictured in Figure 1. Since the inscribed triangle is a right-triangle, the two smaller triangles are similar, and therefore

$$\frac{a}{x} = \frac{x}{b}.$$

Thus x is the *mean proportional* between a and b, that is to say, their geometric mean, $\sqrt{ab}$, and the construction problem has been solved.[4]

---

   [4] The Pythagoreans, in their theory of proportionals, considered no fewer than *ten* different means, of which the arithmetic, geometric, and harmonic means were the most important [62, p. 56]. (The *harmonic mean* of a and b is by definition the reciprocal of the arithmetic mean of $1/a$ and $1/b$.) Having succeeded in constructing the mean proportional, or goemetric mean, the geometers of the day turned to the more general problem of inserting *two* means between two given magnitudes. Thus, given two arbitrary line segments a and b, the goal was to construct (with ruler and compass alone) two other segments x and y such that

$$\frac{a}{x} = \frac{x}{y} = \frac{y}{b}.$$

This construction, however, could not in general be accomplished. For, in the special case when $b = 2a$, elimination of y from the equations leads to the relation

$$x^3 = 2a^3.$$

Thus, the problem is equivalent to that of "duplicating the cube," which is unsolvable in Euclidean geometry (see chapter 6, §1).

But we have proved even more. Since the radius of the semi-circle is $(a + b)/2$, it follows that

$$\sqrt{ab} \leq \frac{a+b}{2}$$

and that equality holds if and only if $a = b$. *The geometric mean of two positive numbers never exceeds their arithmetic mean.*[5]

The inequality just derived can be used as a stepping stone to a more general statement.

Theorem. *The geometric mean of n positive numbers $a_1, a_2, \ldots, a_n$ never exceeds their arithmetic mean:*

$$(a_1 a_2 \cdots a_n)^{1/n} \leq \frac{a_1 + a_2 + \cdots + a_n}{n}.$$

Moreover, equality holds if and only if $a_1 = a_2 = \cdots = a_n$.

This is the *theorem of the arithmetic and geometric means* or simply the *theorem of the means.* It is of widespread applicability in analysis and a great many proofs have been discovered (see [210]). Here we shall single out the ingenious argument given by the great French mathematician Augustin Cauchy (1789–1857). Cauchy's method, based on an important variant of the principle of mathematical induction, is inspired and its usefulness far transcends the immediate application.

Proof. **1. Forward Induction.** Knowing that the theorem is true for two numbers, it would be natural to try to prove it for three. As it turns out, however, it is easier to prove it for four. Since

$$a_1 a_2 \leq \left(\frac{a_1 + a_2}{2}\right)^2 \quad \text{and} \quad a_3 a_4 \leq \left(\frac{a_3 + a_4}{2}\right)^2,$$

---

[5] Elementary algebra provides an even quicker proof. Since

$$(a + b)^2 - (a - b)^2 = 4ab$$

it follows at once that

$$\frac{a+b}{2} \geq \sqrt{ab}.$$

It is surprising how many important inequalities depend on little more than the obvious assertion that the square of a real number is never negative. Additional examples can be found at the end of the section.

it follows that

$$a_1 a_2 a_3 a_4 \leq \left(\frac{a_1 + a_2}{2} \cdot \frac{a_3 + a_4}{2} \right)^2$$

$$\leq \left(\frac{a_1 + a_2 + a_3 + a_4}{4} \right)^4$$

the final inequality being itself an immediate consequence of the theorem of the means applied to the two numbers $(a_1 + a_2)/2$ and $(a_3 + a_4)/2$. Now take the fourth root of both sides. It is clear that equality holds throughout if and only if $a_1 = a_2 = a_3 = a_4$.

We have taken the first step of an obvious inductive procedure. By repeating the argument we can prove the theorem for eight numbers, then sixteen numbers, and, in general, for any set of n positive numbers, provided that n is a power of two.

2. Backward Induction.   The theorem of the means is now known to be true for an infinite sequence of positive integers, $n = 2, 4, 8, 16, \ldots$, tending to infinity. To complete the proof we need only show that *its truth for any $n > 2$ implies its truth for $n - 1$.* (For example, the truth for $n = 16$ will imply the truth for 15, which in turn will imply the truth for 14, and so on for all positive integers less than 16.)

Suppose then that $n - 1$ positive numbers $a_1, a_2, \ldots, a_{n-1}$ are given and let A denote their arithmetic mean. It is to be shown that

$$a_1 a_2 \cdots a_{n-1} \leq A^{n-1}. \tag{1}$$

By the inductive hypothesis, we are entitled to apply the theorem to the n numbers $a_1, a_2, \ldots, a_{n-1}$ and A. Thus,

$$a_1 a_2 \cdots a_{n-1} A \leq \left(\frac{a_1 + a_2 + \cdots + a_{n-1} + A}{n} \right)^n .$$

But $a_1 + a_2 + \cdots + a_{n-1} = (n-1)A$, so the term in parentheses equals A, and (1) follows immediately. Once again, equality holds throughout if and only if all the a's are equal.

Example 1.   Maxima and Minima without Calculus: An Isoperimetric Problem.   Describing the unusual appeal of problems involving maxima and minima, George Pólya writes:

> Problems concerned with greatest and least values, or maximum and minimum problems, are more attractive, perhaps, than other mathematical problems of comparable difficulty, and this may be due to a

quite primitive reason. Everyone of us has his personal problems. We may observe that these problems are very often maximum or minimum problems of a sort. We wish to obtain a certain object at the lowest possible price, or the greatest possible effect with a certain effort, or the maximum work done within a given time and, of course, we wish to run the minimum risk. Mathematical problems on maxima and minima appeal to us, I think, because they idealize our everyday problems.[6]

The differential calculus provides us with a powerful tool for solving optimization problems. In many instances, however, the use of the calculus may be impractical—or even impossible—and we may then find that the more elementary method is also the more effective. Among the many alternatives, the inequality between the arithmetic and geometric means deserves special attention. It can be formulated in two different ways:

The product of n positive numbers with a given sum is greatest when these numbers are all equal.

The sum of n positive numbers with a given product is smallest when these numbers are all equal.

These two logically equivalent statements are called *dual statements.* They are both consequences of the theorem of the means, it is only a matter of deciding which side of the inequality is to be regarded as given. The reciprocal relation that exists between the statements is by no means fortuitous. In the study of variational problems of this type, it is usually the case that the solution of one problem automatically yields the solution of the other one as well.[7]

As an illustration, we shall discuss a simple problem taken from the calculus of variations, an area of mathematics that deals with optimal forms in geometry and nature.

Problem.   Among all triangles with a given perimeter, which one has the greatest area?

The solution is readily conjectured and, once conjectured, just as readily derived from the theorem of the means. Suppose then that a, b, and c are the lengths of the sides of a triangle whose perimeter P has been specified. To

---

[6] *Induction and Analogy in Mathematics* [338], p. 121.

[7] N. Kazarinoff, *Geometric Inequalities* [245], p. 43.

determine the area of the triangle, as a function of a, b, and c, we make use of Heron's famous formula:[8]

Heron's Formula.   *If A is the area of a triangle with sides of length a, b, and c, then*

$$A = \sqrt{s(s-a)(s-b)(s-c)}$$

where $s = (a+b+c)/2$ is the semi-perimeter.

Thus to maximize A it is sufficient to maximize the product

$$(s-a)(s-b)(s-c).$$

But the sum of these three factors is a *constant,*

$$(s-a) + (s-b) + (s-c) = 3s - (a+b+c) = s = P/2,$$

and therefore their product becomes a maximum when all the factors are equal, that is, when

$$s - a = s - b = s - c.$$

Accordingly,

$$a = b = c$$

and the triangle is equilateral.

The result and its dual can be expressed in the following way:

Of all triangles with the same perimeter, the equilateral triangle has the largest area.

Of all triangles with the same area, the equilateral triangle has the smallest perimeter.

The problem just solved is only one of a number of problems involving maxima and minima that were studied by the Greek geometers. The most general result of this type—and justifiably the most famous—states that *among all plane figures with the same perimeter, the circle has the greatest area.* This is the **isoperimetric theorem** (isoperimetric means having the same perimeter). It was known to Archimedes, but it would take two thousand years before a rigorous proof could be found. The difficulty lies in demonstrating the existence of a maximum. Once it is known that there actually is a plane figure of maximum

---

[8] Elementary proofs can be found in [320] and [354].

area among those having a given perimeter, it is easy to show that it must be a circle. This was done by the Swiss geometer Jacob Steiner (1796–1863). That such a figure exists, however, lies much deeper. The issue was finally settled by the great German mathematician Karl Weierstrass (1815–1897), who was the first to point out that a solution to an extremal problem in geometry may not exist. (For example, there can be no smooth curve of minimum length that passes through three noncollinear points.) Weierstrass developed the calculus of variations on a rigorous basis, and he played a central role in the efforts to secure the foundations of analysis during the last half of the nineteenth century.

It is worth pointing out that, in our treatment of the isoperimetric problem for triangles, the existence of a maximum was never at issue. The theorem of the means provides both an optimal solution and the guarantee that it is unique.

Example 2.   Computing Square Roots by the Method of Successive Approximations.   This well-known numerical algorithm for approximating the square root of a positive number c illustrates the usefulness of the theorem of the means even in the simplest case. It is a consequence of Newton's method, applied to the equation $x^2 - c = 0$, but it can be traced back at least as far as the Babylonians.[9]

If x were the positive root of the equation $x^2 = c$, then it would follow that $x = c/x$. But if x differed slightly from this root, say x were an underestimate, then c/x would be an overestimate. It would then be entirely reasonable to choose the number halfway between the underestimate and the overestimate as a better approximation than either x or c/x. Formally, then, we define a sequence of real numbers $x_0, x_1, \ldots$ by taking successively

$$x_{n+1} = \frac{1}{2} \left(x_n + \frac{c}{x_n} \right) \qquad n = 0, 1, 2, \ldots.$$

If x_0 is any positive number, the sequence $x_0, x_1, \ldots$ converges to $\sqrt{c}$ with astonishing speed.

Observe to begin with that $\sqrt{c}$ is the geometric mean of the numbers x_n and c/x_n. Thus the method consists in substituting the arithmetic mean for the geometric mean of these numbers at every stage of the iteration. Observe also that, since the geometric mean never exceeds the arithmetic mean, all approximations after the initial one are excessive approximations.

Assertion: With every iteration after the first the magnitude of the error is at most half its previous value. Reason: Let ϵ_n denote the error at the nth

---

[9]   See, for example, Van der Waerden, *Science Awakening* [426], p. 45.

stage of the approximation, so that

$$\epsilon_n = x_n - \sqrt{c}.$$

A straightforward calculation shows that

$$\epsilon_{n+1} = \frac{\epsilon_n^2}{2x_n}. \tag{2}$$

But ϵ_n/x_n always lies between 0 and 1 (provided n is greater than zero), and therefore

$$|\epsilon_{n+1}| \leq \frac{1}{2}|\epsilon_n|. \tag{3}$$

Conclusion: $\epsilon_n \to 0$ and $x_n \to \sqrt{c}$.

Notice that the choice of the initial approximation x_0 has no influence whatsoever on the final result. Even if we choose x_0 badly, the errors of subsequent approximations still tend to zero. A picture makes this plausible.

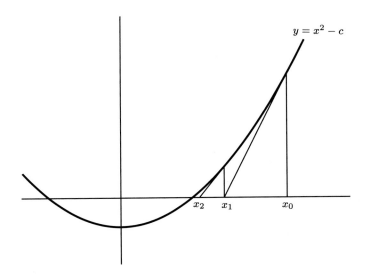

FIGURE 2

As an illustration, let us see how well the algorithm approximates $\sqrt{2}$. The table below gives the results of the first three iterations, beginning with $x_0 = 3/2$. The results are extraordinary. Three iterations alone will produce an accuracy of twelve significant digits.

n	x_n	error
1	17/12	0.002
2	577/408	0.000002
3	665857/470832	0.000000000002

The rapid rate of convergence of the algorithm in this special case far exceeds any expectations we may have drawn from inequality (3). The reason is clear. The true power of the method derives not from (3) but from the recursive formula (2), which shows that, for large values of n, $|\epsilon_{n+1}|$ is roughly proportional to $|\epsilon_n|^2$. Appropriately, this type of convergence is called *quadratic convergence*. Thus, the accuracy of the approximations improve at an ever increasing rate. In particular, when $c > 1$, (2) shows that

$$|\epsilon_{n+1}| \leq \frac{1}{2}|\epsilon_n|^2$$

and therefore, when we are quite close to the root, every iteration doubles the number of correct significant digits. Newton's method, when it converges, always converges quadratically ([101], p. 101).

Example 3.   The Mysterious Number e.   The second most important real number in mathematics, after the Archimedean constant π, is the number e, the base of the natural logarithms. The origins of e, like those of π, are purely geometric—it arises naturally in connection with the quadrature of the hyperbola—but its simplest definition is based on the positive integers: e is the limit of the sequence

$$\left(1+\frac{1}{2}\right)^2, \left(1+\frac{1}{3}\right)^3, \left(1+\frac{1}{4}\right)^4, \ldots \tag{4}$$

With arithmetic as the starting point, let us show directly—without any appeal to geometry—that this remarkable limit does indeed exist. Among the many known proofs, the one based on the theorem of the means is perhaps the most elegant.

We begin by introducing the companion sequence

$$\left(1+\frac{1}{2}\right)^3, \left(1+\frac{1}{3}\right)^4, \left(1+\frac{1}{4}\right)^5, \ldots \tag{5}$$

It is clear that both sequences converge or diverge together since corresponding terms differ by the factor $1 + \frac{1}{n}$ which tends to unity as $n \to \infty$. Since every bounded monotonic sequence has a limit, it is sufficient to show that (5) decreases.

Consider, for a fixed positive integer $n > 1$, the $n + 1$ numbers

$$1 - \frac{1}{n}, 1 - \frac{1}{n}, \ldots, 1 - \frac{1}{n}, \quad \text{and} \quad 1.$$

By the theorem of the means, their product is less than their arithmetic mean raised to the power $n + 1$. Since their product is $(1 - \frac{1}{n})^n$ and their sum is n, it follows that

$$\left(1 - \frac{1}{n}\right)^n < \left(\frac{n}{n+1}\right)^{n+1}.$$

Or, taking reciprocals,

$$\left(1 + \frac{1}{n-1}\right)^n > \left(1 + \frac{1}{n}\right)^{n+1}.$$

This shows that (5) is decreasing and the proof is complete. (A similar argument shows that (4) is increasing [299].)

Voicing his concern over the purely arithmetic approach, Felix Klein writes in his celebrated treatise *Elementary Mathematics from an Advanced Standpoint* [256], p. 146:

> This definition of e is usually, in imitation of the French models, placed at the very beginning of the great text books of analysis, and entirely unmotivated, whereby the really valuable element is missed, the one which mediates the understanding, namely, an explanation why precisely this remarkable limit is used as base and why the resulting logarithms are called natural.

There is perhaps no better way to mediate that understanding than by examining Euler's remarkable treatment of the exponential. The following development, taken from the *Introductio in analysin infinitorum* (1748)—the first of Euler's three great treatises on the differential and integral calculus and the prototype of all modern textbooks on calculus and differential equations—illustrates Euler's extraordinary analytic skills.[10] While the proof is lacking by modern standards of rigor, it exemplifies the 18th century investigations, when mathematicians trusted symbols far more than logic.

To begin with, "Euler unhesitatingly accepts the existence of both infinitely small and infinitely large numbers, and uses them to such remarkable advan-

---

[10] For an English translation of the *Introductio*, see [149].

tage that the modern reader's own hesitation must be tinged with envy."[11] Suppose then that ω is an infinitely small positive number. Then

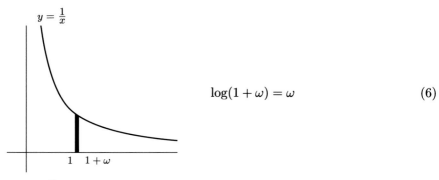

$$\log(1 + \omega) = \omega \tag{6}$$

since $\log(1 + \omega)$ may be viewed as the area of an infinitely thin rectangle of height 1 and base ω (see Figure 3; the *Introductio* employs not a single figure). Euler had long used the letter e to denote the base of the natural logarithms and therefore (6) is equivalent to the assertion that

$$e^{\omega} = 1 + \omega.$$

Now for any nonzero real number x, let $n = x/\omega$. Then n is infinitely large and

$$e^x = e^{n\omega} = (e^{\omega})^n$$
$$= (1 + \omega)^n$$
$$= \left(1 + \frac{x}{n}\right)^n.$$

In modern terminology, e^x is the *limit* of the sequence $(1 + x/2)^2$, $(1 + x/3)^3$, $(1 + x/4)^4, \ldots$. When $x = 1$, we obtain e.

Euler now applies the binomial theorem (!) and concludes that

$$e^x = \left(1 + \frac{x}{n}\right)^n$$
$$= \binom{n}{0} + \binom{n}{1}\frac{x}{n} + \binom{n}{2}\left(\frac{x}{n}\right)^2 + \binom{n}{3}\left(\frac{x}{n}\right)^3 + \cdots$$
$$= 1 + x + \frac{n-1}{n} \cdot \frac{x^2}{2!} + \frac{n-1}{n} \cdot \frac{n-2}{n} \cdot \frac{x^3}{3!} + \cdots.$$

[11] C. H. Edwards, Jr., *The Historical Development of the Calculus* [145], p. 272.

But because n is infinite all of the quotients $\frac{n-1}{n}, \frac{n-2}{n}, \ldots$ may be taken to be 1, and, lo and behold, we arrive at the exponential series

$$e^x = 1 + x + \frac{x^2}{2!} + \frac{x^3}{3!} + \cdots.$$

Euler's proof shows no trace of the rigor that we demand today, yet, clearly, here is mathematics of the highest caliber. In the nineteenth century, when analysis was placed on a firm foundation, infinitely large and infinitely small numbers—ever a source of confusion and controversy—were finally banished from the mathematical landscape. Those sound but sterile ϵ's and δ's took their place, and with them the various steps of Euler's argument could all be made secure. Yet, "when a concept or technique proves to be useful even though the logic of it is confused or even nonexistent, persistent research will uncover a logical justification, which is truly an afterthought."[12] In 1960, Abraham Robinson proved that infinitesimals can be defined in such a way as to provide a rigorous framework for the calculus. The modern theory of *nonstandard analysis,* as the new field is called, has the same internal consistency as the calculus based on real numbers and limits. While the subject is still in its infancy, and its future far from clear, yet it strives to bridge the gap between the finite and the infinite, between the continuous and the discrete. Within this new system, Euler's inspired proof can be fully justified.[13]

> A man of genius makes no mistakes. His errors are volitional and are the portals of discovery.
>
> —James Joyce, *Ulysses*

PROBLEMS

1. Consider the fixed straight line $x + y = 2m$, together with the family of hyperbolas $xy = c$, where c is a constant for each of these curves and variable from curve to curve. Why is it evident from Figure 4 that, for all positive x and y,

$$xy \leq \left(\frac{x+y}{2}\right)^2 ?$$

---

[12] M. Kline, *Mathematical Thought from Ancient to Modern Times* [260], p. 1120.

[13] For a nonstandard approach to elementary calculus, see Keisler's textbook [248]. For a more advanced treatment see [217].

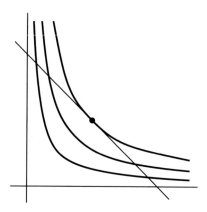

FIGURE 4

2. We know from simple algebra that *the product of two positive quantities with given sum becomes a maximum when these two quantities are equal.* Critique the following argument, due to Colin Maclaurin, which purports to show that the product of *three* positive quantities with given sum s becomes a maximum when they are all equal. Call the three quantities x, y, and z and suppose that z has already been determined. Then x and y are variables, but z is a constant. To maximize the product xyz we must therefore maximize xy. But x and y have a constant sum,

$$x + y = s - z,$$

and therefore their product xy becomes a maximum when $x = y$. It is clear that the argument remains unchanged no matter which quantity x, y, or z plays the constant role. Therefore the desired maximum is attained when $x = y = z$.

Is this proof entirely satisfactory? Can it be extended to more than three quantities?

3. Let $a_1, a_2, \ldots, a_n$ be nonnegative numbers with arithmetic mean A and geometric mean G. Consider the following algorithm:

If a_i is the smallest of the a's and a_j is the largest, we replace a_i by $\bar{a}_i = A$ and a_j by $\bar{a}_j = a_i + a_j - A$.

Show that

$$\bar{a}_i \bar{a}_j \geq a_i a_j.$$

Why does applying the algorithm enough times prove that $G \leq A$?

4. Show that if $0 < b \leq a$, then

$$\frac{1}{8}\frac{(a-b)^2}{a} \leq \frac{a+b}{2} - \sqrt{ab} \leq \frac{1}{8}\frac{(a-b)^2}{b}.$$

5. Prove: Of all triangles with given base and perimeter, the isosceles triangle has the maximum area. What is the dual statement?

6. Given the surface area of a box (rectangular parallelepiped), find the maximum of its volume. What is the dual statement?

7. Find the largest volume of an open-top box that can be constructed from a square sheet of cardboard by cutting a small square from each corner and then folding up the flaps to form the sides.

8. Given the sum of the areas of the five faces of an open-top box, find the maximum of its volume.

9. A post office problem. Find the maximum of the volume of a box if the sum of the length and girth is not to exceed ℓ inches.

10. Of all tetrahedra inscribed in a given sphere, which one has the maximum volume?

11. A function f defined on an interval is said to be **convex** if the portion of its graph in every subinterval lies on or below its chord.

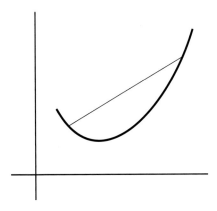

FIGURE 5

a. Show that a continuous function is convex if we assume merely that the *midpoint* of each chord is above (or on) the graph of the function.

Hint. The hypothesis can be formulated analytically as follows:

$$f\left(\frac{a+b}{2}\right) \le \frac{f(a)+f(b)}{2} \tag{7}$$

for every a and b in the domain. Deduce from this that

$$f\left(\frac{a_1+\cdots+a_n}{n}\right) \le \frac{f(a_1)+\cdots+f(a_n)}{n}$$

whenever $a_1,\ldots,a_n$ lie in the domain.

b. Show that every convex function is continuous.

12. Show that the usual calculus criterion for convexity is a sufficient condition: *If $f''(x)$ exists and is nonnegative at every point of an interval, then f is convex.*

Hint. Establish the midpoint condition (7) for all $a < b$ by applying the mean value theorem to f on the intervals $[a, (a+b)/2]$ and $[(a+b)/2, b]$.

13. Use the convexity of $-\log x$ for positive x to give another proof of the inequality between the arithmetic and geometric means.

14. a. Show that if $a_1,\ldots,a_n$ are positive numbers whose sum is π, then

$$\sin a_1 + \cdots + \sin a_n \le n \sin \frac{\pi}{n}.$$

Hint. If we try to argue by induction, then it is not at all clear that we can make the transition to the inequality

$$\sin a_1 + \cdots + \sin a_{n+1} \le (n+1) \sin \frac{\pi}{n+1}.$$

The requirement that $\sum a_i = \pi$ appears to be too restrictive. Prove that the stronger inequality

$$\sin a_1 + \cdots + \sin a_n \le n \sin \left(\frac{a_1+\cdots+a_n}{n}\right)$$

is valid whenever $0 < a_i \le \pi$ $(i = 1,\ldots,n)$.

b. Show that among all polygons with a given number of sides that can be inscribed in a circle, the regular polygon encloses the greatest area.

15. Let $0 < x_i < \pi$ $(i = 1, 2, \ldots, n)$ and put $x = (x_1 + \cdots + x_n)/n$. Show that

$$\frac{\sin x_1}{x_1} \frac{\sin x_2}{x_2} \cdots \frac{\sin x_n}{x_n} \leq \left(\frac{\sin x}{x} \right)^n.$$

16. If a and b are positive numbers, then

$$\left(\frac{\sqrt[n]{a} + \sqrt[n]{b}}{2} \right)^n \to \sqrt{ab} \qquad \text{as } n \to \infty.$$

17. Let A_n denote the arithmetic mean and G_n the geometric mean of the binomial coefficients

$$\binom{n}{0}, \binom{n}{1}, \binom{n}{2}, \ldots, \binom{n}{n}.$$

Show that

$$\sqrt[n]{A_n} \to 2$$

and

$$\sqrt[n]{G_n} \to \sqrt{e}.$$

18. The **harmonic mean** of two positive numbers a and b is defined to be the reciprocal of the arithmetic mean of $1/a$ and $1/b$, namely,

$$\frac{2ab}{a+b}.$$

It is an average that occurs quite naturally in the study of optics and electrical networks.

a. Show that if the two bases of a trapezoid have lengths a and b, then the harmonic mean of a and b is the length of the segment PQ parallel to the bases and passing through the intersection of the two diagonals.

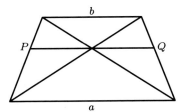

FIGURE 6

b.　Show that the harmonic mean of a and b never exceeds their geometric mean:

$$\frac{2ab}{a+b} \le \sqrt{ab}.$$

When does equality hold?

c.　Formulate a definition of the harmonic mean of n positive numbers. Is it still true that the harmonic mean never exceeds the geometric mean?

19.　The **quadratic mean** or **root-mean-square** is an important average in statistics. For two positive numbers a and b, it is defined to be the square root of the arithmetic mean of a^2 and b^2, namely,

$$\sqrt{\frac{a^2 + b^2}{2}}.$$

a.　Show that the arithmetic mean of a and b never exceeds their quadratic mean:

$$\frac{a+b}{2} \le \sqrt{\frac{a^2 + b^2}{2}}.$$

When does equality hold?

b.　Formulate a definition of the quadratic mean of n positive numbers. Is it still true that the arithmetic mean never exceeds the quadratic mean?

20.　Prove the **Cauchy-Schwarz inequality:** If $a_1, \ldots, a_n$ and $b_1, \ldots, b_n$ are real numbers, then

$$(a_1 b_1 + \cdots + a_n b_n)^2 \le (a_1^2 + \cdots + a_n^2)(b_1^2 + \cdots + b_n^2).$$

When does equality hold?

Hint. Begin with the obvious inequality $(x - y)^2 \ge 0$ in the form $2xy \le x^2 + y^2$. For each $i = 1, 2, \ldots, n$, take

$$x = \frac{a_i}{(\sum a_i^2)^{1/2}} \quad \text{and} \quad y = \frac{b_i}{(\sum b_i^2)^{1/2}}.$$

21. Establish the inequalities

$$(a+b)\left(\frac{1}{a}+\frac{1}{b}\right) \geq 4$$

$$(a+b+c)\left(\frac{1}{a}+\frac{1}{b}+\frac{1}{c}\right) \geq 9$$

$$(a+b+c+d)\left(\frac{1}{a}+\frac{1}{b}+\frac{1}{c}+\frac{1}{d}\right) \geq 16$$

$$\cdots$$

where $a, b, c, d, \ldots$ are positive numbers. What is the general pattern? When does equality hold?

22. If a, b, c are positive numbers whose sum is 1, what is the smallest possible value of

$$\left(\frac{1}{a}-1\right)\left(\frac{1}{b}-1\right)\left(\frac{1}{c}-1\right)?$$

23. If a, b, c are positive numbers whose sum is 1, then

$$\frac{1}{a}+\frac{1}{b}+\frac{1}{c} \geq 9.$$

Prove, generalize, and prove again.

24. If a and b are positive numbers whose sum is 1, then

$$\left(1+\frac{1}{a}\right)\left(1+\frac{1}{b}\right) \geq 9.$$

Prove, generalize, and prove again.

25. Show that if $a < b < c < d$, then

$$(a+b+c+d)^2 > 8(ac+bd).$$

26. If a and b are positive numbers whose sum is 1, then

$$\left(a+\frac{1}{a}\right)^2+\left(b+\frac{1}{b}\right)^2 \geq \frac{25}{2}.$$

27. Show that if $b_1, \ldots, b_n$ is any rearrangement of the positive numbers $a_1, \ldots, a_n$, then

$$\frac{a_1}{b_1} + \cdots + \frac{a_n}{b_n} \geq n.$$

28. If a and b are positive, prove that

$$a^b + b^a > 1.$$

29. Prove **Chebyshev's inequality:** If $a_1 \leq a_2 \leq \cdots \leq a_n$ and $b_1 \leq b_2 \leq \cdots \leq b_n$, then

$$\left\{ \frac{1}{n} \sum_1^n a_k \right\} \left\{ \frac{1}{n} \sum_1^n b_k \right\} \leq \frac{1}{n} \sum_1^n a_k b_k.$$

When does equality hold?

30. Estimate the rate of convergence of the sequence

$$\left(1 + \frac{1}{n} \right)^n \qquad n = 1, 2, 3, \ldots$$

by showing that, for all n,

$$0 < e - \left(1 + \frac{1}{n} \right)^n < \frac{3}{n}.$$

Remark. The evidence given in the table below suggests that, for large values of n, $e - (1 + \frac{1}{n})^n$ is approximately equal to $e/2n$. The evidence does not mislead for a more careful analysis reveals that

$$\frac{e}{2n + 2} < e - \left(1 + \frac{1}{n} \right)^n < \frac{e}{2n + 1}.$$

(See *Problems and Theorems in Analysis* [345], volume I, p. 38.)

n	$(1 + \frac{1}{n})^n$	$e - (1 + \frac{1}{n})^n$
100	2.7048	.0135
1,000	2.7169	.00135
10,000	2.7181	.000135
100,000	2.7182	.0000135

31. Show that

$$\frac{\log n!}{\log n^n} \to 1 \qquad \text{as } n \to \infty.$$

Hint. Apply the inequalities

$$\left(1 + \frac{1}{m}\right)^m < e < \left(1 + \frac{1}{m}\right)^{m+1} \qquad \text{for } m = 1, 2, \ldots, n.$$

32. a.   Let a and d be positive numbers and call A_n the arithmetic, and G_n the geometric, mean of the numbers $a, a + d, a + 2d, \ldots, a + (n - 1)d$. Show that

$$\lim_{n \to \infty} \frac{G_n}{A_n} = \frac{2}{e}.$$

b.   Deduce that

$$\lim_{n \to \infty} \frac{\sqrt[n]{n!}}{n} = \frac{1}{e}.$$

33. a.   Show that

$$\lim_{n \to \infty} n \sin(2\pi e n!) = 2\pi.$$

b.   Conclude that e is irrational. (The usual proof is based on the infinite series expansion $e = 1 + 1/1! + 1/2! + 1/3! + \cdots$; see, for example, [387], p. 391.)

34. Show that

$$1 + x + \frac{x^2}{2!} + \frac{x^3}{3!} + \cdots + \frac{x^{2n}}{(2n)!} = 0$$

has no real roots.

35. The graph of the equation $x^y = y^x$ in the first quadrant consists of a straight line and a curve (see Figure 7). Find the coordinates of the intersection point of the line and the curve.

36. Find all positive numbers x for which the sequence $x, x^x, x^{x^x}, \ldots$ is convergent.

37. *Prove or disprove:* The number $N = e^{\pi\sqrt{163}}$ is an integer.

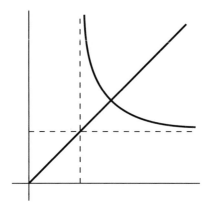

FIGURE 7

38. Show that

$$\int_0^1 \frac{dx}{x^x} = \sum_1^\infty \frac{1}{n^n}.$$

2. The Law of Errors

> Everybody believes in it, the experimenters because they think it is a mathematical theorem, the mathematicians because they think it is an experimental fact.
> —G. Lippmann (French physicist, 1845–1921)

We begin with a problem encountered by every experimental scientist.

No physical quantity can ever be measured with perfect accuracy—all observations are subject to error. And, if several measurements of the same magnitude are made, they will invariably differ from each other. The imperfections of our senses or of our best instruments imply an uncertainty that is said to be accidental, or random, and which tends to occur however great the skill and conscientiousness of the observer. A physicist, for example, who makes several determinations of the speed of light observes that they do not agree exactly. He rejects the notion that the speed is changing and attributes the variations in his results to unavoidable errors of observation. We do not know the origin of random errors—"we attribute them to chance because their causes are too complicated and too numerous." [14] Like the random noise that pervades an electronic communication channel, random errors are inevitable.

The problem was taken up by the great German mathematician Carl Friedrich Gauss. Gauss devoted much of his life to astronomy, and in his famous treatise of 1809, *Theoria Motus Corporum Coelestium in Sectionibus Conics Solum Ambientium* (Theory of the Motion of the Heavenly Bodies Moving about the Sun in Conic Sections)—a masterful investigation of the mathematics of planetary orbits—he put forth a theory of errors of observation that survives to this day.

In that work, Gauss formulates a principle upon which his entire theory is to be built.

Principle of the Arithmetic Mean. *When any number of equally good observations have given*

$$x_1, x_2, \ldots, x_n$$

as the values of a certain magnitude, the most probable value is their arithmetic mean.

What could be a simpler or more natural way of arriving at an equitable balance, an equilibrium, among a given set of observations? In fact, when the x's are points in the plane or in space, their arithmetic mean is precisely the balancing point, or centroid, of the system formed by placing equal masses at each point. But Gauss's approach to the theory of errors is to be based, not on a mechanical equilibrium, but on probability. Without the quantification of uncertainty, an answer to the question "How good is best?" was not possible. [15] Starting from the principle of the arithmetic mean, Gauss derived his famous law of errors, an elegant law that governs the probability that a single measurement x will lie between two given limits.

If μ is the true value of the magnitude being observed, then the observational error is the deviation

$$\text{error} = \mu - x.$$

The problem is to determine the form of the *error function,* a positive function $\Phi(x)$ having the property that, for any given measurement,

---

[14] Henri Poincaré, *Science and Method* (Dover, 1952), p. 75. For an elementary discussion of random error in the context of precision weighing done at the National Bureau of Standards in Washington, see D. Freedman, R. Pisani, and R. Purves, *Statistics* [160], chapters 6, 24.

[15] S. M. Stigler, *The History of Statistics* [403], p. 140.

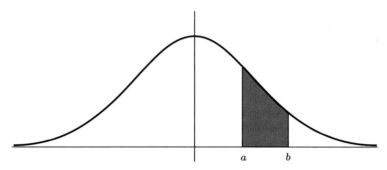

FIGURE 8

$$\text{Probability } [a < \text{ error } < b] = \int_a^b \Phi(x)\, dx.$$

Properly speaking, $\Phi(x)$ ought to be a discontinuous function, taking on only a finite number of distinct values. For any measuring device, no matter how sensitive it may be, cannot be accurate beyond a certain limit. But there is a more serious objection.

> As our mental eye penetrates into smaller and smaller distances and shorter and shorter times, we find nature behaving so entirely differently from what we observe in visible and palpable bodies of our surrounding that *no* model shaped after our large-scale experiences can ever be 'true'. . . . The idea of a *continuous range,* so familiar to mathematicians in our days, is something quite exorbitant, an enormous extrapolation of what is really accessible to us. The idea that you should *really* indicate the exact values of any physical quantity—temperature, density, potential, field strength, or whatever it might be—for *all* the points of a continuous range, say between zero and 1, is a bold extrapolation. We *never* do anything else than determine the quantity approximately for a very limited number of points and then 'draw a smooth curve through them'. This serves us well for many practical purposes, but from the epistemological point of view, from the point of view of the theory of knowledge, it is totally different from a supposed exact continual description.[16]

---

[16] E. Schrödinger, *Science and Humanism* [371], pp. 25ff.

To simplify the analysis, it is convenient to disregard these practical difficulties and to consider an ideal case in which measurements may range over all real values. The discrete error function can then be replaced by a continuous one.[17]

There are three simple properties that $\Phi(x)$ should satisfy:

i. Since errors that are of the same magnitude but of opposite sign are equally likely,

$$\Phi(x) = \Phi(-x).$$

ii. Since small errors are more likely than larger ones and extremely large errors are negligible, $\Phi(x)$ should decrease rapidly to zero as x approaches infinity.

iii. Since it is certain that each error will take on some real value,

$$\int_{-\infty}^{\infty} \Phi(x)\,dx = 1.$$

These conditions certainly do not determine $\Phi(x)$ uniquely, and many mathematicians before Gauss—most notably, Euler, Laplace, and Legendre—had long sought the elusive error function. It was Gauss who finally provided the solution.[18]

Theorem (The Normal Law of Errors). *The only error function that makes the arithmetic mean the most probable value is*

$$\Phi(x) = \frac{h}{\sqrt{\pi}} e^{-h^2 x^2}.$$

The constant h serves as a measure of the accuracy of the observations.

Proof. While the proof is demanding, each step along the way is elementary, and the reader who perseveres will have mastered the ingredients of a most important mathematical model. It is convenient to present the proof in three steps.

---

[17] The idea of using the continuous case to model the discrete case was once a daring novel idea, known as *Boscovich's Hypothesis* (R. J. Boscovich, *Treatise on Natural Philosophy,* Venice, 1758).

[18] The basic result was obtained independently and almost simultaneously by the American mathematician Robert Adrian (1775–1843) [289], p. 149.

Step 1. What exactly is meant by "most probable value"? To motivate the answer, let us assume for a moment that we are dealing with the discrete case in which only a finite number of different measurements are possible. The error function is then discontinuous and $\Phi(x)$ represents the probability that a single measurement differs from the true value μ by the amount x.

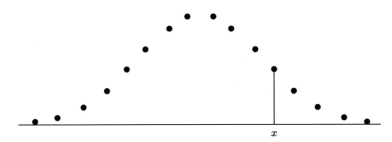

FIGURE 9

Now, it is entirely reasonable to suppose that repeated observations are *independent*. (Recall that two events are said to be independent if the probability of their joint occurrence is the product of the probabilities of their individual occurrences. It is natural to regard events which seem unrelated as being independent of each other. The mathematical rule of multiplication of probabilities is a way of formalizing this vague and intuitive notion.)

In view of the preceding comments, if n equally good observations have produced the values

$$x_1, x_2, \ldots, x_n$$

(with corresponding errors $\mu - x_1, \mu - x_2, \ldots, \mu - x_n$), then the product

$$\Phi(\mu - x_1)\Phi(\mu - x_2)\cdots\Phi(\mu - x_n)$$

represents the likelihood that all these values were observed.

But μ is unknown—it can never be known—so Gauss asserts that its most probable value is that for which the likelihood of the observations is a maximum. [19] By adopting the principle of the arithmetic mean, we are stipulating

---

[19] For a more complete explanation of why this choice of μ is the "most probable value" see Whittaker and Robinson, *The Calculus of Observations* [448], §112. In §110, it is shown how the principle of the arithmetic mean can be deduced from other axioms of a more elementary nature.

that this maximum must occur when

$$\mu = (x_1 + \cdots + x_n)/n.$$

To pass to the continuous case involves no essential change in the argument and we are thereby led to the same maximum problem (see, for example, [448], §112).

Step 2. Question: What form must the error function take if, given any three observations a, b, and c, the product

$$\Phi(x - a)\Phi(x - b)\Phi(x - c) \tag{1}$$

assumes its largest value when

$$x = \frac{a + b + c}{3}? \tag{2}$$

Let us suppose, for simplicity, that Φ is continuously differentiable. Setting the logarithmic derivative of (1) equal to zero to obtain the maximum, we find

$$\frac{\Phi'(x - a)}{\Phi(x - a)} + \frac{\Phi'(x - b)}{\Phi(x - b)} + \frac{\Phi'(x - c)}{\Phi(x - c)} = 0.$$

Let

$$F(x) = \frac{\Phi'(x)}{\Phi(x)},$$

which is then defined and continuous for all real values of x, and observe that (2) holds if and only if $(x - a) + (x - b) + (x - c) = 0$. Since a, b, and c are arbitrary, it follows that $F(x) + F(y) + F(z) = 0$ whenever $x + y + z = 0$. Equivalently,

$$F(x + y) = F(x) + F(y) \tag{3}$$

for all x and y.

This functional equation is quite reminiscent of another that we encountered in connection with the self-similarity of the logarithmic spiral (see chapter 3, §2). Arguing as before, we find that the only continuous solutions of (3) are the scalar multiples of the identity function. Thus

$$\frac{\Phi'(x)}{\Phi(x)} = Ax$$

and

$$\Phi(x) = Be^{Ax^2/2}.$$

But the relation

$$\int_{-\infty}^{\infty} \Phi(x)\, dx = 1$$

shows that A must be negative, say $-2h^2$, so that

$$\frac{1}{B} = \int_{-\infty}^{\infty} e^{-h^2 x^2}\, dx = \frac{\sqrt{\pi}}{h}.^{20}$$

Thus the only error function that makes the arithmetic mean the most probable value is

$$\Phi(x) = \frac{h}{\sqrt{\pi}} e^{-h^2 x^2}. \tag{4}$$

Step 3. It remains only to verify the correctness of the solution, namely, to show that if Φ is given by (4) then the principle of the arithmetic mean is valid, not just for three observations, but for any finite number as well. If the values $x_1, x_2, \ldots, x_n$ have been observed, then the function to be maximized is

$$\Phi(x - x_1)\Phi(x - x_2) \cdots \Phi(x - x_n).$$

By combining the exponentials, we see that this entails *minimizing the sum of the squares of the errors*[21]

$$(x - x_1)^2 + (x - x_2)^2 + \cdots + (x - x_n)^2.$$

But the minimum is obtained when

$$(x - x_1) + (x - x_2) + \cdots + (x - x_n) = 0,$$

---

[20] The second equality follows from the beautiful and well-known result

$$\int_{-\infty}^{\infty} e^{-x^2}\, dx = \sqrt{\pi}.$$

For a proof see chapter 5, §2.

[21] This is an example of the famous *method of least squares,* introduced by Legendre in 1805 in his *Nouvelles méthodes pour la détermination des orbites des comètes.* Commenting on the method, Stigler writes:

> The method of least squares was the dominant theme—the leitmotif—of nineteenth-century mathematical statistics. In several respects it was to statistics what the calculus had been to mathematics a century earlier. "Proofs" of the method gave direction to the development of statistical theory, handbooks explaining its use guided the application of the higher methods, and disputes on the priority of its discovery signaled the intellectual community's recognition of the method's value. ([403], p. 11)

that is, when x is the arithmetic mean of the observations. This completes the proof.

The curve

$$y = \frac{h}{\sqrt{\pi}} e^{-h^2 x^2}$$

is known generally as the **normal curve**, or the **Gaussian curve**, although it was introduced in probability theory nearly a century earlier by De Moivre (1718). It is perhaps the most ubiquitous of curves, appearing throughout the physical, biological, and social sciences. Contained within it are two of the most famous constants in mathematics, π and e, which seem to appear mysteriously in so many settings involving randomness and chance.

For each choice of the constant h, the graph is a symmetric, bell-shaped curve, as pictured below. Gauss called h the *modulus of precision,* and it reflects the accuracy of the observations. For large values of h, most of the area under the curve is concentrated near the y-axis, indicating that in this case most of the observations fall very close to the true value. For small values of h, the curve flattens out and the bulk of the observations are diffuse.

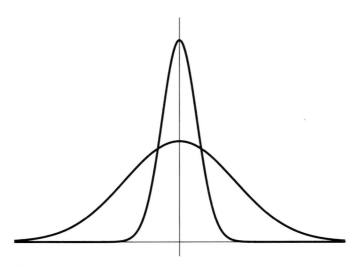

FIGURE 10

Gauss's derivation of the normal law rests on the principle of the arithmetic mean, yet, almost immediately, Laplace saw that a better rationale could be provided. If each deviation from the average measure were itself the cumulative effect of a very large number of independent causes, each one of them having only a very slight influence on the whole, then the normal law would follow.

Here, certainly, is one of the finest examples of order out of chaos in all of probability theory. A single component of an individual measurement can be totally unpredictable, and yet the cumulative effect of many such components will be governed by a deterministic law.

> We know not to what are due the accidental errors, and precisely be-
> cause we do not know, we are aware they obey the law of Gauss. Such
> is the paradox. The explanation is nearly the same as in the preceding
> cases. We need know only one thing: that the errors are very numer-
> ous, that they are very slight, that each may be as well negative as
> positive. What is the curve of probability of each of them? We do
> not know; we only suppose it is symmetric. We prove then that the
> resultant error will follow Gauss's law, and this resulting law is inde-
> pendent of the particular laws which we do not know. Here again the
> simplicity of the result is born of the very complexity of the data.[22]

The central limit theorem, as Laplace's result came to be known, placed Gauss's principle of the arithmetic mean on a firm logical foundation. "Only after this work of Laplace did the widespread applications of probability theory become feasible as a scientifically justified method."[23]

While a precise formulation and proof of the central limit theorem lies well beyond our scope, it is instructive to illustrate it by mentioning a famous piece of apparatus that was devised by the English statistician and natural scientist Sir Francis Galton (1822–1911). Galton called it the *quincunx*,[24] and it illustrates the principle of the law of errors from the joint effect of a large number of small and independent deviations.

The quincunx had a glass face and a funnel at the top. Small ball bearings poured through the funnel cascaded through a triangular array of pins and ul-

---

[22] Henri Poincaré, op. cit. This translation comes from *The World of Mathematics* [316], volume 2, p. 1389.

[23] L. E. Maistrov, *Probability Theory: A Historical Sketch* [289], p. 148. With Laplace's theorem as a starting point, Adolphe Quetelet (1796–1874), the Belgian astronomer and statistician, succeeded in showing that the normal curve could be fitted to a large variety of empirical data taken from all corners of science. Heights and weights of individuals, sizes of skulls, and today even IQ scores appear to be normally distributed. Nature, it seems, can be counted on to obey the laws of probability.

[24] The word is used in *Natural Inheritance*, p. 64, to describe the arrangement of the pins. The description given here is adapted from Stigler [403], p. 276.

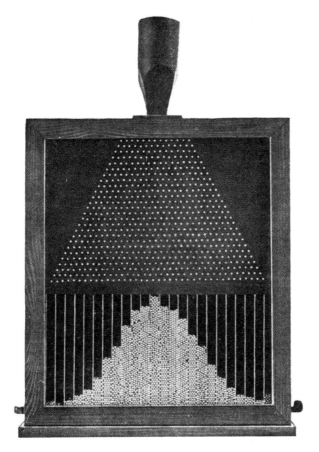

FIGURE 11
Galton's apparatus
From G. Weber, *Belysningsteknik,* 2nd edition, p. 132.

timately collected in compartments at the bottom. The construction was such that each ball bearing would strike one pin in every row and, at least in principle, fall either left or right with equal probabilities. What should we observe? Since each encounter with a pin subjects a ball bearing to a small displacement, equally likely to be left or right, the total displacement—as measured from the compartment directly below the funnel—was the sum of as many independent displacements as there were rows of pins (twenty-six in Figure 11). The resulting outline after many ball bearings were dropped should resemble a normal curve.

By examining the configuration of the pins and asking the obvious combinatorial question—In how many ways can a ball bearing reach a given pin?—
the reader will discover almost at once the reappearance of Pascal's celebrated
arithmetical triangle (problem 8). Here begins another fascinating excursion
into the unexpected connection between the binomial theorem and the normal law (see, for example, [156], volume I, chapter VII).

PROBLEMS

1. In memoirs of 1777 and 1781, Laplace gave an extremely complex argument to show why the function

$$y = \frac{1}{2a} \log\left(\frac{a}{|x|}\right), \qquad |x| \le a$$

should be taken as an error distribution. Here, a represents the upper limit
of the possible errors. Does Laplace's function satisfy all three properties required of an error function?

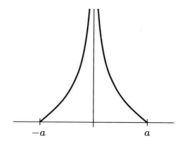

$-a$ $\qquad\qquad$ a

FIGURE 12

2. Given a triangle ABC, we form a new triangle $A'B'C'$ by connecting the
midpoints of the sides (see Figure 13). When this process is repeated indefinitely the resulting sequence of similar triangles converges to a common intersection point P. Identify P in terms of A, B, and C.

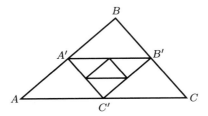

FIGURE 13

3. Show that every continuous solution of the functional equation

$$f\left(\frac{x+y}{2}\right) = \frac{f(x) + f(y)}{2} \qquad \text{for all } x \text{ and } y$$

is of the form $f(x) = ax + b$.

4. Given n real numbers $x_1, \ldots, x_n$, how should x be chosen so that the sum of the absolute values of the deviations

$$\sum_{i=1}^{n} |x - x_i|$$

is as small as possible?

5. Show that, for any equilateral triangle, the sum of the distances from a point in the interior to the opposite sides is always a constant.

6. In the plane of a given triangle, locate a point whose distances from the vertices have the smallest possible sum of squares.

7. Given n data points in the plane,

$$(x_1, y_1), (x_2, y_2), \ldots, (x_n, y_n),$$

it is desired to find a straight line $y = ax + b$ that "best fits" the data in the sense of the method of least squares, i.e., such that the sum of the squares of the vertical deviations

$$\sum_{i=1}^{n} (ax_i + b - y_i)^2$$

is a minimum (see Figure 14). Find a and b.

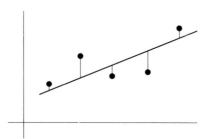

FIGURE 14

Hint. A function of several variables cannot attain a minimum with respect to all its variables jointly, unless it attains a minimum with respect to each variable separately.

8. Consider the infinite triangular array of points depicted in Figure 15. How many *shortest zigzag paths* are there starting from the top and going downward to any point in the array?

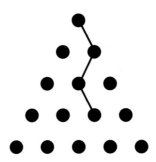

FIGURE 15

9. A Problem of Pólya. Suppose that of an unknown quantity x we know only that

$$a \le x \le b.$$

(The lower bound a is assumed to be positive.) How should we choose p so that the maximum relative error

$$\max_{a \le x \le b} \frac{|p - x|}{x}$$

is as small as possible?

10.   The Weber–Fechner Law. Physiological experiments have revealed that in many cases involving human perception a change in the magnitude of a physical stimulus does not result in a proportional change in the perception of that stimulus. Double the weight of an object, for example, and the senses do not perceive it as being twice as heavy. The Weber–Fechner law asserts that, if r denotes the response due to stimulus s, then *the change in response is proportional to the relative change in the stimulus* [286].

  a.   Show that, under the assumption that r and s are continuous quantities,

$$r = c \log \frac{s}{s_0}, \tag{5}$$

where c is a constant of proportionality and s_0 is the *threshold stimulus* (the value of s for which r is zero).

  b.   Suppose that we wish to determine empirically the true value of a certain stimulus s. We make n equally good observations and obtain the responses $r_1, r_2, \ldots, r_n$. The corresponding stimuli $s_1, s_2, \ldots, s_n$ are obtained from formula (5). Show that, if the errors in response obey the normal law, then the corresponding errors in the stimulus do not.

3. Variations on a Theme

> Unification, the establishment of a relationship between seemingly diverse objects, is at once one of the great motivating forces and one of the great sources of aesthetic satisfaction in mathematics.
>   —Davis and Hersh, *The Mathematical Experience*

The arithmetic and geometric means—the simplest of all averages—have been generalized in many ways and in many different directions. We present here three examples taken from number theory, probability, and analysis. Together they show once again how the essential unity of mathematics may be revealed even at the most elementary level.

a. The Average Value of an Arithmetical Function

Suppose that we wish to study the behavior of an arithmetical function

$$f : \{1, 2, 3, \ldots\} \to \mathbf{R}$$

whose values might well be erratic. It may then be profitable to turn our attention to its average behavior instead, that is, to consider the sequence of arithmetic means

$$\frac{f(1) + \cdots + f(n)}{n} \qquad n = 1, 2, 3, \ldots$$

for in this way, any eccentricities exhibited by the individual terms may be greatly suppressed. If

$$\frac{f(1) + \cdots + f(n)}{n} \to L$$

as n tends to infinity, then we call L the **average value** of f.

As an example, consider the important arithmetical function $r(n)$, defined to be the number of representations of n as a sum of two squares

$$n = x^2 + y^2. \tag{1}$$

It is assumed here that x and y are integers (positive, negative, or zero) and that representations that differ only trivially, that is, in the order of x and y, will be counted as distinct. More precisely, $r(n)$ is the number of ordered pairs (x, y) of integers satisfying (1). For example, $r(10) = 8$ because

$$10 = 1^2 + 3^2 = 1^2 + (-3)^2 = (-1)^2 + 3^2 = (-1)^2 + (-3)^2$$
$$= 3^2 + 1^2 = 3^2 + (-1)^2 = (-3)^2 + 1^2 = (-3)^2 + (-1)^2.$$

We tabulate the first twenty-four values of this function.

n	$r(n)$	n	$r(n)$
1	4	13	8
2	4	14	0
3	0	15	0
4	4	16	4
5	8	17	8
6	0	18	0
7	0	19	0
8	4	20	8
9	4	21	0
10	8	22	0
11	0	23	0
12	0	24	0

The behavior of $r(n)$ is highly irregular. By virtue of Fermat's great theorem, every prime number of the form $4k + 1$ has a unique representation as a sum of two (positive) squares, and therefore $r(n) = 8$ for every number of this form; on the other hand, $r(n) = 0$ for every odd number of the form $4k - 1$. It is not difficult to show that the values of $r(n)$ can be made arbitrarily large ([385], p. 380, example 3).

In light of these observations, the following theorem is striking.

Theorem. *The average number of representations of a natural number as a sum of two squares is π. That is,*

$$\lim_{n \to \infty} \frac{r(1) + \cdots + r(n)}{n} = \pi.$$

The appearance of the number π, initially so unexpected, becomes completely transparent when the theorem is viewed geometrically.

Proof. For a fixed n, let

$$R(n) = r(1) + \cdots + r(n).$$

Interpreting $r(n)$ as the number of *lattice points* (points with integer coordinates) on the circle $x^2 + y^2 = n$, we see that $R(n)$ is one less than the total number of lattice points in the disk $x^2 + y^2 \leq n$.

About each lattice point in the disk place a unit square, centered at the lattice point, with sides parallel to the coordinate axes (see Figure 16). The total area of these squares is equal to $1 + R(n)$. Of course, this is not exactly the area of the disk—some squares project beyond the boundary, while parts of the disk remain uncovered by any square. However, all of the squares are contained in the larger disk with center $(0, 0)$ and radius $\sqrt{n} + 1/\sqrt{2}$. (Reason: The greatest distance from a point inside a square of side 1 to its center is $1/\sqrt{2}$.) By comparing areas we see that

$$1 + R(n) < \pi \left(\sqrt{n} + \frac{1}{\sqrt{2}} \right)^2.$$

A similar argument shows that the disk with center $(0, 0)$ and radius $\sqrt{n} - 1/\sqrt{2}$ is completely covered by the given squares, and hence

$$1 + R(n) > \pi \left(\sqrt{n} - \frac{1}{\sqrt{2}} \right)^2.$$

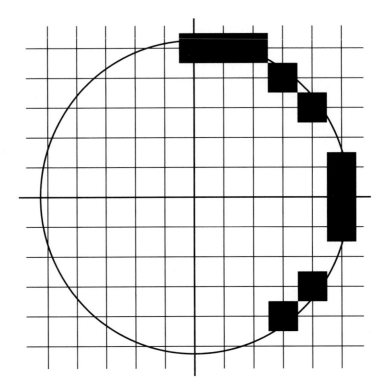

FIGURE 16

These two inequalities may be simplified by observing that $\pi\sqrt{2} < 5$ and $0 < \frac{\pi}{2} - 1 < \sqrt{n}$ $(n = 1, 2, 3, \ldots)$. Therefore,

$$\pi\left(\sqrt{n} + \frac{1}{\sqrt{2}}\right)^2 - 1 = \pi n + \pi\sqrt{2}\sqrt{n} + \frac{\pi}{2} - 1 < \pi n + 6\sqrt{n}$$

and

$$\pi\left(\sqrt{n} - \frac{1}{\sqrt{2}}\right)^2 - 1 = \pi n - \pi\sqrt{2}\sqrt{n} + \frac{\pi}{2} - 1 > \pi n - 6\sqrt{n}.$$

It follows that

$$\pi n - 6\sqrt{n} < R(n) < \pi n + 6\sqrt{n}$$

so that

$$|R(n) - \pi n| < 6\sqrt{n}$$

and hence also that

$$\left| \frac{R(n)}{n} - \pi \right| < \frac{6}{\sqrt{n}}.$$

Thus $R(n)/n \to \pi$ and the proof is complete.

The same considerations arise when we are confronted with the problem of how to assign a meaningful "sum" to a divergent series. If the problem at first seems paradoxical, it is because we have grown accustomed to Cauchy's classical definition of convergence. In his great textbook, the *Cours d'analyse* (1821), in which he sought to establish complete rigor in mathematical analysis, Cauchy set down the simple and nearly transparent meaning that is now commonly attached to the sum of a series—namely, the limit of its partial sums. A divergent series is one for which the limit fails to exist and therefore it has no sum.

What could be a more natural way of summing an infinite series of numbers than to simply add more and more terms? It seems almost inconceivable, then, that to the mathematicians of the seventeenth and eighteenth centuries, the precise nature of convergence and divergence remained elusive. Certainly a sound and irrefutable theory would have to await the construction of the real number system. But, to the mathematicians of the era, the manipulation of series was largely formal, and the issue of rigor was not taken too seriously.[25]

The series

$$1 - 1 + 1 - 1 + \cdots$$

in particular was a source of great confusion and controversy. If its value is denoted by S, then it seemed clear that

$$S = 1 - (1 - 1 + 1 - 1 + \cdots) = 1 - S$$

and therefore

$$S = \frac{1}{2}.$$

This paradoxical result was given by Guido Grandi (1671–1742), a priest and professor at Pisa known for his study of roses ($r = \sin n\theta$) and other curves resembling flowers.[26] Even Leibniz and Euler offered explanations of why the sum of the series should be $1/2$ ([260], pp. 446ff.).

---

[25] For an excellent discussion, see Judith Grabiner, "Is mathematical truth time-dependent" [182].

[26] D. J. Struik, *A Concise History of Mathematics* [407], p. 126. By assuming that the terms of the series $1 - 1 + 1 - 1 + \cdots$ could be grouped in pairs, the first with the second, the third with the

As Hardy explains it:

> This remark is trivial now: it does not occur to a modern mathematician that a collection of mathematical symbols should have a 'meaning' until one has been assigned to it by definition. It was not a triviality even to the greatest mathematicians of the eighteenth century. They had not the habit of definition: it was not natural to them to say, in so many words, 'by X we *mean* Y'. There are reservations to be made ... but it is broadly true to say that mathematicians before Cauchy asked not 'How shall we *define* $1 - 1 + 1 - \ldots$?' but 'What *is* $1 - 1 + 1 - \ldots$?', and that this habit of mind led them into unnecessary perplexities and controversies which were often really verbal.[27]

What is needed is a way of assigning a value to an infinite series that is more widely applicable than the classical definition due to Cauchy. Whereas most mathematicians of the early nineteenth century had rejected divergent series as unsound—Abel called them "the invention of the devil"—yet, by the end of the century they were unreservedly embraced.

The very obvious process of forming the arithmetic mean suggests a way of giving a meaning to Grandi's paradoxical equation $1 - 1 + 1 - 1 + \cdots = \frac{1}{2}$. Since the sequence of partial sums $\{S_n\} : 1, 0, 1, 0, \ldots$ is divergent, we consider instead the sequence of arithmetic means

$$T_n = \frac{S_1 + \cdots + S_n}{n} \qquad n = 1, 2, 3, \ldots.$$

It is easy to see that

$$T_n = \frac{1}{2} + \frac{1 - (-1)^n}{4n},$$

and therefore $T_n \to \frac{1}{2}$. Thus we can reasonably associate with the divergent series $1 - 1 + 1 - 1 + \cdots$ the number $\frac{1}{2}$ as its "value" or "sum."

In the same way, the sum of any divergent series can be defined to be the average value of the sequence of its partial sums (provided of course that the average value exists). This is but one of a number of important and widely applicable methods of modern *summability theory*. (For an excellent account, see Knopp, *Theory and Application of Infinite Series* [262], chapter XIII.)

The utility of the theory, however, is only a part of the story.

---

fourth, and so on, he went on to conclude that $0 + 0 + 0 + \cdots = 1/2$, and he found in this a symbol of the creation of the world from nothing!

[27] G. H. Hardy, *Divergent Series* [204], p. 5.

The construction and acceptance of the theory of divergent series is another striking example of the way in which mathematics has grown. It shows, first of all, that when a concept or technique proves to be useful even though the logic of it is confused or even nonexistent, persistent research will uncover a logical justification, which is truly an afterthought. It also demonstrates how far mathematicians have come to recognize that mathematics is man-made. The definitions of summability are not the natural notion of continually adding more and more terms, the notion which Cauchy merely rigorized; they are artificial. But they serve mathematical purposes, including even the mathematical solution of physical problems; and these are now sufficient grounds for admitting them into the domain of legitimate mathematics. [28]

b. Mathematical Expectation: A Coin Tossing Experiment

How many really basic mathematical objects are there? One of them is surely the "miraculous jar" of the positive integers $1, 2, 3, \ldots$. Another is the concept of a fair coin. Though gambling was rife in the ancient world and although prominent Greeks and Romans sacrificed to Tyche, the goddess of luck, her coin did not arrive on the mathematical scene until the Renaissance. Perhaps one of the things that had delayed this was a metaphysical position which held that God speaks to humans through the action of chance. . . .

The modern theory begins with the expulsion of Tyche from the Pantheon. There emerges the vision of the fair coin, the unbiased coin. This coin exists in some mental universe and all modern writers on probability theory have access to it. They toss it regularly and they speculate about what they "observe."
—Davis and Hersh, *The Mathematical Experience*

The elementary theory of coin tossing begins with two underlying assumptions:

a. The coin is "fair."

---

28 Morris Kline, *Mathematical Thought from Ancient to Modern Times* [260], p. 1120.

b. The successive tosses are *independent.*

The first assumption tells us that in each individual toss the alternatives H (heads) and T (tails) are equally likely, that is, each is assigned probability $1/2$. The second tells us that the probability associated with any given pattern (of length n) of the symbols H and T is obtained by the rule of multiplication of probabilities:

$$\left(\frac{1}{2}\right)\left(\frac{1}{2}\right)\cdots\left(\frac{1}{2}\right) = \frac{1}{2^n}.$$

In other words, all of the 2^n possible patterns of length n are equally likely.

Coin tossing is a far from frivolous pursuit.

In fact, the model may serve as a first approximation to many more complicated chance-dependent processes in physics, economics, and learning theory. Quantities such as the energy of a physical particle, the wealth of an individual, or the accumulated learning of a rat are supposed to vary in consequence of successive collisions or random impulses of some sort. For purposes of a first orientation one assumes that the individual changes are of the same magnitude, and that their sign is regulated by a coin-tossing game. Refined models take into account that the changes and their probabilities vary from trial to trial, but even the simple coin-tossing model leads to surprising, indeed to shocking, results. They are of practical importance because they show that, contrary to generally accepted views, the laws governing a prolonged series of individual observations will show patterns and averages far removed from those derived for a whole population. In other words, currently popular psychological tests would lead one to say that in a population of "normal" coins most individual coins are "maladjusted."[29]

Here, we shall consider only the simplest of patterns and the averages associated with them.

Problem.   *On the average, how many times must an ideal coin be tossed until a run of n consecutive heads appears?*[30]

---

[29] W. Feller, *An Introduction to Probability Theory and Its Applications* [156], volume I, p. 71.

[30] While it is conceptually possible that the game will never terminate, that the desired run of heads will never appear, the probability of this event is zero. In fact, a prolonged tossing of a coin is bound to produce, sooner or later, any given (finite) pattern and to repeat it infinitely often (see, for example, [156], p. 202).

It is natural to think of the average duration of the game intuitively in terms of a conceptual experiment that is performed repeatedly a great many times. Of course, if the theory is to be useful, then the conceptual experiment must conform to the real one, and the reader will find it instructive to actually play the game, either by tossing coins or by enlisting a computer to simulate play.

For computational purposes, however, such an approach is of limited value. What is needed is a means of determining the average based not on a random process but on a precise analytic expression.

The following definition is a natural one.

Definition. Consider a random experiment in which the possible outcomes are the numbers $a_1, a_2, \ldots$ occurring with probabilities $p_1, p_2, \ldots$, respectively. Then the **expected value** of the experiment is the number

$$\mathbf{E} = a_1 p_1 + a_2 p_2 + \cdots$$

provided that the series converges absolutely.[31] If the series fails to converge absolutely, then we say that there is no finite expectation. Thus, the expected value is the **weighted average** of the possible outcomes, each outcome being weighted by the probability with which it occurs.

The concept of expected value originated in connection with games of chance, where the average gain of a player is called that player's *expectation*. The term, however, is misleading and should not be interpreted in the ordinary sense of the word. It does not imply that $\mathbf{E}$ is the most probable value of the experiment, or even a value that will necessarily occur. For example, the expected number of points obtained in a single throw of a fair die is

$$\frac{1 + 2 + 3 + 4 + 5 + 6}{6} = \frac{7}{2}$$

which is different from all of the possible outcomes $1, 2, \ldots, 6$.

When we repeat an experiment with expected value $\mathbf{E}$ a sufficiently large number of times, the "law of large numbers" tells us that the average outcome is likely to be near $\mathbf{E}$. More precisely, if $x_1, x_2, x_3, \ldots$ represent the values that are actually observed, then, with probability 1,

$$\mathbf{E} = \lim_{n \to \infty} \frac{x_1 + \cdots + x_n}{n}.$$

---

[31] The absolute convergence guarantees that every rearrangement of the series is also convergent and has the same sum ([413], p. 586).

This reconciles the definition of expected value with our intuitive interpretation of it.

Let us return now to the coin-tossing problem that was posed earlier. In the simplest case, the game ends as soon as a single head has appeared. The outcomes of the experiment are the positive integers $1, 2, 3, \ldots$ (the possible durations of the game) and they occur with probabilities $1/2, 1/4, 1/8, \ldots$ (the likelihood of the patterns $H, TH, TTH, \ldots$). Thus, the expected number of tosses is given by

$$\mathbf{E} = \sum_{n=1}^{\infty} n/2^n.$$

Since the nth partial sum of the series is equal to $2 - (n + 2)/2^n$ (the proof by induction is straightforward), it follows that $\mathbf{E} = 2$. Two tosses on the average are needed to produce a single head.

For a run of two heads, the problem is far more interesting. We begin by determining the total number S_n of different ways in which the game can end on the nth toss. Clearly, $S_1 = 0$ and $S_2 = 1$. Suppose then that $n > 2$. If the first toss is a tail, then there are S_{n-1} admissible patterns of length $n - 1$ that can follow. If the first toss is a head, then the number of admissible patterns is S_{n-2} (the second toss must have been a tail, otherwise $n = 2$). But any pattern of length n must begin with either a head or a tail. Therefore,

$$S_n = S_{n-1} + S_{n-2},$$

which is the same recursive formula satisfied by the Fibonacci numbers. In view of the definition of S_n, we have $S_n = f_{n-1}$.

Accordingly, if p_n denotes the probability that the game ends on the nth toss, then

$$p_n = \frac{f_{n-1}}{2^n}.$$

To determine the expected duration of the game, we introduce the generating function

$$F(x) = \sum_{n=1}^{\infty} p_n x^n.$$

By formally differentiating the series term-by-term, we see that when $x = 1$,

$$\mathbf{E} = F'(1). \qquad (2)$$

The truth of this assertion will follow as soon as it can be shown that the radius of convergence of the power series is greater than 1. But

$$F(x) = \frac{x}{2} \sum_{n=1}^{\infty} f_n \left(\frac{x}{2}\right)^n = \frac{\left(\frac{x}{2}\right)^2}{1 - \left(\frac{x}{2}\right) - \left(\frac{x}{2}\right)^2} \tag{3}$$

by virtue of the known generating function for the Fibonacci numbers (chapter 3, §3). This tells us three things: first, since (3) is valid whenever $|x| < 2/\tau$, and since $\tau < 2$, it follows that (2) is also valid; second, since $\sum p_n = F(1) = 1$, we need not consider the possibility of an unending game; and third, by differentiating the quotient, we find easily that $\mathbf{E} = F'(1) = 6$. *On the average, six tosses of an ideal coin will produce a run of two consecutive heads.*

We shall not pursue this method further, since the recursive formula for S_n becomes increasingly more complicated for longer runs. Fortunately, however, there is a far more elegant solution.

For simplicity, let us continue to denote by $\mathbf{E}$ the expected number of tosses needed to produce a run of n heads. If at any time during the first n tosses a tail appears, then the game begins from scratch; otherwise it terminates. The table below illustrates these mutually exclusive possibilities leading to success.

Outcome	Probability	Expected Duration of the Game
T	$1/2$	$\mathbf{E}+1$
HT	$1/4$	$\mathbf{E}+2$
HHT	$1/8$	$\mathbf{E}+3$
$HHHT$	$1/16$	$\mathbf{E}+4$
$\cdots$	$\cdots$	$\cdots$
$HHHH\cdots HT$	$1/2^n$	$\mathbf{E}+n$
$HHHH\cdots HH$	$1/2^n$	n

By the law of total expectation,[32]

$$\mathbf{E} = \sum_{i=1}^{n} \frac{\mathbf{E}+i}{2^i} + \frac{n}{2^n},$$

---

[32] Compare the law of total probability (chapter 3, §3, example 3).

and we have reduced the problem to the solution of a single linear equation in a single unknown. Solving for **E**, we find

$$\mathbf{E} = 2^{n+1} - 2.$$

"Seek simplicity, and distrust it," said Alfred North Whitehead. Is this solution quite satisfactory?

In formulating the definition of mathematical expectation, we replaced a probability experiment by an analytic expression. It happens frequently, however, that the tables can be turned and that a deterministic problem in which the obvious calculations seem formidable can be solved by performing a random experiment. A classical example is the famous Buffon needle problem in which the approximate value of π is obtained by throwing needles at random many times (problem 19). Another is the problem of evaluating the definite integral $\int_a^b f(x)\,dx$ of a continuous function f when no elementary antiderivative can be found. If we choose a large set of random numbers $x_1, \ldots, x_n$ on the interval $[a, b]$, and then calculate the arithmetic mean

$$\frac{f(x_1) + \cdots + f(x_n)}{n},$$

then the ratio is approximately equal to the average

$$\frac{1}{b - a} \int_a^b f(x)\,dx$$

(problem 9).

These examples illustrate the famous technique known as the **Monte Carlo method**, introduced around 1945 by Ulam and von Neumann. To solve a deterministic problem that appears intractable, replace it by a probability question that has the same answer and then solve the latter by means of random experimentation.[33]

---

[33] The quest for true randomness remains elusive. No one has ever found a completely satisfactory way of simulating the tosses of a fair coin, or even of deciding what properties such a sequence should possess. Currently popular algorithms for generating "pseudo-random" numbers—numbers that appear to be random—give more or less satisfactory results, but all such methods are intrinsically deterministic. As von Neumann said, "Anyone who considers arithmetical methods of producing random digits is, of course, in a state of sin." For an excellent account of the problem see *The Art of Computer Programming* [263], volume 2, chapter 3.

c. Gauss and the Arithmetic-Geometric Mean

> He thought numerically and algebraically, after the manner of Euler,
> and personified the extension of Euclidean rigor to analysis.
>
> —Kenneth May[34]

The mysterious path of discovery—the tireless experimentation in search of patterns, the veiled connections that suddenly unfold, serendipity—all these elements combine to make mathematics so magical. In this final variation on a theme, we shall turn to one of Gauss's remarkable discoveries in function theory, an unexpected connection between the arc length of the lemniscate and a simple algorithm based on the arithmetic and geometric means.

The lemniscate, a legacy of the ancient Greeks,[35] was rediscovered independently in 1694 by the two Bernoulli brothers, James and his younger brother John. It is a curve in the form of a figure 8 lying on its side, the symbol of infinity. Whereas the ellipse is defined to be the locus of points in the plane for which the sum of the distances to two given points is a constant, the family of curves to which the lemniscate belongs is characterized by the condition that the product of these two distances is a constant. In polar coordinates, the standard equation of the lemniscate is $r^2 = \cos 2\theta$ and its total arc length is

$$L = 4 \int_0^1 \frac{dr}{\sqrt{1 - r^4}}. \tag{4}$$

This special integral was much studied throughout the eighteenth century, most notably by Euler, Lagrange, Legendre, and Gauss, and it played a key role in the subsequent development of a general theory of elliptic integrals.[36] By the end of the century, Gauss had penetrated deeply into the rich properties of the lemniscate, strongly influenced by the analogies with the circular functions.[37]

---

[34] Carl Friedrich Gauss, *Dictionary of Scientific Biography,* volume V, p. 299.

[35] See Stillwell, *Mathematics and Its History* [404], p. 23 ("The Spiric Sections of Perseus (c. 150 B.C.)").

[36] The general elliptic integral is of the form $\int R(x, y)\, dx$, where $R(x, y)$ is a rational function of the two variables x, y and y is the square root of a polynomial of degree three or four in x. The term elliptic integral is used because this type of integral arises in the problem of calculating the arc length of an ellipse.

[37] Two remarkable discoveries that Gauss made during the first three months of 1797 deserve special mention. Only a year after constructing the regular 17-gon by ruler and compass, he found a ruler and compass construction for dividing the lemniscate into five equal parts. (In 1827, Abel solved the general problem, finding all values of n for which the lemniscate could be divided into n equal parts using only ruler and compass. The answer is the same as for the circle! For an excellent modern account of Abel's theorem see M. Rosen, "Abel's Theorem on the Lemniscate," *Amer. Math. Monthly,* 88 (1981), 387–395.) Gauss also recognized that the difficulties encountered in

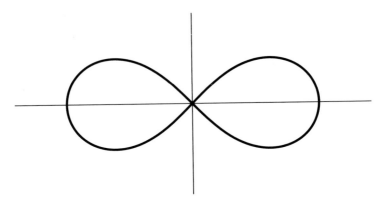

FIGURE 17
The lemniscate.

Notice, for example, the similarity between (4) and the formula

$$2\pi = 4 \int_0^1 \frac{dx}{\sqrt{1 - x^2}}$$

for the circumference of the unit circle.

But (4) lies much deeper. Like other important elliptic integrals, such as the arc length of the ellipse or the hyperbola, or the period of a simple pendulum, the lemniscatic integral cannot be evaluated in terms of the familiar functions of elementary calculus—the algebraic, trigonometric, logarithmic, or exponential functions. Since these integrals are crucial for applications, it was essential that they be transformed into more manageable forms.

The algorithm of the arithmetic and geometric means made its first appearance in a paper of Lagrange published in 1784–1785. However, Gauss claimed to have discovered it independently in 1791—at the age of fourteen!—and it is he who uncovered the true depth of the subject.

---

studying the function $y = \int_0^x dt/\sqrt{1 - t^4}$ are the same as those encountered in studying

$$y = \int_0^x \frac{dt}{\sqrt{1 - t^2}} = \arcsin x.$$

It is simpler to study x as a function of y. In the case of the lemniscate, Gauss proved that the inverse function $x = \varphi(y)$, which he called the *sine of the lemniscate* of y, is a doubly periodic function, with real period L (the arc length of the lemniscate) and imaginary period iL. Analogously, the period of the sine function is 2π, the circumference of the unit circle.

Given two positive numbers a and b, the algorithm transforms them into a pair of sequences $\{a_n\}$ and $\{b_n\}$ defined recursively by $a_0 = a$, $b_0 = b$ and

$$a_1 = (a_0 + b_0)/2, \qquad b_1 = \sqrt{a_0 b_0}$$
$$a_2 = (a_1 + b_1)/2, \qquad b_2 = \sqrt{a_1 b_1}$$
$$a_3 = (a_2 + b_2)/2, \qquad b_3 = \sqrt{a_2 b_2}$$

$$\cdots$$

As an illustration, take $a = \sqrt{2}$ and $b = 1$. The results after only four iterations are striking:

n	a_n	b_n
0	1.41421356237390504880 2	1.00000000000000000000
1	1.20710678118654752440 1	1.18920711500272106671 7
2	1.19815694809463429555 9	1.19812352149312012260 7
3	1.19814023479387720908 3	1.19814023467730720579 8
4	1.19814023473559220744 1	1.19814023473559220743 9

It appears that both sequences converge to the same value and that the rate of convergence is extraordinarily rapid. It is noteworthy that these calculations were performed, not on a computer, but by Gauss himself more than 200 years ago.

Assertion: For any choice of the initial values a and b,

$$\lim_{n \to \infty} a_n = \lim_{n \to \infty} b_n.$$

This common limit is denoted by $\mathrm{agm}(a, b)$ and is called, appropriately, the **arithmetic-geometric mean** of a and b.

For the proof, we may clearly suppose that $a > b$. Then

$$b < b_1 < a_1 < a$$
$$b < b_1 < b_2 < a_2 < a_1 < a$$
$$b < b_1 < b_2 < b_3 < a_3 < a_2 < a_1 < a$$

and so on ad infinitum. This shows that $\{b_n\}$ is increasing and bounded from above and $\{a_n\}$ is decreasing and bounded from below. Both sequences therefore converge. But

$$a_{n+1} - b_{n+1} < a_{n+1} - b_n = \frac{a_n - b_n}{2},$$

and hence, by induction,

$$0 < a_n - b_n < \frac{a - b}{2^n} \tag{5}$$

for every n. Since $(a - b)/2^n \to 0$, the assertion follows.

The rapid rate of convergence of the algorithm in the special case considered cannot be accounted for by the estimate (5). The trick is to observe that

$$a_{n+1}^2 - b_{n+1}^2 = \frac{(a_n - b_n)^2}{4}.$$

If we now divide both sides by $a_{n+1} + b_{n+1} = 2a_{n+2}$ and recall that $\{a_n\}$ decreases to $\mathrm{agm}(a, b)$, then

$$a_{n+1} - b_{n+1} < \frac{(a_n - b_n)^2}{8\,\mathrm{agm}(a, b)}.$$

Thus, $a_n - b_n$ tends to zero quadratically and the number of correct decimal places of $\mathrm{agm}(a, b)$ is approximately doubled with each iteration of the algorithm.

By May 30, 1799 Gauss had made an extraordinary discovery. He recorded the result in his mathematical diary, the 98th entry: "We have established that the arithmetic-geometric mean between 1 and $\sqrt{2}$ is $\pi/\tilde{\omega}$ to the eleventh decimal place; the demonstration of this fact will surely open an entirely new field of analysis."[38]

The words were prophetic. Two seemingly unrelated disciplines would merge into a rich and fruitful whole, with important applications to analytic number theory and computational complexity, a surprising mixture of the classical and the modern, the theoretical and the applied.[39] And the origin was Gauss's observation that two numbers appear to be the same.

By December of that year Gauss had succeeded in demonstrating his empirical discovery. The arithmetic-geometric mean is not nearly so simple as we may have initially supposed.

Theorem (Gauss). *For all positive numbers a and b,*

$$\int_{-\infty}^{\infty} \frac{dx}{\sqrt{(x^2 + a^2)(x^2 + b^2)}} = \frac{\pi}{\mathrm{agm}(a, b)}.$$

---

[38] Gauss, *Werke,* Göttingen-Leipzig, 1868–1927, X.1, p. 542. Gauss used the symbol $\tilde{\omega}$ to denote the *lemniscate constant,* or, in our notation, the number $L/2$ (see formula (4)). Its value, correct to 5 decimal places, is 2.62205. ... For an English translation of Gauss's diary, see J. J. Gray, A commentary on Gauss's mathematical diary, 1796–1814, with an English translation, *Expos. Math.,* 2(1984), 97–130.

[39] The most comprehensive treatment is given by J. M. Borwein and P. B. Borwein, *Pi and the AGM* [53].

Proof. Call the integral $I(a, b)$. The quickest and most elegant proof is based on the inspired observation that the value of the integral remains unchanged when a and b are replaced, respectively, by their arithmetic and geometric means. In other words,

$$I(a, b) = I(\frac{a+b}{2}, \sqrt{ab}). \qquad (6)$$

To prove this, change variables in

$$\int_{-\infty}^{\infty} dt / \sqrt{(t^2 + (\frac{a+b}{2})^2)(t^2 + ab)}$$

by setting $t = \frac{1}{2}(x - \frac{ab}{x})$ and carefully work through the algebra.[40] Of course, it is one thing to verify a known result and quite another to discover it!

By applying formula (6) repeatedly, we find that

$$I(a, b) = I(a_1, b_1) = I(a_2, b_2) = \cdots$$

and hence

$$I(a, b) = \lim_{n \to \infty} I(a_n, b_n).$$

Now it is a simple matter to show that $I(a, b)$ is jointly continuous in a and b. Therefore, since a_n and b_n both approach $\mu = \text{agm}(a, b)$, it follows that

$$\lim_{n \to \infty} I(a_n, b_n) = I(\mu, \mu)$$

$$= \int_{-\infty}^{\infty} \frac{dx}{x^2 + \mu^2}$$

$$= \frac{1}{\mu} \tan^{-1} \frac{x}{\mu} \Big]_{-\infty}^{\infty}$$

$$= \frac{\pi}{\mu}.$$

Thus $I(a, b) = \pi / \text{agm}(a, b)$, as asserted.

The theorem provides us with a powerful tool for approximating a large class of elliptic integrals. In particular, Gauss's formula for the arc length of the lemniscate is now a simple consequence. Indeed,

$$\int_0^{\infty} \frac{dx}{\sqrt{(x^2 + 1)(x^2 + 2)}} = \int_0^1 \frac{dr}{\sqrt{1 - r^4}}.$$

---

[40] Complete details can be found in [315]. For Gauss's own proof see [112].

(For the proof, change variables in the first integral by setting $x = \sqrt{1 - r^2}/r$.) Since the right side is $L/4$ and the left side is one-half $\pi/\operatorname{agm}(\sqrt{2}, 1)$, it follows that

$$2\pi/L = \operatorname{agm}(\sqrt{2}, 1),$$

and this is precisely what Gauss observed.[41]

To pursue these matters further is well beyond the scope of the present work. Yet it seems appropriate to conclude by mentioning one application of considerable import—the high precision calculation of the number π.

Four Hundred Digits of Pi

$\pi = 3.$ 14159 26535 89793 23846
26433 83279 50288 41971
69399 37510 58209 74944
59230 78164 06286 20899
86280 34825 34211 70679
82148 08651 32823 06647
09384 46095 50582 23172
53594 08128 48111 74502
84102 70193 85211 05559
64462 29489 54930 38196
44288 10975 66593 34461
28475 64823 37867 83165
27120 19091 45648 56692
34603 48610 45432 66482
13393 60726 02491 41273
72458 70066 06315 58817
48815 20920 96282 92540
91715 36436 78925 90360
01133 05305 48820 46652
13841 46951 94151 16094

---

[41] For an excellent account of Gauss's many contributions to the arithmetic-geometric mean, see Cox [112].

In the mid-1970's, Richard Brent [63] and Eugene Salamin [370] independently rediscovered a beautiful formula known in an equivalent form to Gauss:

$$\pi = 2\,\text{agm}(\sqrt{2}, 1)^2 / (1 - \sum_1^\infty 2^n \epsilon_n),$$

where $\epsilon_n = a_n^2 - b_n^2$. The dramatic rate of convergence of the arithmetic-geometric mean iteration makes this a highly efficient method for computing π. A decade later, Jonathan and Peter Borwein found an improved method, based in part on work done by the great Indian mathematician Srinivasa Ramanujan (1887–1920).[42] The Borweins' algorithm consists of a simple two-term recursion beginning with the values $\alpha_0 = \sqrt{2} - 1$ and $\beta_0 = 6 - 4\sqrt{2}$:

$$\alpha_1 = \frac{1 - \sqrt[4]{1 - \alpha_0^4}}{1 + \sqrt[4]{1 - \alpha_0^4}}, \quad \beta_1 = (1 + \alpha_1)^4 \beta_0 - 2^3 \alpha_1 (1 + \alpha_1 + \alpha_1^2)$$

$$\alpha_2 = \frac{1 - \sqrt[4]{1 - \alpha_1^4}}{1 + \sqrt[4]{1 - \alpha_1^4}}, \quad \beta_2 = (1 + \alpha_2)^4 \beta_1 - 2^5 \alpha_2 (1 + \alpha_2 + \alpha_2^2)$$

$$\alpha_3 = \frac{1 - \sqrt[4]{1 - \alpha_2^4}}{1 + \sqrt[4]{1 - \alpha_2^4}}, \quad \beta_3 = (1 + \alpha_3)^4 \beta_2 - 2^7 \alpha_3 (1 + \alpha_3 + \alpha_3^2)$$

$$\cdots$$

As n tends to infinity, $1/\beta_n \to \pi$.

This remarkable algorithm—among the most efficient known for the extended precision calculation of π—converges to π *quartically*. This means that the error at each step is approximately proportional to the fourth power of the previous error, thereby quadrupling the number of correct digits with each iteration. As a result, fifteen iterations alone are sufficient to know π to more than 2 billion decimal places![43]

Using a different variant of Ramanujan's ideas, David and Gregory Chudnovsky were the first to reach the 2 billion mark [349].

Of what possible use can it be to know 2 billion digits of π? Are these the concerns only of those who like to break records, who love to compute just for the sake of computing? The challenge of approximating π has been taken up for more than 4000 years—its history is almost as old as the history of civilization

---

[42] For an excellent expository account see [55].

[43] This modern algorithm is, theoretically, close to the fastest possible. By applying the best techniques known for addition, multiplication, and root extraction, the bit complexity of computing the first n digits of π is only marginally greater than the n steps needed to write them down! [437]

itself—yet 39 digits alone will suffice for any conceivable application.[44] What then is the mathematical import of the tireless efforts of the "digit hunters"?

In his efforts to extend Archimedes' calculations of π, Huygens was led to an important technique that has come to be known as the Romberg method of numerical integration. Today the high precision calculation of π finds practical application in testing the "global integrity" of a supercomputer. "A large-scale calculation of pi is entirely unforgiving; it soaks into all parts of the machine and a single bit awry leaves detectable consequences."[45] And the circle of ideas surrounding the arithmetic-geometric mean has led to algorithms for the fast computation of log, exp, and all of the other elementary transcendental functions [52], [63].

Nevertheless, the state of our current ignorance is profound:

> We do not know whether such basic constants as $\pi + e$, π/e, or $\log \pi$ are irrational, let alone transcendental. ... We don't know anything of consequence about the simple continued fraction of pi, except (numerically) the first 17 million terms, which Gosper computed in 1985. ... Likewise, apart from listing the first many millions of digits of π, we know virtually nothing about the decimal expansion of π. It is possible, albeit not a good bet, that all but finitely many of the decimal digits of pi are in fact 0's and 1's. ... Questions concerning the normality of or the distribution of digits of particular transcendentals such as π appear completely beyond the scope of current mathematical techniques.[46]

Thus the search for patterns continues. Just as Gauss was led to an entirely new field of analysis by the observed coincidence of two numbers, so may the next Gauss unlock new mysteries and fashion new and unifying theories based on the empirical work of the digit hunters.

> Perhaps in some far distant century they may say, "Strange that those ingenious investigators into the secrets of the number system had so little conception of the fundamental discoveries that would later develop from them!"
>
> —D.N. Lehmer[47]

---

[44] As many writers have pointed out, 39 digits of π are sufficient to compute the circumference of a circle of radius 2×10^{25} meters (an upper bound for the radius of the known universe) with an error not exceeding 10^{-12} meters (a lower bound for the radius of a hydrogen atom).

[45] See [57], p. 204.

[46] Ibid., pp. 203–204.

[47] "Hunting big game in the theory of numbers" [277].

PROBLEMS

1. Consider the sequence

$$0, 1, \frac{1}{2}, \frac{3}{4}, \frac{5}{8}, \frac{11}{16}, \dots$$

in which the first two terms are 0 and 1 and every term after these is the arithmetic mean of the two terms that precede it. Show that the sequence converges and find its limit.

2. Let $\{x_n\}$ be a sequence of real numbers and let

$$y_n = \frac{x_1 + \cdots + x_n}{n} \qquad n = 1, 2, 3, \dots.$$

Prove that if $\{x_n\}$ converges to L, then $\{y_n\}$ also converges to L.

3. Prove that $\dfrac{1 + \sqrt{2} + \sqrt[3]{3} + \cdots + \sqrt[n]{n}}{n} \to 1.$

4. Let $\{x_n\}$ be a sequence of positive numbers with $\lim_{n \to \infty} x_n = L > 0$. Prove that

$$\lim_{n \to \infty} \sqrt[n]{x_1 x_2 \cdots x_n} = L.$$

5. Prove that

$$\lim_{n \to \infty} \frac{\sqrt[n]{n!}}{n} = \frac{1}{e}.$$

(Compare problem 32(b), §1.)

6. If $\{x_n\}$ is a sequence of positive numbers and if $x_n/x_{n-1} \to L$, prove $\sqrt[n]{x_n} \to L$. What does this say about the relative strength of the ratio test and the root test?

7. *Prove*: If f is continuous on $[0, 1]$, then

$$\int_0^1 f(x)\, dx = \lim_{n \to \infty} \frac{f(1/n) + f(2/n) + \cdots + f(n/n)}{n}.$$

The integral is called the *average value* of f.

8.   Establish the following limiting relations:

a. $\dfrac{1}{n+1} + \dfrac{1}{n+2} + \cdots + \dfrac{1}{2n} \to \log 2$

b. $\dfrac{n}{n^2+1^2} + \dfrac{n}{n^2+2^2} + \cdots + \dfrac{n}{n^2+n^2} \to \dfrac{\pi}{4}$

c. $\dfrac{1}{n}\left(\sin\dfrac{\pi}{n} + \sin\dfrac{2\pi}{n} + \cdots + \sin\dfrac{n\pi}{n}\right) \to \dfrac{2}{\pi}$

d. $\dfrac{1}{\sqrt{n^2+1^2}} + \dfrac{1}{\sqrt{n^2+2^2}} + \cdots + \dfrac{1}{\sqrt{n^2+n^2}} \to \log(1+\sqrt{2})$

e. $\dfrac{\sqrt[n]{(n+1)(n+2)\cdots(n+n)}}{n} \to \dfrac{4}{e}$

9.   Let $\{a_n\}$ be a sequence of numbers in $[0,1]$ and, for $0 \le a < b \le 1$, let $N(n; a, b)$ denote the number of integers $j \le n$ such that a_j is in $[a, b]$. We say that $\{a_n\}$ is **uniformly distributed** in $[0, 1]$ if

$$\lim_{n\to\infty} \frac{N(n; a, b)}{n} = b - a$$

for all a and b.

a.   Show that the sequence

$$\frac{1}{2}, \frac{1}{3}, \frac{2}{3}, \frac{1}{4}, \frac{2}{4}, \frac{3}{4}, \frac{1}{5}, \frac{2}{5}, \frac{3}{5}, \frac{4}{5}, \frac{1}{6}, \frac{2}{6}, \ldots$$

is uniformly distributed in $[0, 1]$.

b.   Prove that if $\{a_n\}$ is uniformly distributed in $[0,1]$ and f is continuous on $[0, 1]$, then

$$\int_0^1 f(x)\, dx = \lim_{n\to\infty} \frac{f(a_1) + \cdots + f(a_n)}{n}.$$

10.   Prove that if $\{x_n\}$ is a sequence of real numbers for which

$$\lim_{n\to\infty} \frac{x_1 + x_2 + \cdots + x_n}{n}$$

exists, then $\lim_{n\to\infty} x_n/n = 0$. Deduce that the series

$$1 - 2 + 3 - 4 + \cdots$$

cannot be "summed" by the method of the text. (For a discussion of other, more general, summability methods, see [175], [262].)

11. Let $f(n)$ denote the number of integral solutions of

$$x^2 + y^2 + z^2 = n$$

(solutions differing only in sign or order are regarded as distinct). Prove that

$$\lim_{n \to \infty} \frac{f(1) + \cdots + f(n)}{n^{3/2}}$$

exists and find its value.

12. Let $f(n)$ be the number of decompositions of a natural number n into a sum of one or more consecutive prime numbers. For example, $f(41) = 3$ because

$$41 = 11 + 13 + 17 = 2 + 3 + 5 + 7 + 11 + 13.$$

Formulate a conjecture concerning the average value of f.

13. Let $P(n)$ denote the number of primitive Pythagorean triangles with hypotenuse equal to n. For example, $P(5) = 1$,

$$3^2 + 4^2 = 5^2,$$

and $P(65) = 2$,

$$33^2 + 56^2 = 63^2 + 16^2 = 65^2.$$

Formulate a conjecture concerning the average value of P.

14. In the game of *craps* a pair of dice is rolled by one of the players. On the first roll, the player wins if the sum of the spots is 7 or 11 and loses if it is 2, 3, or 12. Any other number becomes the "point" and the player continues rolling the dice until either the point or a 7 appears. In the former case, the player wins, and in the latter case he loses. Assuming that the player wins or loses \$1, compute the expected value of the game.

15. A fair coin is tossed until either three consecutive heads or three consecutive tails appear. What is the expected number of tosses?

16. Two players, A and B, toss a fair coin until either HHT or HTT occurs. Player A wins if the pattern HHT comes first, player B wins if HTT comes first. Show that, while the game certainly seems to be fair, A is twice as likely to win as B!

17. In the *jeu de rencontre* (see chapter 1, §3), two identical decks of n different cards are put into random order and matched against each other. A *match* occurs if a card occupies the same position in both decks. Find the expected number of matches.

18. The St. Petersburg Paradox. Two players, A and B, toss a fair coin until it falls heads; if this occurs at the nth toss then A agrees to pay B 2^n dollars. How much should B agree to pay A at the outset to make the game fair? Since the possible outcomes of the game are the values $2^1, 2^2, 2^3, \ldots$ with corresponding probabilities $2^{-1}, 2^{-2}, 2^{-3}, \ldots$, it follows that B's expected gain is *infinite*:

$$\mathbf{E} = \sum_{n=1}^{\infty} 2^n \cdot 2^{-n} = \infty.$$

Does this mean that B should agree to pay A an infinite amount of money for the privilege of playing the game?! How can the paradox be resolved?

19. Buffon's Needle Problem. A thin needle (theoretically, a line segment) of length ℓ is dropped onto a board ruled with equidistant parallel lines, the width of the strip between two consecutive lines being $d \geq \ell$. Show that the probability that the needle crosses one of the lines is $2\ell/\pi d$.

Indication of the proof. The position of the needle can be determined by the distance $x = OP$ from its center to the nearest line, and the smaller angle θ between OP and the needle. The domain of (x, θ) is a rectangle. Find it.

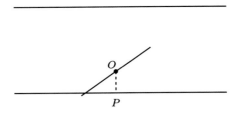

FIGURE 18

Let Ω denote the portion of the rectangle corresponding to the event that the needle crosses one of the lines. Formulate a plausible assumption about the area of Ω and the probability of the desired event.

20. Solve Buffon's problem when the needle is longer than the distance between two consecutive lines.

21. Laplace's Problem. A thin needle is dropped onto a board covered with congruent rectangles, as shown in Figure 19. If the length of the needle is less than the smaller sides of the rectangles, find the probability that the needle will be entirely contained in one of the rectangles of the set.

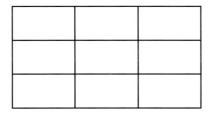

FIGURE 19

22. A machine emits real numbers at random from the unit interval $[0, 1]$, one after another, until their sum exceeds 1. Show that the average number of numbers emitted is e.

23. Define a pair of sequences $\{a_n\}$ and $\{b_n\}$ recursively by the relations

$$a_{n+1} = \frac{a_n + b_n}{2}, \qquad b_{n+1} = \frac{2a_n b_n}{a_n + b_n},$$

where $0 < b_1 < a_1$. Show that for $n = 1, 2, 3, \ldots$,

$$a_n > a_{n+1} > b_{n+1} > b_n > 0$$

and that both sequences $\{a_n\}$ and $\{b_n\}$ tend to the common limit $\sqrt{a_1 b_1}$.

24. The arithmetic-geometric mean algorithm bears a striking resemblance to the ancient algorithm used by Archimedes to calculate π. Starting with the unit circle, circumscribe about it and inscribe in it regular polygons with n sides;

let a_n and b_n be their respective perimeters. Show that

$$a_{2n} = \frac{2a_n b_n}{a_n + b_n} \quad \text{and} \quad b_{2n} = \sqrt{a_{2n} b_n}.$$

Thus the terms of the sequence $a_3, b_3, a_6, b_6, a_{12}, b_{12}, \ldots$ can be generated recursively by taking alternately the harmonic and geometric means of the two preceding terms. In this way, π can be approximated from above and below by numerical bounds whose difference can be made as small as desired.

In the *Measurement of a Circle*, Archimedes was able to show, by passing to polygons with 96 sides, that

$$3\frac{10}{71} < \pi < 3\frac{1}{7}.$$

See *The Works of Archimedes* [15], pp. 91–98. The algorithm converges exceedingly slowly: each iteration decreases the error by a factor of approximately four. (See, for example, [52], p. 352 and [337], p. 166.)

25. (Steinhaus) For every natural number n, there exists in the plane a circle containing exactly n lattice points in its interior.

Hint. A circle with center $(\sqrt{2}, \sqrt{3})$ cannot pass through two or more lattice points.

Remark. A. Schinzel (*L'Enseignement Mathématique* 4 (1958), 71–72) has proved that *for any natural number n there exists a circle containing exactly n lattice points on its circumference.*

26. Bertrand's Paradox. What is the probability that a chord drawn at random on a circle will be longer than the side of an inscribed equilateral triangle? Depending on how we interpret the term "at random," the answer can reasonably be $1/2$, $1/3$, or $1/4$.

27. What is the average straight line distance between two points on a sphere of radius 1?

28. The elements of a determinant of order n are arbitrary integers. What is the probability that the value of the determinant is odd?

29. If $d(n)$ denotes the number of divisors of the natural number n, show that

$$\frac{d(1) + d(2) + \cdots + d(n)}{n} \sim \log n.$$

Here the sign $\sim$ (read "is asymptotically equal to") is used to indicate that the ratio of the two sides tends to unity as $n \to \infty$.

Hint. Interpret $d(n)$ geometrically as the number of lattice points in the plane that lie on the equilateral hyperbola

$$xy = n.$$

Remark. By a simple refinement of the argument, Dirichlet (1849) was able to show that this asymptotic formula can be improved considerably: As $n \to \infty$,

$$\frac{d(1) + d(2) + \cdots + d(n)}{n} - \log n \to 2\gamma - 1,$$

where γ is Euler's constant (see, for example, [211], p. 264). This shows the intimate connection that exists between γ and the theory of numbers.

30. (Unsolved) It has been argued on probabilistic grounds that the expected number of primes p in the octave interval $(x, 2x)$ for which $2^p - 1$ is a prime is e^γ, where γ is Euler's constant. Equivalently: If $M(n)$ is the nth Mersenne prime, then

$$\frac{\log_2 \log_2 M(n)}{n} \to e^{-\gamma}$$

(see, for example, [373], p. 32). Figure 20 shows a plot of $\log_2 \log_2 M(n)$ as a function of n for the first thirty values of n. Using the method of least squares, find the line that best fits the given data and compare its slope with $e^{-\gamma}$.

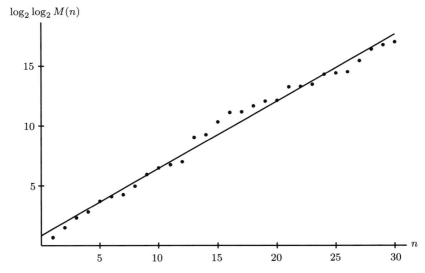

FIGURE 20

CHAPTER **5**

APPROXIMATION:
FROM PI TO THE PRIME NUMBER THEOREM

Primzahlen unter $a \, (= \infty) \, \dfrac{a}{1a}$

—Carl Friedrich Gauss[1]

---

[1] This cryptic entry, on the back page of a copy of a table of logarithms that he had obtained as a boy of fourteen, is a statement of the celebrated prime number theorem: the number of primes less than a given number a is asymptotically equal to $a/\log a$ as $a \to \infty$. See Hans Reichardt, ed., *C. F. Gauss, Leben und Werk,* Haude and Spener, Berlin, 1960.

1. "Luck Runs in Circles"

Steinhaus, with his predilection for metaphors, used to quote a Polish proverb, '*Forturny kolem sie tocza*' (*Luck runs in circles*), to explain why π, so intimately connected with circles, keeps cropping up in probability theory and statistics, the two disciplines which deal with randomness and luck.

—Mark Kac[2]

a. A Delicate Balance

We begin with a problem in ordinary arithmetic.

How does the product of the first n even integers

$$2 \cdot 4 \cdot 6 \cdots 2n$$

compare with the product of the first n odd integers

$$1 \cdot 3 \cdot 5 \cdots 2n - 1$$

when n is a large number?

More precisely, we should like to estimate their ratio

$$\frac{2 \cdot 4 \cdot 6 \cdots 2n}{1 \cdot 3 \cdot 5 \cdots 2n - 1} \qquad \text{as } n \to \infty.$$

There is no simple formula for this expression, so an approximation will have to suffice. What can reasonably be expected? Do the quotients approach a finite limit as $n \to \infty$ or do they increase without bound?

It may help to reformulate the problem in the more descriptive language of coin-tossing. *If a fair coin is tossed 2n times, what is the probability that it lands heads exactly half the time?* The answer is $\binom{2n}{n}/2^{2n}$ (the denominator represents the number of possible outcomes, the numerator the number of favorable ones), and it simplifies readily to

$$\frac{1 \cdot 3 \cdot 5 \cdots 2n - 1}{2 \cdot 4 \cdot 6 \cdots 2n}.$$

"Experience modifies human beliefs," wrote Pólya. Does this information reinforce your original conjecture? Does it weaken it?

---

[2] *Enigmas of Chance* [242], p. 55.

n	$\dfrac{2 \cdot 4 \cdot 6 \cdots 2n}{1 \cdot 3 \cdot 5 \cdots 2n - 1}$
10	6
100	18
1,000	56
10,000	177
100,000	560

TABLE 1

With the aid of a computer we first gather some data (see Table 1).

Note that the entries in the second column of Table 1 have been rounded to the nearest integer. Naturally, they have been computed recursively, as a product of quotients $\frac{2}{1}, \frac{4}{3}, \frac{6}{5}, \dots$, and not as a quotient of two products!

What do the data suggest? Is there a pattern? It appears, for example, that, as n increases from one power of 10 to the next, the value of the corresponding ratio is multiplied by roughly the square root of 10. Thus, for example, 56 is roughly ten times 6, 177 is roughly ten times 18, and 560 is exactly ten times 56.

On the basis of the evidence, it is not difficult to formulate the conjecture

$$3n < \left(\frac{2 \cdot 4 \cdot 6 \cdots 2n}{1 \cdot 3 \cdot 5 \cdots 2n - 1} \right)^2 < 4n$$

for every $n > 1$. This is certainly true when $n = 2$, and we proceed by induction. Call the middle term P_n. Then $P_{n+1} = P_n \cdot (\frac{2n+2}{2n+1})^2$, and we find

$$P_{n+1} < 4n \left(\frac{2n+2}{2n+1} \right)^2 < 4(n+1)$$

(the first inequality is a consequence of the induction hypothesis, while the second follows by elementary algebra). This establishes the upper bound for P_n. For the lower bound, the proof fails—the induction hypothesis guarantees only that $P_{n+1} > 3n(\frac{2n+2}{2n+1})^2$, and the term on the right is *never* larger than $3(n + 1)$. What's wrong? The lower bound is correct, as we shall soon see, so why has induction failed? The explanation lies in an important aspect of this method of proof which is frequently overlooked.

In a proof by mathematical induction, there is often a delicate balance between what needs to be proved and what is needed to prove it. Try to prove too little and the induction hypothesis will be weak, and the transition from n to $n+1$ may not readily follow. Strengthen the proposition and you strengthen the tools needed to prove it.

The remedy therefore is to prove more. Accordingly, we conjecture that

$$P_n > 3n + 1.$$

The induction hypothesis is now equal to the task, and we find that

$$P_{n+1} > (3n + 1) \cdot \left(\frac{2n+2}{2n+1}\right)^2 > 3(n+1) + 1.$$

Since $P_2 > 7$, all is well.

That P_n has upper and lower bounds that are both multiples of n suggests that P_n itself may behave like a multiple of n. Table 2 suggests even more.

n	P_n/n
10	3.22108899
100	3.14945642
1,000	3.14237815
10,000	3.14167119
100,000	3.14160050
1,000,000	3.14159343
10,000,000	3.14159273
100,000,000	3.14159266
1,000,000,000	3.14159265

TABLE 2

The evidence is compelling—it appears that

$$\left(\frac{2 \cdot 4 \cdot 6 \cdots 2n}{1 \cdot 3 \cdot 5 \cdots 2n - 1}\right)^2 \frac{1}{n}$$

approaches π as $n \to \infty$. Or, in the language of infinite products,

$$\frac{\pi}{2} = \left(\frac{2\,2}{1\,3}\right)\left(\frac{4\,4}{3\,5}\right)\left(\frac{6\,6}{5\,7}\right)\cdots.$$

We have been led, by experimentation, to one of the most remarkable formulas of analysis, a representation of π as the limit of a simple sequence of rational numbers. It is called *Wallis's product,* after John Wallis (1616–1703), the leading English mathematician before Newton. While elegant in its simplicity, it is hardly a practical way to compute π.

b. Wallis's Formula

Wallis was led to his celebrated formula in an effort to calculate the area of a circle analytically, and his method took him through an extraordinary process of analogy, intuition, and highly complex interpolations. The argument is put forth in his important book *Arithmetica infinitorum* (*The Arithmetic of Infinites*), published in 1655, which had a decisive influence on Newton's early mathematical work. Starting with Cavalieri's formula

$$\int_0^1 x^n \, dx = \frac{1}{n+1},$$

which he derived empirically (of course the symbol $\int$ was not then in use), he proceeded to compute the areas bounded by the axes, the ordinate at x, and the curves

$$y = (1 - x^2)^0, \quad y = (1 - x^2)^1, \quad y = (1 - x^2)^2, \quad y = (1 - x^2)^3, \ldots$$

(see Figure 1) and arrived at the results

$$x,$$

$$x - \frac{1}{3}x^3,$$

$$x - \frac{2}{3}x^3 + \frac{1}{5}x^5,$$

$$x - \frac{3}{3}x^3 + \frac{3}{5}x^5 - \frac{1}{7}x^7,$$

$$x - \frac{4}{3}x^3 + \frac{6}{5}x^5 - \frac{4}{7}x^7 + \frac{1}{9}x^9,$$

$$\cdots$$

respectively. Now since the equation of the unit circle is $y = (1 - x^2)^{1/2}$, the problem amounts to determining what expression corresponding to the exponent $1/2$ should be interpolated between the known values x and $x - \frac{1}{3}x^3$. This Wallis could not answer for it seemed to him that the interpolated series would have to contain more than one term but less than two! Nevertheless, "he plunged into a maelstrom of numerical work," and, through a highly complicated and difficult analysis, arrived finally at the formula that bears his name ([408], pp. 244ff.).

Ten years later, Newton discovered the underlying pattern, effected the interpolation, and thereby discovered the *binomial series* for fractional and neg-

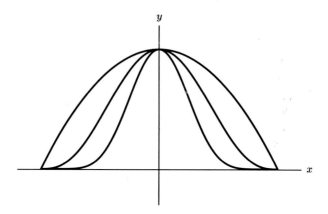

FIGURE 1
The graph of $y = (1 - x^2)^n$ for $n = 1, 2, 3$.

ative exponents, one of the landmarks in the development of mathematical science.

The simplest proof of Wallis's formula is not Wallis's own, but it begins unerringly in the same way. Observe that by taking $x = 1$ in the sequence of area functions displayed above we arrive at the values

$$1 = 1$$

$$1 - \frac{1}{3} = \frac{2}{3}$$

$$1 - \frac{2}{3} + \frac{1}{5} = \frac{2 \cdot 4}{3 \cdot 5}$$

$$1 - \frac{3}{3} + \frac{3}{5} - \frac{1}{7} = \frac{2 \cdot 4 \cdot 6}{3 \cdot 5 \cdot 7}$$

$$1 - \frac{4}{3} + \frac{6}{5} - \frac{4}{7} + \frac{1}{9} = \frac{2 \cdot 4 \cdot 6 \cdot 8}{3 \cdot 5 \cdot 7 \cdot 9}$$

and these are essentially the ratios we wanted to approximate. Consideration of the integrals

$$I(n) = \int_0^1 (1 - x^2)^n \, dx$$

for $n \geq 1$ therefore seems promising.

The preceeding remarks were intended simply as motivation; the proof itself begins now.

The simplest way to evaluate $I(n)$ is to integrate by parts. Setting $f(x) = (1 - x^2)^n$ and $g'(x) = 1$, we find that

$$I(n) = x(1 - x^2)^n \big]_0^1 + 2n \int_0^1 x^2(1 - x^2)^{n-1} \, dx$$

$$= 2n \int_0^1 (1 - (1 - x^2))(1 - x^2)^{n-1} \, dx$$

$$= 2n\, I(n - 1) - 2n\, I(n).$$

Thus, we have the recursive formula

$$I(n) = \frac{2n}{2n + 1} I(n - 1),$$

which is *valid for any real number $n \geq 1$*.

Taking n successively to be $1, 2, 3, \ldots$, we find by repeated application of this formula that

$$I(1) = \frac{2}{3}I(0) = \frac{2}{3}$$

$$I(2) = \frac{4}{5}I(1) = \frac{2}{3}\frac{4}{5}$$

$$I(3) = \frac{6}{7}I(2) = \frac{2}{3}\frac{4}{5}\frac{6}{7}$$

$$\cdots$$

and, in general,

$$I(n) = \frac{2 \cdot 4 \cdot 6 \cdots 2n}{3 \cdot 5 \cdot 7 \cdots 2n + 1}$$

for every natural number n.

But the number π has not yet made its appearance. Since the area of one quarter of the unit circle is

$$\frac{\pi}{4} = \int_0^1 \sqrt{1 - x^2} \, dx,$$

or $I(1/2)$, we now apply the recursive formula with $n = \frac{3}{2}, \frac{5}{2}, \frac{7}{2}, \ldots$ The resulting equations are then

$$I(\tfrac{3}{2}) = \tfrac{3}{4}I(\tfrac{1}{2}) = \frac{3}{4}\frac{\pi}{4}$$

$$I(\tfrac{5}{2}) = \tfrac{5}{6}I(\tfrac{3}{2}) = \frac{3}{4}\frac{5}{6}\frac{\pi}{4}$$

$$I(\tfrac{7}{2}) = \tfrac{7}{8}I(\tfrac{5}{2}) = \frac{3}{4}\frac{5}{6}\frac{7}{8}\frac{\pi}{4}$$

$$\cdots$$

so that, generally,

$$I\left(n - \frac{1}{2}\right) = \frac{1 \cdot 3 \cdot 5 \cdots 2n - 1}{2 \cdot 4 \cdot 6 \cdots 2n} \cdot \frac{\pi}{2}.$$

But

$$I(n) < I\left(n - \frac{1}{2}\right) < I(n - 1)$$

since $0 < 1 - x^2 < 1$ for $0 < x < 1$, and therefore

$$\frac{2 \cdot 4 \cdot 6 \cdots 2n}{3 \cdot 5 \cdot 7 \cdots 2n + 1} < \frac{1 \cdot 3 \cdot 5 \cdots 2n - 1}{2 \cdot 4 \cdot 6 \cdots 2n} \cdot \frac{\pi}{2} < \frac{2 \cdot 4 \cdot 6 \cdots 2n - 2}{3 \cdot 5 \cdot 7 \cdots 2n - 1}.$$

These inequalities can be written in the form

$$\pi < \left(\frac{2 \cdot 4 \cdot 6 \cdots 2n}{1 \cdot 3 \cdot 5 \cdots 2n - 1}\right)^2 \frac{1}{n} < \pi\left(1 + \frac{1}{2n}\right) \tag{1}$$

and Wallis's formula follows at once.

Notice also that (1) implies that

$$\lim_{n \to \infty} \left(\frac{2 \cdot 4 \cdot 6 \cdots 2n}{1 \cdot 3 \cdot 5 \cdots 2n - 1} - \sqrt{\pi n}\right) = 0.$$

Thus, $\sqrt{\pi n}$ is a simple and remarkably accurate approximation for the desired ratio $(2 \cdot 4 \cdot 6 \cdots 2n)/(1 \cdot 3 \cdot 5 \cdots 2n - 1)$. Equivalently, for large values of n, the probability of obtaining exactly n heads in $2n$ tosses of a fair coin is very nearly equal to $1/\sqrt{\pi n}$. Perhaps luck does run in circles after all.

PROBLEMS

1. Wallis's empirical method of discovery is well illustrated by his original derivation of Cavalieri's formula

$$\int_0^1 x^k \, dx = \frac{1}{k+1}$$

which he gives in the *Arithmetica infinitorum* (Proposition 39). Knowing that

$$\int_0^1 x^k \, dx = \lim_{n\to\infty} \frac{0^k + 1^k + \cdots + n^k}{n^k + n^k + \cdots + n^k},$$

he argues as follows in the case $k = 3$:

$$\frac{0^3 + 1^3}{1^3 + 1^3} = \frac{2}{4} = \frac{1}{4} + \frac{1}{4}$$

$$\frac{0^3 + 1^3 + 2^3}{2^3 + 2^3 + 2^3} = \frac{3}{8} = \frac{1}{4} + \frac{1}{8}$$

$$\frac{0^3 + 1^3 + 2^3 + 3^3}{3^3 + 3^3 + 3^3 + 3^3} = \frac{4}{12} = \frac{1}{4} + \frac{1}{12}$$

$$\frac{0^3 + 1^3 + 2^3 + 3^3 + 4^3}{4^3 + 4^3 + 4^3 + 4^3 + 4^3} = \frac{5}{16} = \frac{1}{4} + \frac{1}{16}$$

$$\frac{0^3 + 1^3 + \cdots + 5^3}{5^3 + 5^3 + \cdots + 5^3} = \frac{6}{20} = \frac{1}{4} + \frac{1}{20}$$

$$\frac{0^3 + 1^3 + \cdots + 6^3}{6^3 + 6^3 + \cdots + 6^3} = \frac{7}{24} = \frac{1}{4} + \frac{1}{24}.$$

On the basis of this numerical evidence he concludes that

$$\frac{0^3 + 1^3 + \cdots + n^3}{n^3 + n^3 + \cdots + n^3} = \frac{1}{4} + \frac{1}{4n} \tag{2}$$

for every n, so the limit as $n \to \infty$ is $\frac{1}{4}$. Give a rigorous proof of formula (2). What is the corresponding formula for the general exponent k ($k = 1, 2, \ldots$)?

2. The discovery of the binomial series

And whatever the common Analysis [that is, algebra] performs by Means of Equations of a finite number of Terms (provided that can be done) this new method can always perform the same by Means of infinite Equations. So that I have not made any Question of giving this the Name of Analysis likewise. For the Reasonings in this are no

less certain than in the other; nor the Equations less exact; albeit we
Mortals whose reasoning Powers are confined within narrow Limits,
can neither express, nor so conceive all the Terms of these Equations
as to know exactly from thence the Quantities we want.

—Isaac Newton

The binomial series, Newton's first mathematical discovery of lasting significance, was formulated in 1665 as a result of his reading of Wallis's *Arithmetica
infinitorum.* Strongly influenced by Wallis's ideas about interpolation and ex-
trapolation, he was led ultimately to the following conjecture: *The expansion*

$$(1+x)^m = 1 + mx + \frac{m(m-1)}{2!}x^2 + \frac{m(m-1)(m-2)}{3!}x^3 + \cdots$$

*is valid not only for positive integral values of the exponent m, but for fractional
and negative values as well.*

When m is a positive integer the series terminates; the coefficients in the
expansion are just the ordinary binomial coefficients $\binom{m}{0}, \binom{m}{1}, \binom{m}{2}, \ldots, \binom{m}{m}$.
While the binomial theorem for integral powers had been known in Europe at
least since 1527, the binomial coefficients had not been known in a form that
permitted their natural extension to negative and fractional exponents. Indeed,
Newton discovered the general binomial coefficient as a result of his efforts to
effect the quadrature of the circle!

Newton never gave a formal proof of his conjecture—like Wallis he relied on examples and analogy—but in a famous letter of October 24, 1676,
addressed to the Secretary of the Royal Society, for transmission to Leibniz,
he outlined his method of discovery. He describes it as follows:[3]

At the beginning of my mathematical studies, when I had met with
the works of our celebrated Wallis, on considering the series by the
intercalation of which he himself exhibits the area of the circle and
the hyperbola, the fact that, in the series of curves whose common
base or axis is x and the ordinates

$$(1-x^2)^{0/2}, \quad (1-x^2)^{1/2}, \quad (1-x^2)^{2/2},$$
$$(1-x^2)^{3/2}, \quad (1-x^2)^{4/2}, \quad (1-x^2)^{5/2}, \text{ etc.}$$

---

[3] See *The Correspondence of Isaac Newton,* volume II (1676–1687), edited by H. W. Turnbull,
Cambridge University Press, 1960, pp. 130–131. For a more detailed account of Newton's interpolation procedure, see C. H. Edwards, Jr., *The Historical Development of the Calculus* [145], pp.
178–187.

if the areas of every other of them, namely

$$x, \quad x - \frac{1}{3}x^3, \quad x - \frac{2}{3}x^3 + \frac{1}{5}x^5, \quad x - \frac{3}{3}x^3 + \frac{3}{5}x^5 - \frac{1}{7}x^7, \text{ etc.}$$

could be interpolated, we should have the areas of the intermediate ones, of which the first $(1 - x^2)^{1/2}$ is the circle: in order to interpolate these series I noted that in all of them the first term was x and that the second terms $\frac{0}{3}x^3$, $\frac{1}{3}x^3$, $\frac{2}{3}x^3$, $\frac{3}{3}x^3$, etc. were in arithmetical progression, and hence that the first two terms of the series to be intercalated ought to be $x - \frac{1}{3}(\frac{1}{2}x^3)$, $x - \frac{1}{3}(\frac{3}{2}x^3)$, $x - \frac{1}{3}(\frac{5}{2}x^3)$, etc. To intercalate the rest I began to reflect that the denominators 1, 3, 5, 7, etc. were in arithmetical progression, so that the numerical coefficients of the numerators only were still in need of investigation. But in the alternately given areas these were the figures of powers of the number 11, namely of these, 11^0, 11^1, 11^2, 11^3, 11^4, that is, first 1; then 1, 1; thirdly, 1, 2, 1; fourthly 1, 3, 3, 1; fifthly 1, 4, 6, 4, 1, etc. And so I began to inquire how the remaining figures in these series could be derived from the first two given figures, and I found that on putting m for the second figure, the rest would be produced by continual multiplication of the terms of this series,

$$\frac{m - 0}{1} \times \frac{m - 1}{2} \times \frac{m - 2}{3} \times \frac{m - 3}{4} \times \frac{m - 4}{5}, \text{ etc.}$$

For example, let $m = 4$, and $4 \times \frac{1}{2}(m - 1)$, that is 6 will be the third term, and $6 \times \frac{1}{3}(m - 2)$, that is 4 the fourth, and $4 \times \frac{1}{4}(m - 3)$, that is 1 the fifth, and $1 \times \frac{1}{5}(m - 4)$, that is 0 the sixth, at which term in this case the series stops. Accordingly, I applied this rule for interposing series among series, and since, for the circle, the second term was $\frac{1}{3}(\frac{1}{2}x^3)$, I put $m = \frac{1}{2}$, and the terms arising were

$$\frac{1}{2} \times \frac{\frac{1}{2} - 1}{2} \quad \text{or} \quad -\frac{1}{8},$$

$$-\frac{1}{8} \times \frac{\frac{1}{2} - 2}{3} \quad \text{or} \quad +\frac{1}{16},$$

$$\frac{1}{16} \times \frac{\frac{1}{2} - 3}{4} \quad \text{or} \quad -\frac{5}{128},$$

and so to infinity. Whence I came to understand that the area of the circular segment which I wanted was

$$x - \frac{\frac{1}{2}x^3}{3} - \frac{\frac{1}{8}x^5}{5} - \frac{\frac{1}{16}x^7}{7} - \frac{\frac{5}{128}x^9}{9} \text{ etc.}$$

a.   As Newton observed, the first few powers of the number 11 are

$$(11)^0 = 1$$
$$(11)^1 = 1\ 1$$
$$(11)^2 = 1\ 2\ 1$$
$$(11)^3 = 1\ 3\ 3\ 1$$
$$(11)^4 = 1\ 4\ 6\ 4\ 1$$

and these are precisely the first five rows of the arithmetical triangle. Does the pattern continue? How can succeeding rows be obtained?

b.   Show that Newton's quadrature result for the circle immediately gives rise to the expansion

$$\sqrt{1 - x^2} = 1 - \frac{1}{2}x^2 - \frac{1}{8}x^4 - \frac{1}{16}x^6 - \cdots .$$

Verify this result by multiplying the series by itself; the result should be

$$\sqrt{1 - x^2}\ \sqrt{1 - x^2} = 1 - x^2.$$

Commenting on Newton's remarkable binomial investigations, C. H. Edwards, Jr. writes:

> The final result of this sequence of investigations—the application to infinite series of the familiar procedures of simple arithmetic—was of greater importance than any single example such as the binomial series. As Boyer puts it, Newton "had found that the analysis by infinite series had the same inner consistency and was subject to the same general laws as the algebra of finite quantities. Infinite series were no longer to be regarded as approximating devices only; they were alternative forms of the functions they represented"[4] ....
>
> Thus was banished forever the "horror of the infinite" that had impeded the Greeks, and was set loose the torrent of infinite series expansions that were to play a central role in the development and applications of the new calculus. ([145], p. 187)

3.   Prove that the general binomial expansion

$$(1 + x)^m = 1 + mx + \frac{m(m - 1)}{2!}x^2 + \frac{m(m - 1)(m - 2)}{3!}x^3 + \cdots$$

---

[4] C. B. Boyer, *A History of Mathematics* [62], p. 395.

is valid for every real exponent m, provided that $|x| < 1$, by following these steps.

    a.   First show that the radius of convergence of the series is 1.

    b.   Next, let $f(x)$ denote the sum of the series for $|x| < 1$ and show that

$$(1 + x) f'(x) = m f(x).$$

    c.   Finally, use part (b) to show that the quotient

$$\frac{f(x)}{(1 + x)^m}$$

has a zero derivative for $|x| < 1$, and hence that $f(x) = c(1 + x)^m$ for some constant c. Complete the proof by showing that $c = 1$.

4. Establish the following infinite product formulas:

    a. $\prod_{n=2}^{\infty} (1 - \frac{1}{n^2}) = \frac{1}{2}$

    b. $(1 + x)(1 + x^2)(1 + x^4)(1 + x^8) \cdots = \frac{1}{1-x}$ whenever $|x| < 1$

    c. $\prod_{n=2}^{\infty} \frac{n^3 - 1}{n^3 + 1} = \frac{2}{3}$

5. The sequence of **Catalan numbers**, named after the Belgian mathematician Eugene Charles Catalan (1814–1894), is defined recursively by the formulas $C_0 = 1$ and, for $n > 0$,

$$C_n = C_0 C_{n-1} + C_1 C_{n-2} + C_2 C_{n-3} + \cdots + C_{n-1} C_0.$$

The first few terms of the sequence are found to be:

$$C_1 = C_0 C_0 = 1,$$
$$C_2 = C_0 C_1 + C_1 C_0 = 2,$$
$$C_3 = C_0 C_2 + C_1 C_1 + C_2 C_0 = 5,$$
$$C_4 = C_0 C_3 + C_1 C_2 + C_2 C_1 + C_3 C_0 = 14.$$

Using the method of generating functions, derive the nonrecursive formula

$$C_n = \frac{1}{n + 1} \binom{2n}{n}.$$

Hint. Let

$$F(x) = C_0 x + C_1 x^2 + C_2 x^3 + \cdots.$$

Show that $F(x)$ satisfies a quadratic equation and deduce that

$$F(x) = \frac{1 - \sqrt{1 - 4x}}{2}.$$

(For an elementary solution that does not rely on generating functions, see H. Dörrie, *100 Great Problems of Elementary Mathematics* [137], p. 21.)

6. Euler's Polygon-Triangulation Problem. It was Euler who first discovered the Catalan numbers through a combinatorial problem in plane geometry which he posed to Christian Goldbach in 1751.

> *In how many ways can a convex polygon of n sides be divided into triangles by drawing diagonals that do not intersect?*

Find an explicit formula for the number D_n of different triangulations.

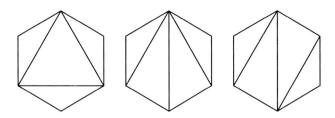

FIGURE 2

a. Calculate D_3, D_4, D_5, D_6.

b. Show that no matter how the n-gon is triangulated, the number of diagonals is always $n - 3$ and the number of triangles is $n - 2$.

c. Define $D_2 = 1$ and show that for $n \geq 3$

$$D_n = D_2 D_{n-1} + D_3 D_{n-2} + D_4 D_{n-3} + \cdots + D_{n-1} D_2.$$

d. Conclude that $\{D_n\}$ is the sequence of Catalan numbers, and hence that

$$D_n = \frac{2 \cdot 6 \cdot 10 \cdots (4n - 10)}{(n - 1)!}.$$

7. Catalan's Problem. In 1838, Catalan solved the following problem:

> *In how many ways can a product of n different factors be calculated by pairs?*

More precisely, given a product of say n letters in a fixed order, in how many ways can we insert parentheses so that inside each pair of left and right parentheses there are exactly two "terms"? A term may consist of any two adjacent letters, a letter and an adjacent parenthetical grouping, or two adjacent groupings. For example, the three-membered product abc has two possible groupings:

$$(ab)c \quad \text{and} \quad a(bc),$$

and the four-membered product $abcd$ has five:

$$((ab)c)d, \quad (a(bc))d, \quad a((bc)d), \quad a(b(cd)), \quad (ab)(cd).$$

Solve Catalan's problem.

(The Catalan numbers arise frequently and unexpectedly in a wide variety of combinatorial problems. For further examples, see Martin Gardner's expository article in *Scientific American,* June 1976, pp. 120–125. Included therein is a simple geometric correspondence between Catalan's problem and Euler's polygon-triangulation problem. For generalizations of the Catalan numbers, see [221].)

8. A one-dimensional random walk. The classical ruin problem described in chapter 3 (§3, example 3) can be rephrased in the language of "random walks." Imagine a random process in which a particle moves along the x-axis in discrete steps, at discrete units of time. Let us suppose that the motion begins at the origin and that at each stage of the process (say at times $t = 1, 2, 3, \ldots$) the particle moves either one unit to the right or one unit to the left, each with probability $1/2$. We say that the particle performs a *random walk.* Find the probability that the particle eventually returns to its initial position.

FIGURE 3

For $n \geq 1$, let u_n denote the probability that the particle returns to the origin at time $t = n$; let p_n denote the probability that the first return to the origin occurs at time $t = n$. It is clear that $p_n = u_n = 0$ whenever n is odd. For convenience we define $p_0 = 0$ and $u_0 = 1$.

a. Show that, for $n \geq 1$, the probabilities u_n and p_n are related in the following way:

$$u_n = p_0 u_n + p_1 u_{n-1} + p_2 u_{n-2} + \cdots + p_n u_0.$$

b.   Introduce the generating functions

$$F(x) = \sum_{n=0}^{\infty} p_n x^n, \qquad U(x) = \sum_{n=0}^{\infty} u_n x^n.$$

Show that the equations in (a) lead to the relation

$$U(x) - 1 = F(x) U(x).$$

c.   Show that

$$u_{2n} = \binom{2n}{n} / 2^{2n}$$

$$= \frac{1 \cdot 3 \cdot 5 \cdots 2n - 1}{2 \cdot 4 \cdot 6 \cdots 2n},$$

and conclude that

$$F(x) = 1 - \sqrt{1 - x^2}.$$

d.   Using part (c), show that

$$p_2 + p_4 + p_6 + \cdots = 1.$$

Thus the probability is 1 that the particle will sooner or later (and therefore infinitely often) return to its initial position.[5]

e.   While the result of part (d) seems reasonable, it comes as a surprise that a great many trials are needed to achieve practical certainty.
Show that the relation

$$U(x) = \frac{1 - F(x)}{1 - x^2}$$

leads to the formula

$$u_{2n} = p_{2n+2} + p_{2n+4} + \cdots.$$

Thus

$$u_{2n} = 1 - (p_2 + p_4 + \cdots + p_{2n}).$$

Theorem.   *The probability that no return to the origin occurs up to and including time $t = 2n$ is the same as the probability that a return occurs at time $t = 2n$.*

---

[5] An interesting result due to Pólya states that in a 2-dimensional random walk, the probability of an ultimate return to the origin is also 1; in three dimensions, however, this probability is only about 0.35 (see [156], volume I, p. 360).

For example, the probability that no return occurs up to time $t = 100$ is $u_{100} \approx 0.08$.

f. Show that the expected waiting time for the first return to the origin is infinite! Equivalently, in the language of coin-tossing games: *If a coin is tossed until for the first time the number of heads equals the number of tails, the game has infinite expected duration.*

9. Prove that

$$\sum_{n=1}^{\infty} \left(\frac{1 \cdot 3 \cdot 5 \cdots 2n - 1}{2 \cdot 4 \cdot 6 \cdots 2n} \right) \frac{1}{n} = \log 4.$$

Hint. Show that

$$\frac{1}{\sqrt{1 - 4x}} = 1 + \sum_{n=1}^{\infty} \binom{2n}{n} x^n$$

and that

$$2 \log \left(\frac{1 - \sqrt{1 - 4x}}{2x} \right) = \sum_{n=1}^{\infty} \frac{1}{n} \binom{2n}{n} x^n.$$

10. Let E denote the arc length of the perimeter of an ellipse with semiaxes a and b. Suppose that $a > b$ and let

$$\varepsilon = \frac{\sqrt{a^2 - b^2}}{a}$$

be the eccentricity of the ellipse ($0 < \varepsilon < 1$).

a. Starting from the parametric representation $x = a \sin t$, $y = b \cos t$, show that

$$E = 4a \int_0^{\pi/2} \sqrt{1 - \varepsilon^2 \sin^2 t} \, dt.$$

The integral is a *complete elliptic integral of the second kind*; it cannot be evaluated in terms of elementary functions, and hence there is no simple formula for E in terms of a and b.

b. By applying the binomial series and then integrating term-by-term, show that

$$E = 2\pi a \left\{ 1 - \sum_{n=1}^{\infty} \left(\frac{1 \cdot 3 \cdot 5 \cdots 2n - 1}{2 \cdot 4 \cdot 6 \cdots 2n} \right)^2 \frac{\varepsilon^{2n}}{2n - 1} \right\}. \tag{3}$$

For very small values of the eccentricity ε the series provides excellent numerical approximations for E.

c.   When $\varepsilon = 0$, $b = a$ and the ellipse becomes a circle; the length of the perimeter is then $E = 2\pi a$. Show that in the other extreme case, $\varepsilon = 1$, $b = 0$ and formula (3) gives rise to the curious relation

$$\sum_{n=1}^{\infty} \left(\frac{1 \cdot 3 \cdot 5 \cdots 2n - 1}{2 \cdot 4 \cdot 6 \cdots 2n} \right)^2 \frac{1}{2n - 1} = 1 - \frac{2}{\pi}.$$

11.   (Gauss) Show that for $-1 < x < 1$,

$$\frac{1}{\mathrm{agm}(1 + x, 1 - x)} = 1 + \sum_{n=1}^{\infty} \left(\frac{1 \cdot 3 \cdot 5 \cdots 2n - 1}{2 \cdot 4 \cdot 6 \cdots 2n} \right)^2 x^{2n}.$$

12.   Prove that

$$\frac{22}{7} - \pi = \int_0^1 \frac{x^4 (1 - x)^4}{1 + x^2} dx.$$

13.   Let $f(x)$ be increasing, differentiable, and nonnegative for $x \geq 1$, and suppose that the graph of $f(x)$ is concave down. Show that

$$\int_1^n f(x)\, dx - \left\{ \frac{1}{2} f(1) + f(2) + \cdots + f(n - 1) + \frac{1}{2} f(n) \right\}$$

approaches a finite limit as $n \to \infty$.

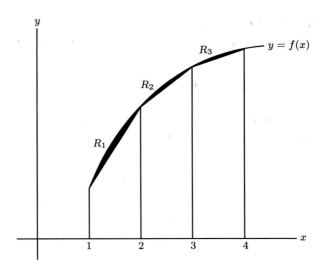

FIGURE 4

Indication of the proof. Geometrically, the problem asserts that the combined areas of the infinitely many shaded regions $R_1, R_2, R_3, \ldots$ pictured in Figure 4 is finite. To approximate the area of the nth region, construct tangents to the curve at $(n, f(n))$ and $(n + 1, f(n + 1))$, respectively. Since the areas of the resulting trapezoids (see Figure 5) both exceed the area under the curve, so does their arithmetic mean. Prove that this leads to the inequalities

$$0 < \text{area}(R_n) < \frac{f'(n) - f'(n + 1)}{4}.$$

Now sum over all n.

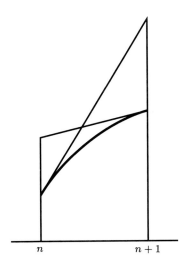

FIGURE 5

14. Stirling's Formula. How large is

$$n! = 1 \cdot 2 \cdot 3 \cdots n?$$

The question is an important one for factorials seem to occur almost everywhere in mathematics, and it is essential that we be able to compute them accurately for large values of n, without laboriously carrying out all the implied multiplications. (The number 100! already has 158 decimal digits and 1000! has more than 2500 digits.)

The question is also vague. Do we mean: Is there a simple, exact formula for the factorial function, a formula in the sense in which $n(n+1)/2$ is a formula for the sum of the first n integers

$$1 + 2 + 3 + \cdots + n?$$

More precisely, is there an *elementary function*—a combination of polynomials, exponentials, logarithms, trigonometric functions and their inverses—which will generate $n!$ for all positive integral values of the variable? The answer is surprising, and it is also deep—there is no, in fact there can be no, elementary formula for $n!$. The product of the first n integers produces a function that is intrinsically far more complex than their sum. (For the fascinating story of this problem, see [123].)

n	2^n	$n!$
1	2	1
2	4	2
3	8	6
4	16	24
5	32	120
6	64	720
7	128	5,040
8	256	40,320
9	512	362,880
10	1,024	3,628,800
11	2,048	39,916,800
12	4,096	479,001,600
13	8,192	6,227,020,800
14	16,384	87,178,291,200
15	32,768	1,307,674,368,000
16	65,536	20,922,789,888,000
17	131,072	355,687,428,096,000
18	262,144	6,402,373,705,728,000
19	524,288	121,645,100,408,832,000
20	1,048,576	2,432,902,008,176,640,000

If, however, we moderate our demands and ask only for an approximation to $n!$, then the problem is entirely reasonable and a solution was found independently in 1730 by James Stirling and Abraham De Moivre. It has come to be known as **Stirling's formula**:

$$n! \sim \sqrt{2\pi n} \left(\frac{n}{e}\right)^n.$$

(The sign $\sim$, read "is asymptotically equal to," is used to indicate that the ratio of the two sides approaches unity as $n \to \infty$.) Thus if we approximate $n!$ by the expression on the right we commit an error which may be large, but which is small compared with $n!$ when n is large. The percentage error approaches zero. For example, Stirling's formula approximates $5! = 120$ by 118 and $10! = 3,628,800$ by 3,598,696. The percentage errors are 1.6 and 0.8, respectively. For $100!$ the error is only 0.08 per cent.

Prove Stirling's formula by following these steps:

a.   By choosing $f(x) = \log x$ in problem 13, show that

$$\log n! = \log 1 + \log 2 + \cdots + \log n$$
$$= \left(n + \frac{1}{2}\right) \log n - n + c_n$$

where

$$\lim_{n \to \infty} c_n = c$$

exists and is finite.

b.   Conclude that

$$\lim_{n \to \infty} \frac{n!}{\sqrt{n}(n/e)^n} = e^c.$$

c.   Complete the proof by applying Wallis's formula, in the form

$$\lim_{n \to \infty} \frac{2 \cdot 4 \cdot 6 \cdots 2n}{1 \cdot 3 \cdot 5 \cdots 2n - 1} \cdot \frac{1}{\sqrt{n}} = \sqrt{\pi},$$

to show that $c = \log \sqrt{2\pi}$.

2. On the Probability Integral

"Do you know what a mathematician is?" Kelvin once asked a class. He stepped to the board and wrote

$$\int_{-\infty}^{+\infty} e^{-x^2} \, dx = \sqrt{\pi}.$$

Putting his finger on what he had written, he turned to the class. "A mathematician is one to whom *that* is as obvious as that twice two makes four is to you."

—E. T. Bell[6]

According to the fundamental theorem of calculus, *every continuous function on a closed interval has an antiderivative.* What is more, the essentially unique solution can even be exhibited explicitly. Thus, if f is continuous on $[a, b]$ then the "area function" F defined by

$$F(x) = \int_a^x f(t)\, dt$$

is differentiable on $[a, b]$ and

$$F'(x) = f(x)$$

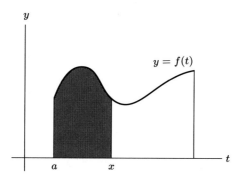

FIGURE 6

throughout the interval. In this way, the fundamental theorem of calculus explicitly states the inverse relationship that exists between area and tangent problems, that is to say, between integration and differentiation.

While the area function is of great geometric appeal, the beginner often finds it lacking, for he is left with the uneasy feeling that he still does not know what F really is. His naive hope, of course, is that F can always be expressed in terms of familiar functions like sin or log or $\tan^{-1}$. Such functions are said to be elementary. More precisely, an **elementary function** is one that can be obtained from polynomials, exponentials, logarithms, trigonometric or inverse

---

6 *Men of Mathematics* [32], p. 452.

trigonometric functions in a finite number of steps by using ordinary algebraic operations (addition, subtraction, multiplication, and division) together with composition.

Unfortunately, even the simplest elementary function may not possess an elementary antiderivative. (By contrast, the derivative of an elementary function is always elementary.) A famous example is provided by the function $f(x) = e^{-x^2}$. The integral

$$F(x) = \int_0^x e^{-t^2}\, dt,$$

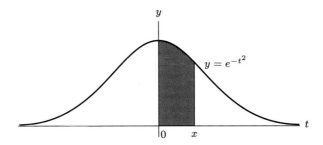

FIGURE 7

which plays a fundamental role in probability theory and in many physical problems, cannot be expressed in terms of the ordinary functions of algebra and trigonometry and calculus. It is not an elementary function. The proof, however, is hard. In general, it is an exceedingly difficult task to decide whether or not a given function can be integrated in elementary terms.[7]

In light of this, it is quite extraordinary that the value of the improper integral $\int_0^\infty e^{-x^2}\, dx$ can be determined precisely. There are many known derivations, but most of them require advanced methods of analysis. The following simple proof is based on an elegant application of Wallis's formula.

---

[7] See, for example, J. F. Ritt, *Integration in Finite Terms,* Columbia University Press, New York, 1948. For a more elementary introduction to the problem, see D. G. Mead's article "Integration," in the *American Mathematical Monthly,* 68 (1961), pp. 152–156. Other examples of functions that do not have elementary antiderivatives are

$$\frac{\sin x}{x}, \quad \sin x^2, \quad \sqrt{1 - k^2 \sin^2 x}$$

where $0 < k < 1$.

Theorem. $\int_0^\infty e^{-x^2}\, dx = \sqrt{\pi}/2$.

Proof. It is convenient to present the proof in three steps.

Step 1. Since the integrand e^{-x^2} cannot be integrated in elementary terms, it may be helpful to approximate it by simpler functions which can be. The simplest approximation to a differentiable function is a linear one—the tangent line to the curve. We begin, accordingly, with the obvious estimate

$$e^x \geq 1 + x \qquad \text{for all } x. \tag{1}$$

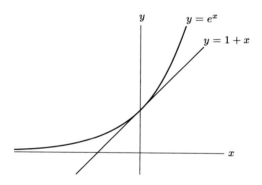

FIGURE 8

(Since the graph of the exponential function is concave up, it lies above every one of its tangent lines, in particular, its tangent line at $x = 0$.) This implies that

$$1 - x^2 \leq e^{-x^2} \leq \frac{1}{1 + x^2} \qquad \text{for all } x. \tag{2}$$

(For the lower bound, simply replace x by $-x^2$ in (1); for the upper bound, replace x by x^2 and then take reciprocals).

Step 2. The tangent line is a good approximation to a differentiable curve *only in the vicinity of the point of tangency.* Therefore, the inequalities just obtained are of value only when x is close to zero. However, if we can transform the function $f(x) = e^{-x^2}$ into another exponential function, which bounds the same area, and yet concentrates most of that area near $x = 0$, then the estimates (2) may yet be used to advantage.

For this purpose we define

$$g(x) = \lambda f(\lambda x)$$

where λ is a large positive number. Geometrically, we have *compressed* the graph of f horizontally, in the direction of the origin, by a factor λ, and then *expanded* it vertically, away from the origin, by the same factor λ. Examining Figure 9, we see that for large values of λ, the graph of g rises to a sharp peak at $x = 0$ and that most of its area is concentrated near the y-axis.

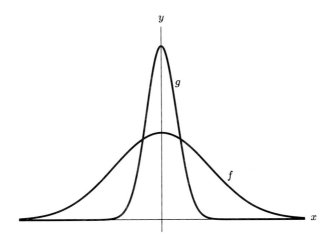

FIGURE 9

Notice that g is still an exponential function, since $g(x) = \lambda e^{-(\lambda x)^2}$, and that f and g bound the same area between 0 and ∞, since

$$\int_0^\infty g(x)\, dx = \int_0^\infty \lambda f(\lambda x)\, dx = \int_0^\infty f(t)\, dt.$$

We are now ready to approximate the integral $I = \int_0^\infty g(x)\, dx$.

Step 3.   For convenience, we take $\lambda = \sqrt{n}$, where n is a large positive integer. Then

$$I = \sqrt{n} \int_0^\infty e^{-nx^2}\, dx.$$

Raising each term in (2) to the nth power, we find that

$$\sqrt{n} \int_0^1 (1 - x^2)^n\, dx \leq I \leq \sqrt{n} \int_0^\infty \frac{1}{(1 + x^2)^n}\, dx.$$

Now from section 1 we know that the values of the integral on the left are given by

$$\int_0^1 (1 - x^2)^n \, dx = \frac{2 \cdot 4 \cdot 6 \cdots 2n}{3 \cdot 5 \cdot 7 \cdots 2n + 1}.$$

The integral on the right is handled in much the same way—by means of a suitable reduction formula—and we find without difficulty that

$$\int_0^\infty \frac{1}{(1 + x^2)^n} \, dx = \frac{\pi}{2} \cdot \frac{1 \cdot 3 \cdot 5 \cdots 2n - 3}{2 \cdot 4 \cdot 6 \cdots 2n - 2}.$$

Thus, the inequalities

$$\sqrt{n} \cdot \frac{2 \cdot 4 \cdot 6 \cdots 2n}{3 \cdot 5 \cdot 7 \cdots 2n + 1} \le I \le \frac{\pi \sqrt{n}}{2} \cdot \frac{1 \cdot 3 \cdot 5 \cdots 2n - 3}{2 \cdot 4 \cdot 6 \cdots 2n - 2}$$

are valid for every positive integer n.

The proof of the theorem is now at hand. By invoking Wallis's formula, we see that the products

$$\frac{1}{\sqrt{n}} \cdot \frac{2 \cdot 4 \cdot 6 \cdots 2n}{1 \cdot 3 \cdot 5 \cdots 2n - 1}$$

can be made as close to $\sqrt{\pi}$ as desired, and therefore that both the upper and lower bounds for I approach $\sqrt{\pi}/2$ as $n \to \infty$. This shows that

$$I = \int_0^\infty e^{-x^2} \, dx = \frac{\sqrt{\pi}}{2}.$$

PROBLEMS

1. Here is another elegant proof of the formula

$$\int_{-\infty}^\infty e^{-x^2} \, dx = \sqrt{\pi}$$

which was discovered by Stieltjes ([402], pp. 263–264).

    a.   Let

$$I_n = \int_0^\infty x^n e^{-x^2} \, dx \qquad n = 0, 1, 2, \ldots.$$

Establish the recursive relation

$$I_n = \frac{n - 1}{2} I_{n-2}$$

by integrating by parts.

    b.   Deduce that

$$I_{2k} = \frac{1 \cdot 3 \cdot 5 \cdots 2k - 1}{2^k} I_0$$

and

$$I_{2k+1} = \frac{1 \cdot 2 \cdot 3 \cdots k}{2}.$$

    c.   Show that for all real values of λ, the expression

$$\lambda^2 I_{n-1} + 2\lambda I_n + I_{n+1} = \int_0^\infty x^{n-1}(\lambda + x)^2 e^{-x^2}\, dx$$

is positive, and conclude that

$$I_n^2 < I_{n-1} I_{n+1}.$$

    d.   By taking $n = 2k$ and $n = 2k + 1$, show that

$$\frac{2 \cdot 4 \cdot 6 \cdots 2k}{1 \cdot 3 \cdot 5 \cdots 2k - 1} \frac{1}{\sqrt{4k + 2}} < I_0 < \frac{2 \cdot 4 \cdot 6 \cdots 2k}{1 \cdot 3 \cdot 5 \cdots 2k - 1} \frac{1}{\sqrt{4k}}.$$

Now apply Wallis's formula.

2.   Give a third proof of the theorem by introducing the (extraordinary!) function

$$F(x) = \left\{ \int_0^x e^{-t^2}\, dt \right\}^2 + \int_0^1 \frac{e^{-x^2(t^2+1)}}{t^2 + 1}\, dt \tag{3}$$

and verifying that $F'(x)$ is identically zero. For the second integral on the right, apply *Leibniz's rule* for differentiating under the integral sign.

3.   Evaluate the **Fresnel integrals**:

$$\int_0^\infty \sin x^2\, dx = \int_0^\infty \cos x^2\, dx = \sqrt{\frac{\pi}{8}}.$$

    Hint. Replace the auxiliary function in (3) by its complex counterpart

$$G(x) = \left\{ \int_0^x e^{it^2}\, dt \right\}^2 + i \int_0^1 \frac{e^{ix^2(t^2+1)}}{t^2 + 1}\, dt$$

and show that $G'(x)$ is identically zero.

4.   Prove:

$$\int_0^\pi \log \sin x\, dx = -\pi \log 2.$$

Hint. $\int_0^{\pi/2} \log \sin x \, dx = \int_0^{\pi/2} \log \cos x \, dx = \frac{1}{2} \int_0^{\pi/2} \log(\frac{\sin 2x}{2}) \, dx.$

5. Find the value of
$$\int_0^1 \frac{\log(1 + x)}{1 + x^2} \, dx.$$
Hint. Set $x = \tan \theta$ and break up the transformed integral into three parts.

3. Polynomial Approximation and the Dirac Delta Function

a. A Theorem of Weierstrass

The evaluation of the probability integral $\int_0^\infty e^{-x^2} \, dx$ in the previous section was based on three ideas of fundamental importance:

1. *The possibility of approximating a given continuous function by a polynomial.* There, the given function e^{-x^2} was differentiable and the approximating polynomial was linear.

2. *The utility of the transformation*
$$f(x) \to \lambda f(\lambda x).$$
By means of this transformation, e^{-x^2} could be replaced by another exponential function whose relevant behavior was centered near $x = 0$.

3. *Wallis's formula:*
$$\lim_{n \to \infty} \left(\frac{2 \cdot 4 \cdot 6 \cdots 2n}{1 \cdot 3 \cdot 5 \cdots 2n - 1} \right)^2 \frac{1}{n} = \pi.$$

We encountered Wallis's formula again in the derivation of Stirling's famous approximation for $n!$ (§1, problem 14). For now, let us explore in somewhat greater detail the intriguing relation that exists between polynomial approximation and the so-called "peaking functions" $\lambda f(\lambda x)$.

Polynomials are among the simplest and most tractable functions in analysis. Their values can always be computed by the same elementary operations of addition and multiplication; fast and efficient algorithms exist for approximating their roots with any desired degree of accuracy; and they can be integrated and differentiated with ease. Because of their simplicity, it is of great importance to know that they can be used to approximate any continuous function

uniformly on a closed and bounded interval. This is the content of the celebrated Weierstrass approximation theorem (1885).

Weierstrass Approximation Theorem. *If f is continuous on the closed interval*
$[a, b]$, then for any positive number ε there exists a polynomial p for which

$$|f(x) - p(x)| < \varepsilon \qquad whenever \ a \leq x \leq b.$$

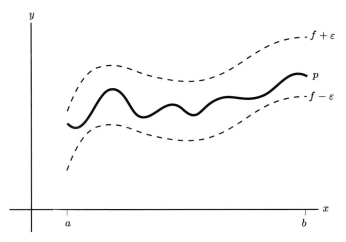

FIGURE 10

Geometrically, the theorem asserts that in any arbitrarily narrow band
bounded by the curves $y = f(x) - \varepsilon$ and $y = f(x) + \varepsilon$, for $a \leq x \leq b$, there
always lies at least one graph of a polynomial. This is what is meant by *uniform
approximation.*

The result is extraordinary—polynomials are infinitely smooth but a continuous function may fail to have a derivative at even a single point.

Many proofs of this remarkable theorem are now known, some of them
providing explicit constructions of the approximating polynomials. One proof
relies on Fejér's famous method of averaging the partial sums of Fourier series. Another proof, discovered by Bernstein, has its background in probability
theory (see problem 2), while a third, due to Lebesgue, reduces the problem
to the approximation of a single continuous function, $|x|$, on $[-1, 1]$. (For an
excellent account of each of these three methods, see: J. D. Pryce, *Basic Methods of Linear Functional Analysis* [350], pp. 196–197; N. D. Kazarinoff, *Analytic
Inequalities* [246], pp. 52–59; and B. Sz.-Nagy, *Introduction to Real Functions*

and Orthogonal Expansions [410], pp. 76–80, respectively. While some of the arguments are elementary, all are demanding, and each requires at least some knowledge of advanced calculus. Nevertheless, to quote Kazarinoff, "If you take up the challenge and master the proof, you will have won a great prize.")

In this section we shall present an elegant argument, based on ideas we have already encountered, which makes Weierstrass's theorem seem plausible. To motivate the approach, we first introduce a famous function of mathematical physics.

b. The Delta Function

In his celebrated treatise *The Principles of Quantum Mechanics* (1930), P. Dirac introduced a revolutionary new function which he called the *δ-function* and which he described as follows:

> Our work ... [has] led us to consider quantities involving a certain kind of infinity. To get a precise notation for dealing with these infinities, we introduce a quantity $\delta(x)$ depending on a parameter x satisfying the conditions
>
> $$\int_{-\infty}^{\infty} \delta(x)\, dx = 1$$
> $$\delta(x) = 0 \qquad \text{for } x \neq 0.$$
>
> $\hspace{10cm}$ (1)
>
> To get a picture of $\delta(x)$, take a function of the real variable x which vanishes everywhere except inside a small domain, of length ε say, surrounding the origin $x = 0$, and which is so large inside this domain that its integral over this domain is unity. The exact shape of the function inside this domain does not matter, provided there are no unnecessarily wild variations (for example provided the function is always of order ε^{-1}). Then in the limit $\varepsilon \to 0$ this function will go over into $\delta(x)$.
>
> $\delta(x)$ is not a function of x according to the usual mathematical definition of a function, which requires a function to have a definite value for each point in its domain, but is something more general, which we may call an 'improper function' to show up its difference from a function defined by the usual definition. Thus $\delta(x)$ is not a quantity which can be generally used in mathematical analysis like an ordinary function, but its use must be confined to certain simple types of expression for which it is obvious that no inconsistency can arise.

The most important property of $\delta(x)$ is exemplified by the following equation,

$$\int_{-\infty}^{\infty} f(x)\delta(x)\,dx = f(0), \tag{2}$$

where $f(x)$ is any continuous function of x. We can easily see the validity of this equation from the above picture of $\delta(x)$. The left-hand side of (2) can depend only on the values of $f(x)$ very close to the origin, so that we may replace $f(x)$ by its value at the origin, $f(0)$, without essential error. Equation (2) then follows from the first of equations (1). By making a change of origin in (2), we can deduce the formula

$$\int_{-\infty}^{\infty} f(x)\delta(x - a)\,dx = f(a),$$

where a is any real number. Thus *the process of multiplying a function of x by $\delta(x - a)$ and integrating over all x is equivalent to the process of substituting a for x.*[8]

That such a function cannot exist is obvious. Dirac himself makes this clear when he calls the δ-function an "improper function" and goes on to say that it is "merely a convenient notation." It is, however, far more than that. For good notation, in addition to illuminating many a difficult argument, reflects the profound relation that exists between content and form.

New results have often become possible only because of a new mode of writing. The introduction of Hindu-Arabic numerals is one example; Leibniz' notation for the calculus is another one. An adequate notation reflects reality better than a poor one, and as such appears endowed with a life of its own which in turn creates new life.[9]

What is more, the efforts to put the δ-function on a firm mathematical basis have given rise to important new areas of analysis, among which Laurent Schwartz's *theory of distributions* is perhaps the single most important example.

While the δ-function may not exist as a true mathematical function, it is possible to approximate it to any degree of accuracy.[10]

---

[8] *The Principles of Quantum Mechanics,* fourth edition, Oxford University Press, Oxford, 1958, pp. 58–59.

[9] D. J. Struik, *A Concise History of Mathematics* [407], p. 94.

[10] In a sharp critique of the mathematician's "literal-mindedness," Jacob Schwartz writes in his essay "The Pernicious Influence of Mathematics on Science," in *Discrete Thoughts, Essays on Mathematics, Science, and Philosophy,* Birkhäuser, Boston, 1986:

Definition. A sequence $K_1, K_2, K_3, \ldots$ of continuous functions on $(-\infty, \infty)$ is called a **peaking kernel** if the following three properties are satisfied:

1. Every $K_n(x) \geq 0$ for all x;
2. $\int_{-\infty}^{\infty} K_n(x)\, dx = 1$ for all $n = 1, 2, 3, \ldots$;
3. For every $\varepsilon > 0$, $\lim_{n \to \infty} \int_{-\varepsilon}^{\varepsilon} K_n(x)\, dx = 1$.

Property 2 asserts that the area between every curve $y = K_n(x)$ and the x-axis is always equal to unity, while property 3 assures us that most of that area is concentrated near $x = 0$ (provided that n is sufficiently large). Thus the δ-function may be viewed qualitatively as the "limiting value" of any peaking kernel. (For a brief history of the use of peaking kernels in pure mathematics long before the appearance of Dirac's δ-function, see Van der Pol and Bremmer, *Operational Calculus Based on the Two-Sided Laplace Integral* [424], chapter V.)

Examples

1. A particularly simple way of generating a peaking kernel $\{K_1, K_2, \ldots\}$ is to start off with an arbitrary continuous function $K(x) \geq 0$, satisfying

$$\int_{-\infty}^{\infty} K(x)\, dx = 1,$$

and then define

$$K_n(x) = nK(nx) \qquad \text{for } n = 1, 2, 3, \ldots,$$

just as we did in the previous section. Properties 1, 2, and 3 are then obvious.

In this way, mathematics has often succeeded in proving, for instance, that the fundamental objects of the scientist's calculations do not exist. The sorry history of the δ-function should teach us the pitfalls of rigor. Used repeatedly by Heaviside in the last century, used constantly and systematically by physicists since the 1920's, this function remained for mathematicians a monstrosity and an amusing example of the physicists' naiveté—until it was realized that the δ-function was not literally a function but a generalized function. It is not hard to surmise that this history will be repeated for many of the notions of mathematical physics which are currently regarded as mathematically questionable. The physicist rightly dreads precise argument, since an argument which is only convincing if precise loses all its force if the assumptions upon which it is based are slightly changed, while an argument which is convincing though imprecise may well be stable under small perturbations of its underlying axioms.

Important examples of kernels generated in this way include

$$K(x) = \frac{1}{\pi}\frac{1}{1+x^2} \qquad \text{(Cauchy kernel)}$$

$$K(x) = \frac{1}{\sqrt{\pi}}e^{-x^2} \qquad \text{(Weierstrass kernel)}$$

$$K(x) = \frac{1}{2}e^{-|x|} \qquad \text{(Picard kernel)}.$$

Figure 11 illustrates the Cauchy kernel for the values $n = 1, 3$, and 6.

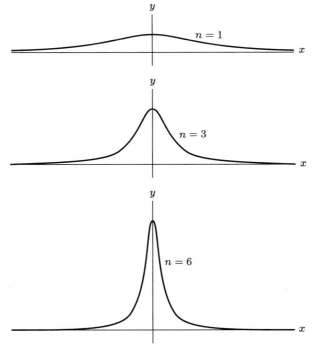

FIGURE 11

Approximations to the δ-function: $K_n(x) = \frac{1}{\pi}\frac{n}{1+n^2x^2}$

2. An important example of a peaking kernel which is not obtainable from a single function $K(x)$ is derived from the polynomials $(1-x^2)^n$, $n = 1, 2, 3, \ldots$, which played so important a role in the proof of Wallis's formula for π. Let

$$K_n(x) = \frac{(1 - x^2)^n}{\int_{-1}^{1} (1 - x^2)^n \, dx} \qquad \text{for } -1 \le x \le 1,$$

and $K_n(x) = 0$ for $|x| > 1$. Properties 1 and 2 are obvious, but 3 requires a proof. If ε is a fixed, but arbitrary, number in the interval $0 < \varepsilon < 1$, then

$$\int_{\varepsilon}^{1} (1 - x^2)^n \, dx < \int_{0}^{1} (1 - \varepsilon^2)^n \, dx = (1 - \varepsilon^2)^n$$

and

$$\int_{0}^{1} (1 - x^2)^n \, dx > \int_{0}^{1} (1 - x)^n \, dx = \frac{1}{n + 1}$$

so that

$$\frac{\int_{\varepsilon}^{1} (1 - x^2)^n \, dx}{\int_{0}^{1} (1 - x^2)^n \, dx} < (n + 1)(1 - \varepsilon^2)^n.$$

Therefore

$$\lim_{n \to \infty} \int_{\varepsilon}^{\infty} K_n(x) \, dx = 0.$$

Since every K_n is an even function, this proves property 3.

c. The Sifting Property

Let us now return to the most important property of the δ-function:

$$\int_{-\infty}^{\infty} f(x)\delta(x - a) \, dx = f(a).$$

This symbolic relation is known as the *sifting property,* for it asserts that $\delta(x)$ acts as if it were a "sieve"—by integrating any continuous function f against the translate $\delta(x - a)$, we just pick out the value of f at a. To make this relation precise, we have only to approximate $\delta(x)$ by a sequence of peaking functions and then pass to the limit.

Theorem.   *Let* $\{K_1, K_2, \ldots\}$ *be a peaking kernel. Then for any function f, continuous and bounded on* $(-\infty, \infty)$,

$$\lim_{n \to \infty} \int_{-\infty}^{\infty} f(t)K_n(t - x) \, dt = f(x).$$

Furthermore, on any finite interval, the functions defined by the integrals approximate f uniformly, within any $\varepsilon > 0$, provided n is sufficiently large.

This theorem represents one of the most powerful and general known methods for approximating functions [377]. A picture makes it plausible (see Figure 12). Since f is bounded, for large values of n the dominant part of the integral $\int_{-\infty}^{\infty} f(t)K_n(t-x)\,dt$ occurs near $t = x$. Since f is continuous, its values in a small neighborhood of x are all roughly the same (and equal to $f(x)$). Therefore,

$$\int_{-\infty}^{\infty} f(t)K_n(t-x)\,dt \approx f(x) \int_{-\infty}^{\infty} K_n(t-x)\,dt = f(x)$$

which is what the theorem asserts.

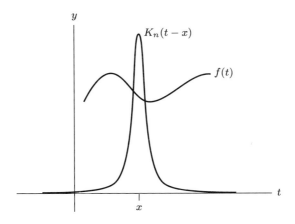

FIGURE 12

As an application we shall now prove Weierstrass's famous theorem.

d. Landau's Proof of the Weierstrass Theorem[11]

Let f be continuous on $[a, b]$ and suppose, merely to simplify the formulas, that $0 < a < b < 1$. Suppose, furthermore, that f has been extended continuously

---

[11] E. Landau (1908). The details of the proof can be found in R. P. Boas's extraordinary introduction to the foundations of real analysis, *A Primer of Real Functions* [46]. See also D. Jackson's paper "A Proof of Weierstrass's Theorem," in *Selected Papers on Calculus* [8], pp. 227–231.

to the entire real line and that $f(x) = 0$ outside $(0,1)$. Define $P_n(x)$ on $[a, b]$ by

$$P_n(x) = c_n \int_0^1 f(t)[1 - (t - x)^2]^n \, dt$$

where

$$1/c_n = \int_{-1}^1 (1 - t^2)^n \, dt.$$

Clearly, each P_n is a polynomial of degree $2n$ at most. But $P_n(x)$ is just the value of the integral $\int_{-\infty}^\infty f(t)K_n(t - x) \, dt$, where $\{K_n\}$ is the peaking kernel defined in Example 2. Therefore, the theorem applies and P_n approximates f uniformly on $[a, b]$, within any $\varepsilon > 0$, provided n is sufficiently large. This proves Weierstrass's theorem.

PROBLEMS

1. The approximation of $|x|$ on $[-1, 1]$ by polynomials played a significant role in the early development of approximation theory. In 1898, Lebesgue reduced the problem of uniform approximation of any continuous function to that of $|x|$. In 1911, Bernstein showed that $|x|$ could be approximated on $[-1, 1]$ by polynomials of degree $2n$ with an error not exceeding $\frac{1}{2n+1}$. Show that uniform approximation of $|x|$ is possible by the following method of successive approximation.

a.   First show that the function $\sqrt{x}$ can be approximated to any desired degree of accuracy on the interval $[0, 1]$ by polynomials. Let $P_0(x) \equiv 0$ and for $n \geq 0$ define

$$P_{n+1}(x) = P_n(x) + \frac{1}{2} \left(x - P_n^2(x) \right).$$

Prove that for all x in $[0, 1]$,

$$0 \leq \sqrt{x} - P_n(x) \leq \frac{2}{n}.$$

Hint. If we try to argue by induction we find that the induction hypothesis is too weak. Prove instead that the slightly stronger inequality

$$0 \leq \sqrt{x} - P_n(x) \leq \frac{2\sqrt{x}}{2 + n\sqrt{x}}$$

is valid for $0 \leq x \leq 1$ and all $n = 0, 1, 2, \ldots$. For this purpose the following identity is crucial:

$$\frac{\sqrt{x} - P_{n+1}(x)}{\sqrt{x} - P_n(x)} = 1 - \frac{1}{2} \left(\sqrt{x} + P_n(x) \right).$$

b. Show that $\{P_n\}$ is an increasing sequence, i.e., $P_0(x) \leq P_1(x) \leq P_2(x) \leq \cdots$ for all x in $[0, 1]$.

c. Deduce from part (a) that $|x|$ can be uniformly approximated on $[-1, 1]$, to within any $\varepsilon > 0$, by polynomials.

2. Bernstein Polynomials. Perhaps the most elegant proof of the Weierstrass approximation theorem is based on a result due to Bernstein (1912). Let us suppose that $f(x)$ is defined and continuous on the closed interval $[0, 1]$. The *n*th **Bernstein polynomial** associated with $f(x)$ is

$$B_n(x) = \sum_{k=0}^{n} f(k/n) \binom{n}{k} x^k (1 - x)^{n-k}.$$

Bernstein's theorem asserts that, as $n \to \infty$,

$$B_n(x) \to f(x)$$

uniformly on $[0, 1]$.

Figure 13 illustrates the approximation properties of the Bernstein polynomials of a concave function. As the figure reflects, the Bernstein approximants mimic the behavior of the function to a remarkable degree. It is true generally, for example, that monotonic and concave functions yield monotonic and concave approximants, respectively ([124], pp. 114ff.). A simple probability experiment makes Bernstein's theorem plausible. Imagine a game in which a player tosses a "biased" coin for which the probability of a head is x and the probability of a tail is $1 - x$. The successive tosses are assumed to be independent and the number x is assumed to lie strictly between 0 and 1. The coin is tossed n times and the number of heads is recorded.

a. Show that the probability of obtaining exactly k heads is given by the expression

$$\binom{n}{k} x^k (1 - x)^{n-k}.$$

b. If the player wins $f(k/n)$ dollars whenever exactly k heads appear (a negative win is regarded as a loss), then the player's expected winnings will be $B_n(x)$ dollars.

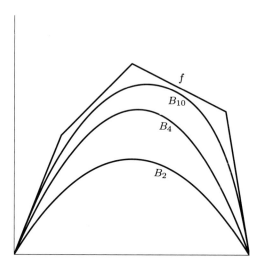

FIGURE 13
Illustrating the approximation properties of the Bernstein polynomials
$B_2(x) = \frac{8}{5}(x - x^2)$
$B_4(x) = \frac{11}{5}x - \frac{9}{5}x^2 - \frac{3}{10}x^3 - \frac{1}{10}x^4$
$B_{10}(x) = \frac{5}{2}x - 18x^3 + 63x^4 - \frac{567}{5}x^5 + \frac{189}{2}x^6 - \frac{243}{4}x^8 + 42x^9 - \frac{197}{20}x^{10}$
The graph of f is a polygonal arc joining $(0,0)$, $(.2,.5)$, $(.5,.8)$, $(.9,.6)$, $(1,0)$.

c.   Conclude that, for each x,

$$B_n(x) \to f(x).$$

(For a proof that the convergence is uniform on $[0, 1]$, see [420], pp. 116–117.)

3.   For a collection of points $\mathbf{P}_0, \mathbf{P}_1, \ldots, \mathbf{P}_n$ in the plane, the **Bézier spline of degree n** is the parametric curve defined by

$$\mathbf{X}(t) = \sum_{k=0}^{n} \mathbf{P}_k \binom{n}{k} t^k (1 - t)^{n-k}, \qquad 0 \le t \le 1.$$

The curve is represented here in vector form, but the separate components of $\mathbf{X}(t) = (x(t), y(t))$ are readily found by substituting the appropriate components of each $\mathbf{P}_k$. These natural generalizations of the Bernstein polynomials (see problem 2) were discovered around 1963 by Bézier, who used them as a tool in computer-aided design while working at Renault. Today Bézier curves (and surfaces) continue to play a fundamental role in computer graphics and modeling when approximations in the large are more important than closeness of approximation. Figure 14 shows the cubic Bézier splines for various locations of the defining points $\mathbf{P}_0, \mathbf{P}_1, \mathbf{P}_2, \mathbf{P}_3$. Notice that the first and last points always lie on the curve, while the others "pull" the curve toward themselves.

a. Show that the equation of the cubic spline can be written compactly in matrix notation

$$\mathbf{X}(t) = (t^3 \ t^2 \ t \ 1) \begin{pmatrix} -1 & 3 & -3 & 1 \\ 3 & -6 & 3 & 0 \\ -3 & 3 & 0 & 0 \\ 1 & 0 & 0 & 0 \end{pmatrix} \begin{pmatrix} \mathbf{P}_0 \\ \mathbf{P}_1 \\ \mathbf{P}_2 \\ \mathbf{P}_3 \end{pmatrix}.$$

b. Show that

$$\mathbf{X}'(t) = (3t^2 \ 2t \ 1 \ 0) \begin{pmatrix} -1 & 3 & -3 & 1 \\ 3 & -6 & 3 & 0 \\ -3 & 3 & 0 & 0 \\ 1 & 0 & 0 & 0 \end{pmatrix} \begin{pmatrix} \mathbf{P}_0 \\ \mathbf{P}_1 \\ \mathbf{P}_2 \\ \mathbf{P}_3 \end{pmatrix}.$$

Conclude that $\mathbf{X}'(0) = 3(\mathbf{P}_1 - \mathbf{P}_0)$ and $\mathbf{X}'(1) = 3(\mathbf{P}_3 - \mathbf{P}_2)$. What does this say about the shape of the curve at the endpoints?

(For further information see R. H. Bartels, J. C. Beatty, and B. A. Barsky, *An Introduction to Splines for Use in Computer Graphics and Geometric Modeling,* M. Kaufmann Publishers, Los Altos, 1987; G. E. Farin, *Curves and Surfaces for Computer Aided Geometric Design,* Academic Press, Boston, 1988; J. D. Foley and A. van Dam, *Fundamentals of Interactive Computer Graphics,* Addison-Wesley, Reading, 1983; and D. F. Rogers and J. A. Adams, *Mathematical Elements for Computer Graphics,* McGraw Hill, New York, 1990.)

4. Lagrange Interpolation. Given $n + 1$ distinct points $x_0, x_1, \ldots, x_n$ and $n + 1$ arbitrary values $y_0, y_1, \ldots, y_n$, show that there is a unique polynomial P of degree $\leq n$ for which

$$P(x_i) = y_i \qquad i = 0, 1, \ldots, n.$$

It is called the *Lagrange polynomial* determined by the conditions.

Hint. For the existence of $P(x)$, make use of the product $(x - x_0)$ $(x - x_1) \cdots (x - x_n)$ to show that for each $i = 0, 1, \ldots, n$ there is a polynomial $P_i(x)$ of degree not exceeding n for which

$$P_i(x_j) = \delta_{ij}.$$

(The symbol δ_{ij} is *Kronecker's delta*: $\delta_{ij} = 0$ when $i \neq j$ and $\delta_{ii} = 1$.)

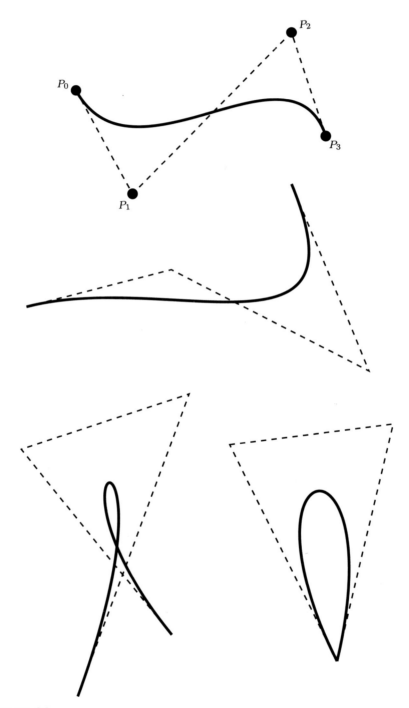

FIGURE 14
Cubic Bézier splines

5. Prove: If $P_1, P_2, \ldots$ is a sequence of polynomials and $P_n \to f$ uniformly on the entire real line, then f must also be a polynomial.

6. If f is continuous on $[a, b]$, then the *nth moment* of f is defined to be

$$c_n = \int_a^b x^n f(x) \, dx \qquad n = 0, 1, 2, \ldots .$$

Show that if $c_n = 0$ for every n then f must be identically zero.

Hint. Begin by showing that $\int_a^b p(x) f(x) \, dx = 0$ for every polynomial $p(x)$.

7. Prove: Every function defined and continuous on $[0, 1]$ can be uniformly approximated by polynomials containing only terms of even degree. This is a special case of the remarkable Müntz–Szász theorem, which asserts that every continuous function on $[0, 1]$ can be uniformly approximated by finite linear combinations from the set $\{1, x^{\lambda_1}, x^{\lambda_2}, \ldots\}$, where $0 < \lambda_1 < \lambda_2 < \cdots$, if and only if the series $1/\lambda_1 + 1/\lambda_2 + \cdots$ is divergent [174].

Hint. Use problem 1 to show that the identity function can be uniformly approximated on $[0, 1]$ by polynomials in x^2.

8. Simultaneous Interpolation and Approximation. Let f be continuous on the closed interval $[a, b]$ and let $x_1, x_2, \ldots, x_n$ be distinct points in $[a, b]$. Show that for each $\varepsilon > 0$ there is a polynomial p such that

$$|f(x) - p(x)| < \varepsilon \qquad \text{for all } x \text{ in } [a, b]$$

and

$$f(x_i) = p(x_i) \qquad \text{for all } i = 1, 2, \ldots, n.$$

4. Euler's Proof of the Infinitude of the Primes

The methods fashioned by the Greeks in their study of perfect numbers were reexamined 2000 years later by Euler, who used them to produce another proof of the infinitude of the primes. Euclid's proof is unparalleled in its simplicity and its elegance—it proves only what it needs to and it does this with the greatest economy. Euler's proof lies deeper, for it is based on the divergence of the harmonic series, but it offers far more in return. By combining the discrete methods of classical number theory with the continuous methods of the calculus—methods that seem far removed from ordinary arithmetic—Euler laid the foundations for a new branch of mathematics that is now known as *analytic number theory.*

What Euler discovered was that the product

$$\frac{2}{1}\frac{3}{2}\frac{5}{4}\frac{7}{6} \cdots \frac{P}{P-1}$$

can be made arbitrarily large and hence there can be no largest prime. (This is quite reminiscent of the product $\frac{2}{1}\frac{4}{3}\frac{6}{5} \cdots \frac{2n}{2n-1}$ which figured so prominently in the proof of Wallis's formula for π.) Here the numerators are successive primes while each denominator is one less than the corresponding numerator. To make matters precise, let N be an arbitrary positive integer and P the largest prime not exceeding N. It is to be shown that

$$\frac{2}{1}\frac{3}{2}\frac{5}{4}\frac{7}{6} \cdots \frac{P}{P-1} > 1 + \frac{1}{2} + \frac{1}{3} + \cdots + \frac{1}{N}. \tag{1}$$

This striking inequality implies at once that there can be no largest prime: since the right side tends to infinity with N, so must the left.

To establish (1) we start with the formula for the sum of a geometric progression:

$$1 + x + x^2 + \cdots + x^N = \frac{1 - x^{N+1}}{1 - x}$$

$$< \frac{1}{1 - x} \qquad \text{provided } 0 < x < 1.$$

Choosing x successively to be $\frac{1}{2}, \frac{1}{3}, \frac{1}{5}, \frac{1}{7}, \ldots, \frac{1}{P}$, and then multiplying the results, we find

$$\frac{2}{1}\frac{3}{2}\frac{5}{4}\frac{7}{6} \cdots \frac{P}{P-1} = \frac{1}{1 - \frac{1}{2}} \frac{1}{1 - \frac{1}{3}} \frac{1}{1 - \frac{1}{5}} \cdots \frac{1}{1 - \frac{1}{P}}$$

$$> \left(1 + \frac{1}{2} + \frac{1}{2^2} + \cdots + \frac{1}{2^N}\right) \left(1 + \frac{1}{3} + \frac{1}{3^2} + \cdots + \frac{1}{3^N}\right)$$

$$\cdots \left(1 + \frac{1}{P} + \frac{1}{P^2} + \cdots + \frac{1}{P^N}\right).$$

Call the last product A_N. When multiplied out, A_N is the sum of all the terms

$$\frac{1}{2^\alpha \cdot 3^\beta \cdot 5^\gamma \cdots P^\mu}$$

where the numbers $\alpha, \beta, \gamma, \ldots, \mu$ range over all integers between 0 and N. In other words, A_N is the *sum of the reciprocals* of all the divisors of

$$A = (2 \cdot 3 \cdot 5 \cdots P)^N.$$

But the divisors of A certainly include all the numbers $1, 2, 3, \ldots, N$. Reason: Every positive integer up to N can be expressed as a product involving only the primes $2, 3, 5, \ldots, P$; since 2^N is already larger than N, none of these primes can appear to a power higher than the Nth. Accordingly,

$$A_N > 1 + \frac{1}{2} + \frac{1}{3} + \cdots + \frac{1}{N}$$

and inequality (1) follows. This proves that the number of primes is infinite.

We now know that the product

$$\prod_{p \leq N} \left(1 - \frac{1}{p} \right)$$

(where p ranges over all the primes up to N) tends to zero as $N \to \infty$, and this in turn provides information about the sum

$$\sum_{p \leq N} \frac{1}{p}.$$

Theorem (Euler).   *The series $\sum \frac{1}{p}$, extended over all the primes, is divergent, i.e.,*

$$\frac{1}{2} + \frac{1}{3} + \frac{1}{5} + \cdots = \infty.$$

This is really quite remarkable—the primes are not known individually and yet we now know something about all of their reciprocals.

Proof.   As before, let N be an arbitrary positive integer and $2, 3, 5, \ldots, P$ the primes up to N. Set

$$B_N = \frac{1}{1 - \frac{1}{2}} \frac{1}{1 - \frac{1}{3}} \frac{1}{1 - \frac{1}{5}} \cdots \frac{1}{1 - \frac{1}{P}}.$$

We already know that

$$B_N > 1 + \frac{1}{2} + \frac{1}{3} + \cdots + \frac{1}{N}.$$

It is natural to modify this inequality in two ways. First, it is clear from Figure 15 that the area under the curve $y = \frac{1}{x}$ from 1 to N is less than $1 + \frac{1}{2} + \cdots + \frac{1}{N-1}$, and therefore

$$B_N > \log N. \qquad (2)$$

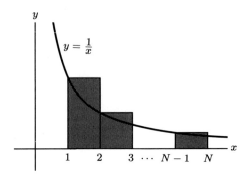

FIGURE 15

Next, recognizing that sums are usually more manageable than products, we apply the logarithm to both sides of (2) and obtain $\log B_N > \log\log N$, or,

$$\log \frac{1}{1-\frac{1}{2}} + \log \frac{1}{1-\frac{1}{3}} + \cdots + \log \frac{1}{1-\frac{1}{P}} > \log\log N. \qquad (3)$$

What has been gained? The new inequality certainly does not appear more advantageous than the original. To see that it is, we have only to choose a suitable upper bound for the logarithm. Figure 16 shows that, since the graph of $\log \frac{1}{1-x}$ is concave up and has slope 1 when $x = 0$, any linear function $y = mx$ will do, provided that $m > 1$ and x is sufficiently small (and positive). For simplicity, we may take $m = 2$, so that

$$\log \frac{1}{1-x} < 2x \qquad (4)$$

which the reader will have no difficulty in verifying for all $0 < x \le \frac{1}{2}$. By combining (3) and (4) we obtain

$$2\left(\frac{1}{2} + \frac{1}{3} + \frac{1}{5} + \cdots + \frac{1}{P} \right) > \log\log N.$$

The proof is over: since $\log\log N$ approaches infinity with N, so must the sum on the left.

Remark. It can be shown by more advanced methods that $\log\log N$ is a remarkably good approximation to $\sum_{p \le N} \frac{1}{p}$, in the sense that the difference

$$\sum_{p \le N} \frac{1}{p} - \log\log N$$

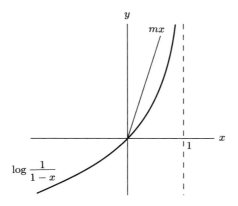

FIGURE 16

tends to a finite limit as $N \to \infty$ ([211], p. 351). Thus, the series $\sum \frac{1}{p}$ diverges to infinity exceedingly slowly. For example, if $N = 1000$, then $\log N \approx 6.9$ and $\log \log N$ is almost 2. When $N = 10^{25}$, $\log \log N$ is only a little larger than 4. Put another way, if a hypothetical computer were set to work on the task of summing the series $\sum \frac{1}{p}$, and if it could process one million terms every second—first testing p for primality and then discarding it from the sum when composite—then three billion centuries would elapse before it reached four!

The product

$$\prod_{p \le N} \left(1 - \frac{1}{p}\right)$$

deserves closer scrutiny for it admits an interesting probabilistic interpretation.

First we need a definition. If A is a set of positive integers, we denote by $A(N)$ the number of its elements among the first N integers. The **density** of A is defined to be

$$\mathbf{D}[A] = \lim_{N \to \infty} \frac{A(N)}{N},$$

provided that the limit exists. Thus the density may be thought of, loosely, as *the probability that a randomly chosen integer will belong to the set A.* The set of even integers has density $1/2$, in agreement with our intuitive belief that "half" of all integers are even. Similarly, the density of all integers divisible by 10 is $1/10$. A set may be quite irregular and still have a density. For example the set of all *square-free* integers (integers which are not divisible by any square greater than 1) has density $6/\pi^2$ ([211], p. 269). The odds are nearly 2 to 1 that an integer chosen at random will not have a repeated prime factor.

Not every set A has a density. A striking and unexpected example is given by the set of those integers whose leading digit is a 1. Intuition suggests that any of the 9 possible first digits is equally likely, and that therefore the density should be 1/9. In fact, the set has no density at all. (For an account of this fascinating problem, see R. P. Boas, "Some Remarkable Sequences of Integers," in *Mathematical Plums* [230], chapter 3.) We shall not be concerned with such irregular looking sets here.

Let us consider the set of integers divisible by a prime p. The density of this set is clearly $1/p$. If q is another prime, then the density of the set of those integers divisible by both p and q is just $1/pq$ (since, by unique factorization, divisibility by p and q is equivalent to divisibility by pq). But

$$\frac{1}{pq} = \frac{1}{p} \cdot \frac{1}{q} \tag{5}$$

and the multiplication rule for independent events comes to mind. (Recall that two events are said to be *independent* if the probability of their joint occurrence is equal to the product of the probabilities of their individual occurrences.) Thus equation (5) can be interpreted by saying that the "events" of being divisible by p and q are independent.[12]

Now, if two events A_1 and A_2 are independent, then their complementary events $\tilde{A}_1$ and $\tilde{A}_2$ are also independent. Indeed, we have the following sequence of obvious assertions:

$$\mathbf{D}[\tilde{A}_1]\mathbf{D}[\tilde{A}_2] = (1 - \mathbf{D}[A_1])(1 - \mathbf{D}[A_2])$$
$$= 1 - \mathbf{D}[A_1] - \mathbf{D}[A_2] + \mathbf{D}[A_1]\mathbf{D}[A_2]$$
$$= 1 - \mathbf{D}[A_1] - \mathbf{D}[A_2] + \mathbf{D}[A_1 \cap A_2]$$
$$= 1 - \mathbf{D}[A_1 \cup A_2]$$
$$= \mathbf{D}[\tilde{A}_1 \cap \tilde{A}_2].$$

Thus the product $\prod_{p \leq N}(1 - \frac{1}{p})$ accurately reflects the probability that a randomly chosen integer is not divisible by any prime less than or equal to N.

---

[12] These ideas are set forth and elaborated by Mark Kac, one of the founders of modern probability theory, in his beautiful monograph *Statistical Independence in Probability, Analysis and Number Theory* [241]. Referring to equation (5), Kac writes: "This holds, of course, for any number of primes, and we can say, using a picturesque but not a very precise language, that the primes play a game of chance! This simple, nearly trivial, observation is the beginning of a new development which links in a significant way number theory on the one hand and probability theory on the other." (p. 54) For a more informal account of the origins of this important connection, the reader is urged to consult Kac's remarkable autobiography *Enigmas of Chance* [242].

Table 3 gives the value of this product for various values of N. It shows for example that more than 75% of all integers are divisible by a prime less than 10, while nearly 95% are divisible by a prime less than 10,000. *Most numbers have at least one small factor.*

N	$\prod_{p \leq N}(1 - \frac{1}{p})$
10	.2286
10^2	.1203
10^3	.0810
10^4	.0609
10^5	.0488
10^6	.0407

TABLE 3

PROBLEMS

1. Euler deduced the infinitude of the primes from the incorrect identity

$$\prod_p \frac{1}{1 - 1/p} = \sum_1^\infty \frac{1}{n}$$

(both the sum and the product are divergent). Derive formally the convergent expansions

$$\prod_p \frac{1}{1 - 1/p^2} = \sum_1^\infty \frac{1}{n^2}$$

$$\prod_p \frac{1}{1 - 1/p^3} = \sum_1^\infty \frac{1}{n^3}$$

$$\prod_p \frac{1}{1 - 1/p^4} = \sum_1^\infty \frac{1}{n^4}$$

$$\cdots$$

(In chapter 6, §2, problem 18, we will show how Euler obtained values for all the sums $\sum 1/n^2, \sum 1/n^4, \sum 1/n^6, \ldots$)

2. Let $\{a_n\}$ be a sequence of positive numbers. Prove that the product $\prod_n (1 + a_n)$ converges if and only if the series $\sum_n a_n$ converges.

3. Consider Figure 17. Show that when the construction is continued indefinitely, the figure remains bounded.

FIGURE 17

4. Let N be a positive integer and $2, 3, 5, \ldots, P$ the primes not exceeding N. Show that

$$\frac{1}{2} + \frac{1}{3} + \frac{1}{5} + \cdots + \frac{1}{P} > \log \log N - \frac{1}{2}.$$

Hint. Use the power series expansion of $\log(1+x)$ to approximate $\log \frac{1}{1-1/p}$.

5. Show that if $\mathbf{D}[A]$ and $\mathbf{D}[B]$ exist, then $\mathbf{D}[A \cup B]$ and $\mathbf{D}[A \cap B]$ need not exist.

6. Show that the set of positive integers with leading digit 1 does not have a density.

7. Show that a set of positive integers $A = \{a_n\}$, with $a_1 < a_2 < a_3 < \cdots$, has density **D** if and only if

$$\lim_{n \to \infty} \frac{n}{a_n} = \mathbf{D}.$$

8. Show that if a set of positive integers $A = \{a_n\}$ has the property that $\sum 1/a_n$ converges, then A has density zero. Is the converse true?

9. Egyptian Fractions. The ancient Egyptians endeavored to represent all fractions (except $2/3$) as the sum of a finite number of distinct unit fractions

$$\frac{m}{n} = \frac{1}{x_1} + \frac{1}{x_2} + \cdots + \frac{1}{x_k}.$$

Using the divergence of the harmonic series, show that every positive rational number can be represented in this way.

    Remarks. For an excellent historical acount of the problem, see Van der Waerden, *Science Awakening* [426]. Egyptian fractions remain a rich source of problems. For example, R. L. Graham has proved the striking result that every positive rational number less that $\frac{\pi^2}{6} - 1$ can be expressed as a finite sum of inverses of distinct squares ("On finite sums of unit fractions," *Proc. London Math. Soc.* 14 (1964), 193–207). Erdős and Sierpiński have conjectured, respectively, that every rational number of the form $4/n$ or $5/n$, where $n > 1$, is expressible as the sum of three unit fractions. (For these and other results, together with an extensive bibliography, see Guy, *Unsolved Problems in Number Theory* [191], pp. 87ff.)

10. One of the best known applications of the harmonic series is the problem of determining how long a cantilevered arch can (theoretically) be made by piling identical books (or bricks) at the edge of a table so that the top book

FIGURE 18

projects as far as possible. Show that under ideal conditions the total overhang obtainable with n books, each of unit length, is half the nth partial sum of the harmonic series, and so can be arbitrarily large if n is large enough.

11.   Thinning Out the Harmonic Series.   Show that if we drop from the harmonic series all the reciprocals of integers whose decimal representations contain one or more zeros, the remaining series converges. Try to find its sum correct to the nearest integer.

   Hint. In summing a slowly convergent series, it is often best not to try to make the remainder *small,* but to determine as accurately as possible what the remainder *is.*

12.   Let $H_n = 1 + \frac{1}{2} + \frac{1}{3} + \cdots + \frac{1}{n}$ denote the nth partial sum of the harmonic series and define $S(N)$ to be the smallest integer n for which $H_n \geq N$. There is no simple formula for $S(N)$, but the first few of its values are readily computed:

N	$S(N)$
1	1
2	4
3	11
4	31
5	83
6	227
7	616
8	1,674
9	4,550
10	12,367

Prove that

$$\lim_{N \to \infty} \frac{S(N+1)}{S(N)} = e.$$

5. The Prime Number Theorem

Even a cursory examination of a table of primes will reveal that the distribution of the primes among the integers is extremely irregular, seemingly without pattern. "They grow like weeds among the natural numbers, seeming to obey no other law than that of chance, and nobody can predict where the next one will sprout."[13] For example, among the 100 numbers immediately preceding 10 million there are 9 primes,

9,999,901	9,999,931	9,999,971
9,999,907	9,999,937	9,999,973
9,999,929	9,999,943	9,999,991

while among the 100 numbers immediately following 10 million there are only 2,

$$10,000,019 \quad \text{and} \quad 10,000,079.$$

"Upon looking at these numbers, one has the feeling of being in the presence of one of the inexplicable secrets of creation," writes Don Zagier, echoing the number mysticism of the ancients.[14]

The number of primes is infinite and yet no one has ever found infinitely many of them. No one has ever found a formula for the nth prime, nor a formula for the prime that immediately follows a given one. In fact, no one has ever found a simple rule for finding any prime larger than a given one, and all such questions appear to be hopelessly out of reach. There is just no apparent reason why one number is prime while another is not.

But whatever eccentricities the primes may exhibit "in the small," their aggregate behavior is orderly and predictable. As we ascend to higher ranges of numbers, the primes certainly grow fewer, and in the far reaches of the number system, their frequency shows a steady but slow decline. Tables 4 and 5 show the diminishing frequency of the primes in various intervals of length 100 and 1000, respectively.

In order to investigate the distribution of the primes we shall make use of the prime counting function $\pi(x)$, the number of primes not exceeding x. We already know that the number of primes is infinite so that $\pi(x)$ tends to infinity with x. It is only natural to compute the ratio $\pi(x)/x$ which measures the proportion of primes among the integers up to x. The last entry in Table 6 shows that among the integers up to 10^{10} only slightly more than four and a half percent are primes. The data suggest that

$$\frac{\pi(x)}{x} \to 0 \qquad \text{as } x \to \infty, \tag{1}$$

the density of the primes is zero. The evidence is not misleading.

---

[13] D. Zagier, "The first 50 million prime numbers" [461], p. 7.

[14] Ibid., p. 8.

Interval	Number of Primes	Interval	Number of Primes
1–100	25	10000–10100	11
100–200	21	10100–10200	12
200–300	16	10200–10300	10
300–400	16	10300–10400	12
400–500	17	10400–10500	10
1000–1100	16	100000–100100	6
1100–1200	12	100100–100200	9
1200–1300	15	100200–100300	8
1300–1400	11	100300–100400	9
1400–1500	17	100400–100500	8

TABLE 4
Numbers of primes in various intervals of length 100

Interval	Number of Primes
1–1,000	168
1,000–2,000	135
10,000–11,000	106
100,000–101,000	81
1,000,000–1,001,000	75
10,000,000–10,001,000	61
100,000,000–100,001,000	54
1,000,000,000–1,000,001,000	49
10,000,000,000–10,000,001,000	44
100,000,000,000–100,000,001,000	47
1,000,000,000,000–1,000,000,001,000	37
10,000,000,000,000–10,000,000,001,000	34
100,000,000,000,000–100,000,000,001,000	30
1,000,000,000,000,000–1,000,000,000,001,000	24

TABLE 5
Numbers of primes in various intervals of length 1000

x	$\dfrac{\pi(x)}{x}$
10	.4000
10^2	.2500
10^3	.1680
10^4	.1229
10^5	.0959
10^6	.0785
10^7	.0665
10^8	.0576
10^9	.0508
10^{10}	.0455

TABLE 6

Theorem (Euler). *The set of primes has density zero.*

Proof. The proof is based on a simple modification of the sieve of Eratosthenes. According to this ancient principle, if we wish to determine all the
primes up to a prescribed limit x, we need only identify those primes up to
$\sqrt{x}$ and then strike out all their multiples; the remaining numbers will then be
prime.

We shall modify this idea slightly. Let $A(x, n)$ denote the number of positive integers not greater than x and not divisible by any of the first n primes
$p_1, \ldots, p_n$. Obviously

$$\pi(x) \leq n + A(x, n) \tag{2}$$

since any prime is either one of $p_1, \ldots, p_n$ or a positive integer not divisible by
any of these. The value of this inequality lies in the fact that n may be treated
as a parameter and its best value prescribed later.

We first estimate $A(x, n)$. If p is a prime, then the number of integers not
greater than x and divisible by p is

$$\left[\frac{x}{p}\right]$$

where $[x]$ denotes the largest integer not exceeding x. (Reason: $[x/p] = m$ if
and only if $mp \leq x < (m+1)p$.) If q is any other prime, then the number of

integers not greater than x and divisible by both p and q is

$$\left[\frac{x}{pq}\right] ;$$

and so on. Thus, by the principle of inclusion and exclusion (see chapter 1, §3),

$$A(x, n) = [x] - \sum\left[\frac{x}{p_i}\right] + \sum\left[\frac{x}{p_i p_j}\right] - \cdots \pm \left[\frac{x}{p_1 \cdots p_n}\right],$$

where $i, j, \ldots$ are unequal and run from 1 to n. The total number of square brackets is

$$1 + \binom{n}{1} + \binom{n}{2} + \cdots + \binom{n}{n} = 2^n$$

and the error introduced by removing any one of them is less than 1. Therefore, if we approximate $A(x, n)$ by

$$x - \sum\frac{x}{p_i} + \sum\frac{x}{p_i p_j} - \cdots \pm \frac{x}{p_1 \cdots p_n}$$

$$= x\left(1 - \frac{1}{p_1}\right)\left(1 - \frac{1}{p_2}\right)\cdots\left(1 - \frac{1}{p_n}\right)$$

we will have introduced an error of at most 2^n. Thus

$$A(x, n) \leq x\prod_{i=1}^{n}\left(1 - \frac{1}{p_i}\right) + 2^n. \tag{3}$$

This is the desired estimate for $A(x, n)$.

By combining (2) and (3), we find that

$$\pi(x) \leq n + x\prod_{i=1}^{n}\left(1 - \frac{1}{p_i}\right) + 2^n$$

and hence that

$$\frac{\pi(x)}{x} \leq \frac{2^{n+1}}{x} + \prod_{i=1}^{n}\left(1 - \frac{1}{p_i}\right)$$

since $n < 2^n$. Let ε be an arbitrary positive number and choose n so large that the product on the right is less than $\varepsilon/2$ (in §4 we proved that the product tends to zero). If x is sufficiently large, $2^{n+1}/x < \varepsilon/2$, and so

$$\frac{\pi(x)}{x} < \frac{\varepsilon}{2} + \frac{\varepsilon}{2} = \varepsilon.$$

This proves Euler's theorem.

It is curious that both of Euler's theorems are consequences of the same fact, namely, that the product

$$\left(1 - \frac{1}{2}\right)\left(1 - \frac{1}{3}\right)\left(1 - \frac{1}{5}\right)\cdots\left(1 - \frac{1}{p}\right)$$

tends to zero as $p \to \infty$, and yet they point in opposite directions. The import of the second theorem is that the primes are rather sparsely distributed among the integers; in fact it implies that *the nth prime p_n has an order of magnitude greater than n*, or

$$\frac{p_n}{n} \to \infty \tag{4}$$

(just take $x = p_n$ in (1)). On the other hand, Euler's first theorem tells us that

$$\sum \frac{1}{p_n} = \infty \tag{5}$$

and therefore the order of magnitude of p_n cannot be too much greater than n; if it were, for example, n^2 or $n^{3/2}$ or even $n(\log n)^2$, then the series $\sum 1/p_n$ would converge.

In 1751, Euler wrote "Mathematicians have tried in vain to this day to discover some order in the sequence of prime numbers, and we have reason to believe that it is a mystery into which the human mind will never penetrate. To convince ourselves, we have only to cast a glance at tables of primes, which some have taken the trouble to compute beyond a hundred thousand, and we should perceive at once that there reigns neither order nor rule."[15] If what Euler despaired of finding was an explicit formula for p_n in terms of n, then his pessimism was justified, for no such formula is known. At the same time, Euler thought it likely that a recursive formula would be found for p_n in terms of $p_1, p_2, \ldots, p_{n-1}$. He had discovered such a formula for $\sigma(n)$, the sum of the divisors of n, and he believed that the irregularities of $\sigma(n)$ must be at least as great as those of p_n since $\sigma(n) = n + 1$ if and only if n is prime. Unfortunately, no satisfactory formula for p_n has ever been discovered. Real progress was made only when mathematicians gave up futile attempts to find an exact formula for the nth prime and looked instead for an approximation.

---

[15] The quotation is from Euler's memoir "Découverte d'une loi tout extraordinaire des nombres par rapport à la somme de leurs diviseurs," in which he presents the celebrated pentagonal number theorem (see his *Collected Works*, ser. 1, vol. 2, p. 241). In chapter 6, we shall give Pólya's complete translation of this famous work.

Euler came ever so close to the truth. If it is possible to approximate p_n by a simple function $f(n)$ then, by virtue of (4) and (5), that function must satisfy

$$\frac{f(n)}{n} \to \infty \quad \text{and} \quad \sum \frac{1}{f(n)} = \infty.$$

The most obvious function that satisfies these requirements is $n \log n$, and we are led to speculate that p_n may behave like a multiple of $n \log n$. The empirical evidence suggests that this is so and that the correct multiple is unity.

n	p_n	$n \log n$	$\dfrac{p_n}{n \log n}$
10	29	23	1.2595
100	541	461	1.1745
1,000	7,919	6,908	1.1464
10,000	104,729	92,103	1.1371
100,000	1,299,709	1,151,293	1.1289

TABLE 7
An approximation for the nth prime

We have been led to an astonishing conjecture: *the nth prime number is approximately equal to $n \log n$*—not in the sense that the difference $p_n - n \log n$ approaches zero (or even remains bounded), but rather in the sense that the ratio $p_n / n \log n$ approaches 1. This is the content of the celebrated prime number theorem.

Prime Number Theorem.

$$p_n \sim n \log n, \tag{6}$$

where the symbol $\sim$ (read "is asymptotically equal to") is used to indicate that the ratio of the two sides tends to unity as $n \to \infty$.

This simple law governing the behavior of the primes is one of the most remarkable facts in all of mathematics. Here is order extracted out of apparent chaos—a simple approximation for the elusive primes. It is true that the difference between p_n and $n \log n$ increases beyond all bounds, but it is the percentage error which really matters. It tends steadily (but not all that quickly) to zero. Thus, for example, the right side of (6) approximates $p_{100} = 541$ by

461 and $p_{1000} = 7919$ by 6908. The percentage errors are nearly 15 and 13, respectively. For $p_{6,000,000} = 104,395,301$ the error is still almost 10 percent.

The prime number theorem is the central theorem in the analytic theory of numbers. It is no less remarkable for the order it brings to the apparent chaos of the primes than for the fact that it establishes that order through the simplicity of the logarithmic function. In so doing it links two seemingly unrelated mathematical concepts—the continuous and the discrete. Euler certainly could have discovered it but the question of approximation did not seem to occur to him.

The conjecture was first formulated not in the form given by (6) but in the logically equivalent form

$$\pi(x) \sim \frac{x}{\log x} \tag{7}$$

and it is this relation that is commonly referred to as the prime number theorem. (The reader will find it instructive to show that each of the two asymptotic formulas (6) and (7) can be derived from the other. *Hint*: p_n as a function of n and $\pi(x)$ as a function of x are inverses of each other.) In this form it asserts that, for large values of x, the proportion $\pi(x)/x$ of the primes up to x will be "about" $1/\log x$.

The empirical evidence in support of relation (7) had long been overwhelming. If we go back to Table 6 and invert the entries in the second column, thereby obtaining not the fraction of the primes up to x but their reciprocals, $x/\pi(x)$, then we obtain Table 8. The pattern is now almost unmistakable. As we pass from one power of ten to the next, the ratio $x/\pi(x)$ increases by roughly

x	$\dfrac{x}{\pi(x)}$
10	2.5
10^2	4.0
10^3	5.9
10^4	8.1
10^5	10.4
10^6	12.7
10^7	15.0
10^8	17.3

TABLE 8

2.3 (at least when x is large). At this point "any mathematician worth his salt" recognizes 2.3 as the approximate value of log 10, and from here it is a simple matter to formulate the conjecture (7).[16]

The first published statement of the prime number theorem is due to Legendre, in 1808, in his well-known treatise "Essai sur la théorie des nombres" (2nd edition). On the basis of empirical evidence, Legendre conjectured that $\pi(x)$ is approximately equal to

$$\frac{x}{\log x - A}$$

where he gave 1.08 as the approximate value of the constant A. So simple was Legendre's law that Abel in 1823 called it "the most remarkable in all mathematics." The formula holds with astonishing accuracy for relatively small values of x (see Table 9), but ultimately the best value for the constant is $A = 1$.

x	$\pi(x)$	$\dfrac{x}{\log x - 1.08} - \pi(x)$	$\mathrm{Li}\,(x) - \pi(x)$
10^2	25	3	5
10^3	168	4	10
10^4	1,229	1	17
10^5	9,592	-7	38
10^6	78,498	23	130
10^7	664,579	399	339
10^8	5,761,455	5,331	754
10^9	50,847,534	60,497	1,701
10^{10}	455,052,511	614,486	3,104

TABLE 9

Independently, during the period 1792–1793 (when he was only about 15 years old), Gauss also made extensive calculations on the distribution of prime numbers. By studying prime number tables contained in a book of logarithms that had been given to him as a present the year before, Gauss compiled an extraordinary set of data on the number of primes in various intervals of length 1000 (also known as *chiliads*). On the basis of that evidence he con-

---

[16] P. J. Davis and R. Hersh, *The Mathematical Experience* [126], p. 214.

jectured that *the average density of the primes near a large number x is approximately* $1/\log x$. In other words, an interval of length Δx about x should contain roughly $\Delta x/\log x$ primes, provided of course that Δx is small compared to x and yet large enough to make statistics meaningful. For example, according to Gauss's formula we should expect to find about 62 primes in the interval between 10 million and 10 million plus 1,000, because $10^3/\log 10^6 \approx 62$. The actual number of primes in this interval is 61. Knowing the "local density" at x, Gauss then concluded—what good calculus student would not!—that $\pi(x)$ ought to be approximately equal to

$$\mathrm{Li}\,(x) = \int_2^x \frac{dt}{\log t}.$$

The function $\mathrm{Li}\,(x)$ is known as the **logarithmic integral**. It is not an elementary function but its values are readily calculated. Table 9 compares the accuracy of both Gauss's and Legendre's approximations. It suggests that, of the two, Gauss's is ultimately the better one. This can be accounted for by the theory. Nevertheless, it is a simple matter to show that both approximations are asymptotically equivalent.

Theorem. *The two functions* $\mathrm{Li}\,(x)$ *and* $x/\log x$ *are asymptotically equal.*

Proof. If we integrate $\mathrm{Li}\,(x)$ by parts, we find

$$\mathrm{Li}\,(x) = \int_2^x \frac{dt}{\log t} = \frac{x}{\log x} - \frac{2}{\log 2} + \int_2^x \frac{dt}{(\log t)^2}.$$

Therefore it suffices to show that

$$\lim_{x \to \infty} \frac{\int_2^x dt/(\log t)^2}{x/\log x} = 0. \tag{8}$$

Since $1/(\log t)^2$ is positive and decreasing for $t > 1$, we have for every $x \geq 4$,

$$0 < \int_2^x \frac{dt}{(\log t)^2} = \int_2^{\sqrt{x}} \frac{dt}{(\log t)^2} + \int_{\sqrt{x}}^x \frac{dt}{(\log t)^2}$$

$$< \frac{\sqrt{x} - 2}{(\log 2)^2} + \frac{x - \sqrt{x}}{(\log \sqrt{x})^2}$$

$$< \frac{\sqrt{x}}{(\log 2)^2} + \frac{4x}{(\log x)^2}.$$

Accordingly,

$$0 < \frac{\int_2^x dt/(\log t)^2}{x/\log x} < \frac{\log x}{\sqrt{x}(\log 2)^2} + \frac{4}{\log x}$$

and (8) follows at once. This proves the theorem.

The proof reminds us again that the real power of the method of integration by parts is not to integrate but to approximate.

Gauss never published his conjecture but there is little reason to doubt his claim that he undertook the work long before Legendre's memoir was written. Certainly he had anticipated other great discoveries, such as the elliptic functions of Jacobi and the non-Euclidean geometry of Bolyai and Lobachevsky, both of which he acknowledged as his own years after the original work had been done. He returned to his formula for $\pi(x)$ many times throughout his life, checking it against newly acquired data. Writing to Encke in 1849 he described how very often he "spent an idle quarter of an hour to count through another chiliad here and there" until finally all the primes up to 3 million had been tabulated and their distribution compared with his remarkable conjecture.

Neither Legendre nor Gauss would ever prove his conjecture. It would take nearly a century and the efforts of many of the world's best mathematicians —among them Dirichlet, Chebyshev, and Riemann—each building on the work of those who came before, until mathematical analysis had developed to the point where a rigorous proof of the prime number theorem could finally be given. That proof came in 1896, independently by Hadamard and de la Vallée Poussin. But the story of their work, its ultimate simplifications and modifications, is the beginning of another and far more difficult story in the theory of numbers. This story is over.

PROBLEMS

1. Prove that for natural numbers $n > 1$ the inequality

$$\frac{\pi(n-1)}{n-1} < \frac{\pi(n)}{n}$$

holds if and only if n is a prime.

2. Bertrand's postulate asserts that for every $n > 1$ there is a prime number p satisfying $n < p < 2n$. Bertrand verified this for $n < 3,000,000$ and, in 1850, Chebyshev provided the proof. Prove the following extension: *If a and b are two positive real numbers and a < b, then for sufficiently large real numbers x there is always at least one prime number between ax and bx.*

Hint. Use the prime number theorem to show that

$$\lim_{x \to \infty} \frac{\pi(ax)}{\pi(bx)} = \frac{a}{b}.$$

3. Show that

$$\frac{\pi(2x) - \pi(x)}{\pi(x)} \to 1 \qquad \text{as } x \to \infty.$$

Thus, for large x, the number of primes between x and $2x$ is roughly the same as the number less than x. Is this consistent with the fact that the density of the primes near x decreases as x increases?

4. Sierpiński's Conjecture (unsolved). If the numbers $1, 2, 3, \ldots, n^2$ with $n > 1$ are arranged in n rows each containing n numbers:

1	2	3	$\cdots$	n
$n+1$	$n+2$	$n+3$	$\cdots$	$2n$
$2n+1$	$2n+2$	$2n+3$	$\cdots$	$3n$
$\cdots$	$\cdots$	$\cdots$	$\cdots$	$\cdots$
$(n-1)n+1$	$(n-1)n+2$	$(n-1)n+3$	$\cdots$	n^2

then each row contains at least one prime number.

5. (Continuation) Show that each of the following statements is a consequence of Sierpiński's conjecture:

a. (Bertrand's postulate) For every natural number $n > 1$ there is at least one prime number between n and $2n$.

b. (Unsolved) Between the squares of any two consecutive natural numbers there are at least two prime numbers.

c. (Unsolved) Between the cubes of any two consecutive natural numbers there are at least two prime numbers.

d. (Unsolved) If the natural numbers are arranged in rows in such a manner that the nth row contains n consecutive natural numbers,

1				
2	3			
4	5	6		
7	8	9	10	
11	12	13	14	15

$\cdots$

then each but the first row contains a prime number. In other words, between any two triangular numbers there is at least one prime number.

6.   Gilbreath's Conjecture (unsolved).   A table of natural numbers is formed in the following manner: the first row contains the differences of consecutive prime numbers (i.e., the numbers $p_{n+1} - p_n$, $n = 1, 2, \ldots$); the next row contains the absolute values of the differences of the consecutive numbers in the first row; each of the following rows contains the absolute differences of the consecutive terms of the preceding row. Gilbreath's conjecture is that the first term in each row is 1.

The initial terms of the first 10 rows are as follows:

$$
\begin{array}{ccccccccccc}
1 & 2 & 2 & 4 & 2 & 4 & 2 & 4 & 6 & 2 \\
 & 1 & 0 & 2 & 2 & 2 & 2 & 2 & 2 & 4 \\
 & & 1 & 2 & 0 & 0 & 0 & 0 & 0 & 2 \\
 & & & 1 & 2 & 0 & 0 & 0 & 0 & 2 \\
 & & & & 1 & 2 & 0 & 0 & 0 & 2 \\
 & & & & & 1 & 2 & 0 & 0 & 2 \\
 & & & & & & 1 & 2 & 0 & 2 \\
 & & & & & & & 1 & 2 & 2 \\
 & & & & & & & & 1 & 0 \\
 & & & & & & & & & 1 \\
\end{array}
$$

CHAPTER 6

INFINITE SUMS: A POTPOURRI

"What's one and one and one and one and one and one and one and one and one and one?"

"I don't know," said Alice, "I lost count."

"She can't do Addition," the Red Queen interrupted.

—Lewis Carroll, *Through the Looking Glass*

1. Geometry and the Geometric Series

a. The Quadrature of the Parabola

Many mathematicians have endeavored to square the circle, the ellipse, or the segment of a circle, of an ellipse or of a hyperbola; and the edifice of lemmas which they erected for this purpose has generally been found open to objection. No one, however, seems to have thought of attempting the quadrature of the segment of a parabola, which is precisely the one which can be carried out.

—Archimedes, from the Preface to his treatise
on the *Quadrature of the Parabola*[1]

Within the rich geometric legacy of the ancient Greeks lie three famous construction problems which have captivated mathematicians for more than 2000 years: the quadrature of the circle, the duplication of the cube, and the trisection of the angle.

1. *The quadrature of the circle*: To construct a square equal in area to a given circle.

2. *The duplication of the cube*: To construct a cube having twice the volume of a given cube.

3. *The trisection of the angle*: To construct an angle one-third as large as a given angle.

These are the "three famous problems" of antiquity, all requiring solutions using only Euclidean tools, that is, straightedge and compass alone.

The restricted rules of construction render the problems of purely theoretical interest and reflect the Greek view of the world of mathematics.

"Here we see a type of mathematics that is quite unlike that of the Egyptians and Babylonians. It is not the practical application of a science of number to a facet of life experience, but a theoretical question involving a nice distinction between accuracy in approximation and exactitude in thought. . . . In the

---

[1] The somewhat free translation given here is from O. Toeplitz, *The Calculus: A Genetic Approach* [417], p. 43. A more literal translation is given by Heath in *The Works of Archimedes* [15].

Greek world mathematics was more closely related to philosophy than to practical affairs, and this kinship has persisted to the present day."[2]

The problems cannot be solved—at least not with straightedge and compass alone—and it is perhaps for this reason that they have always held so great an allure. It was not until the late 19th century that the impossibility of all three constructions was finally established. But the search for solutions would exercise a profound influence on Greek mathematics and on much later mathematical thought. It would lead, in antiquity, to the discovery of the conic sections, to curves of higher degree, and to several transcendental curves, among them the *quadratrix* and the *spiral of Archimedes*. Much later, it would influence the modern theory of equations, involving domains of rationality, algebraic numbers, and group theory.[3]

Here, we shall indicate briefly how Archimedes, using the "method of exhaustion," the ancient forerunner of the modern theory of integration, arrived at the area of a parabolic segment.

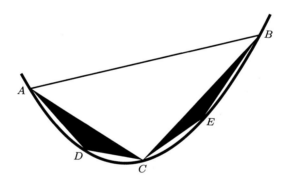

FIGURE 1

First, Archimedes exhausted the given region by means of a cleverly chosen sequence of inscribed triangles (see Figure 1). His procedure, which is recursive, begins with the approximating triangle ABC, where the vertex C is chosen so that the tangent to the parabola at C is parallel to AB. The next approximation is obtained by adding two new triangles ACD and BCE, the vertices D and E being chosen in the same way, so that the tangent at D is

---

[2] C. B. Boyer, *A History of Mathematics* [62], p. 64.

[3] See, for example, E.W. Hobson, *Squaring the Circle* [223] and F. Klein, *Famous Problems of Elementary Geometry* [257].

parallel to AC and the tangent at E is parallel to BC. At the next stage four new triangles are added, one in each of the remaining regions, and after that eight more, and so on ad infinitum.

To prove that this method really does exhaust the entire parabolic segment, we begin by observing that the area of each triangle formed in the construction is more than half that of the corresponding segment in which it is inscribed. To see this, in the case of the initial triangle ABC, we need only observe that the area of the triangle is half that of the circumscribed rectangle $ABB'A'$ (see Figure 2) whose base $A'B'$ is parallel to AB and tangent to the parabola at C.

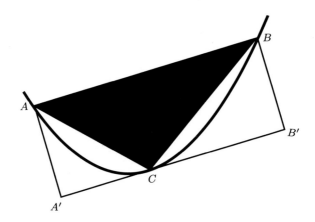

FIGURE 2

Now, the *Principle of Eudoxus*[4] asserts that:

> Two unequal magnitudes being set out, if from the greater there be subtracted a magnitude greater than its half, and from that which is left a magnitude greater than its half, and if this process be repeated continually, there will be left some magnitude which will be less than the lesser magnitude set out.

Application of this principle shows that, *for any $\epsilon > 0$, we obtain after a finite number of steps an inscribed polygon whose area differs from that of the original parabolic segment by less than ϵ.*

---

[4] Euclid, Book X, Proposition 1. The principle is a corollary of the well-known *Axiom of Archimedes*.

Next, by an elementary geometric argument, Archimedes was able to show that *the area of each of the triangles ACD and BCE is exactly one-eighth the area of triangle ABC* (see, for example, [145], pp. 35–40). It follows that

$$\text{area}(\triangle ACD) + \text{area}(\triangle BCE) = \frac{1}{4}\text{area}(\triangle ABC),$$

and hence by induction that *the sum of the areas of the inscribed triangles added at each stage of the process is equal to $\frac{1}{4}$ of the sum of the areas of the triangles added at the previous stage.* If we write

$$\alpha = \text{area}(\triangle ABC)$$

then, by the nature of the exhaustion process, it is clear that the area of the parabolic segment is equal to

$$\alpha + \frac{\alpha}{4} + \frac{\alpha}{4^2} + \frac{\alpha}{4^3} + \cdots .$$

Using the formula for the sum of a geometric series, we obtain

$$1 + \frac{1}{4} + \frac{1}{4^2} + \frac{1}{4^3} + \cdots = \frac{1}{1 - \frac{1}{4}} = \frac{4}{3}$$

and therefore

$$\text{Area of the parabolic segment} = \frac{4}{3}\text{area}(\triangle ABC),$$

which is Archimedes' own result.[5]

When the result is applied to the segment bounded by the parabola $y = x^2$ and the horizontal line $y = 1$, we find that the area is $\frac{4}{3}$. Or, in modern integral notation,

$$\int_{-1}^{1} \left(1 - x^2\right) \, dx = \frac{4}{3} .$$

---

[5] Archimedes never summed the infinite series $1 + \frac{1}{4} + \frac{1}{4^2} + \cdots$ explicitly. Sharing the Greek "horror of the infinite," he first derived the elementary identity

$$1 + \frac{1}{4} + \frac{1}{4^2} + \cdots + \frac{1}{4^n} = \frac{4}{3} - \frac{1}{3} \cdot \left(\frac{1}{4}\right)^n .$$

Then, to avoid "taking the limit," he completed the proof with a typical double reductio ad absurdum argument, characteristic of Greek geometry, in which he showed that either of the inequalities

$$\text{area(parabolic segment)} < \frac{4}{3}\text{area}(\triangle ABC)$$

or

$$\text{area(parabolic segment)} > \frac{4}{3}\text{area}(\triangle ABC)$$

leads to a contradiction. In this way, his rigor was far superior to that which is found in the works of Newton and Leibniz.

We have thus shown that the areas of the parabolic segment and of the triangle ABC are "commensurable" with each other, and it is now a simple matter of elementary geometry to construct a square equal in area to that of the segment using only Euclidean tools. Thus the parabolic segment has been squared.

b. The Leibniz Series

The unsuccessful attempts to effect the quadrature of the circle through finite geometric means led ultimately to the highly successful arithmetical quadratures which relied on infinite processes—as, of course, they had to since π is transcendental.[6] Thus, as we have already seen, both Viète and Wallis found elegant representations for π in terms of infinite products, while Brouncker found one in terms of an infinite continued fraction.

But the most astonishing formula—and one of the most beautiful mathematical discoveries of the 17th century—is the famous *Leibniz series* for the number π:

$$\frac{\pi}{4} = 1 - \frac{1}{3} + \frac{1}{5} - \frac{1}{7} + \cdots. \tag{1}$$

Through this remarkable alternating series, with its simple law of formation, we see π linked with the integers in a far more striking way than through its decimal expansion, which displays no apparent order or regularity in the succession of its digits. Its beauty, however, is not matched by its practicality: it would require about 100,000 terms to compute π even to the accuracy obtained by Archimedes.

Many proofs of Leibniz's formula are now known. Perhaps the quickest proof is based on the formula

$$\frac{\pi}{4} = \tan^{-1} 1 = \int_0^1 \frac{1}{1 + x^2}\, dx.$$

It is tempting to expand the integrand into a geometric series, so that

$$\frac{1}{1 + x^2} = 1 - x^2 + x^4 - x^6 + \cdots,$$

---

[6] The proof that π is transcendental—and hence that the circle cannot be squared using only Euclidean tools—was first given by Lindemann in 1882.

and then integrate term by term. Formally, then,

$$\frac{\pi}{4} = \int_0^1 \frac{1}{1+x^2}\,dx = \int_0^1 dx - \int_0^1 x^2\,dx + \int_0^1 x^4\,dx - \cdots$$

$$= 1 - \frac{1}{3} + \frac{1}{5} - \frac{1}{7} + \cdots$$

which is Leibniz's formula.[7]

The trouble with this argument is its assumption that term by term integration is permissible, even though the series $1 - x^2 + x^4 - x^6 + \cdots$ is divergent when $x = 1$. But, to the mathematicians of the period, the manipulation of series was largely formal, and questions of convergence and divergence were not taken too seriously. Infinite processes, once shunned by the ancient Greeks, were now embraced with an enthusiasm that left little concern for rigor. It was not until the 19th century, through the work of Cauchy, Abel, and their contemporaries, that the logical foundations of the subject were finally secured and the results of such formal manipulations rigorously justified.

It is true that if formal integration of a power series produces a convergent result, then it must necessarily be the correct result. But in the present case it is simpler to verify the formal calculation directly. Starting with the formula for the sum of a finite geometric progression, we obtain $1 - x^2 + x^4 - \cdots - x^{2n-2} = (1 - x^{2n})/(1 + x^2)$, where, for simplicity, we have assumed that n is *even*. Consequently, it follows by *finite* termwise integration that

$$1 - \frac{1}{3} + \frac{1}{5} - \cdots - \frac{1}{2n-1} = \int_0^1 \frac{1 - x^{2n}}{1 + x^2}\,dx$$

$$= \int_0^1 \frac{1}{1+x^2}\,dx - \int_0^1 \frac{x^{2n}}{1+x^2}\,dx,$$

and hence that

$$\left| 1 - \frac{1}{3} + \frac{1}{5} - \cdots - \frac{1}{2n-1} - \frac{\pi}{4} \right| \le \int_0^1 \frac{x^{2n}}{1+x^2}\,dx$$

$$\le \int_0^1 x^{2n}\,dx$$

$$= \frac{1}{2n+1} \to 0.$$

---

[7] Contrary to common assumption, this argument is not Leibniz's own. Rather, Leibniz derived (1) as an application of his general "transmutation method" for the transformation of integrals (see [145], pp. 245ff.).

The calculation shows both that the series $1 - \frac{1}{3} + \frac{1}{5} - \frac{1}{7} + \cdots$ converges and that its sum is $\frac{\pi}{4}$.

This wonderful formula, of which Huygens wrote "that it would be a discovery always to be remembered among mathematicians," was one of the earliest results obtained by Leibniz (1673) from his newly discovered calculus.[8] It is, however, only a special case of the infinite series expansion for $\tan^{-1} x$ which had been given earlier (1670) by James Gregory (1638–1675), the great young Scottish mathematician who had anticipated some of the key discoveries of Newton and Leibniz. Using the method of long division followed by termwise integration, Gregory showed that

$$\int_0^x \frac{1}{1+t^2}\, dt = \tan^{-1} x = x - \frac{x^3}{3} + \frac{x^5}{5} - \frac{x^7}{7} + \cdots.$$

The result is still known as *Gregory's series.*[9] Leibniz's formula for π now follows at once by taking $x = 1$.

Formula (1) reduces the mysterious number π to the integers with a simplicity and an elegance—and an element of surprise—that is a feature of the greatest art. And yet the result still seems mysterious, for the derivation involving integration and the geometric series does not reveal the number-theoretic relation that exists between π and the odd numbers. Just why is a circle related to odd numbers? The answer is provided by an important extension of Fermat's great theorem. It is due to Jacobi and it can be proved in an elementary way ([385], p. 467).

Theorem 1. *The number of ways of representing a natural number n as the sum of two integral squares*

$$n = a^2 + b^2$$

---

[8] See Leibniz's *Historia et Origo Calculi Differentialis* (History and Origin of the Differential Calculus) in *The Early Mathematical Manuscripts of Leibniz* [93], p. 46.

[9] In the subsequent bitter priority controversy concerning the calculus, Newton's supporters accused Leibniz of plagiarism, claiming in particular that, through communications with Oldenberg, at that time secretary of the Royal Society of London, he had already learned of similar series that had been discovered by Newton and Gregory. (For Leibniz's defense, see the *Historia et Origo Calculi Differentialis* in *The Early Mathematical Manuscripts of Leibniz* [93], pp. 46–48.) It is ironic that, in the end neither side was justified in claiming priority for formula (1). For it is now known that Gregory's series for the arctangent was discovered by the 15th century Hindu mathematician Nilakantha—using the same technique of deriving an infinite series by straight division followed by term by term integration. See: C. T. Rajagopal and T. V. Vedamurthi Aiyar, "On the Hindu Proof of Gregory's Series," *Scripta Mathematica* 17 (1951), 65–74. See also [369] and [447], p. 255, footnote 13.

is equal to four times the difference between the number of the divisors of n of the form 4t + 1 and the number of divisors of the form 4t − 1.

As in chapter 4, §3, we denote by $r(n)$ the number of representations of n as the sum of two integral squares. We count representations as distinct even when they differ only trivially, that is, with respect to the sign or order of a and b. Thus, by virtue of the theorem, $r(n) = 8$ whenever n is a prime of the form $4t + 1$; the representation is unique apart from its eight trivial variations. And this, of course, is Fermat's great theorem. On the other hand, $r(n) = 0$ when n is of the form $4t − 1$.

Jacobi's theorem now enables us to give an elegant and explicit formula for the sums $R(N) = r(1) + r(2) + \cdots + r(N)$, namely,

$$R(N) = 4 \left\{ \left[\frac{N}{1} \right] - \left[\frac{N}{3} \right] + \left[\frac{N}{5} \right] - \left[\frac{N}{7} \right] + \cdots \right\}. \tag{2}$$

(The symbol [ ] denotes the bracket function.) Notice that the terms of the series are equal to zero as soon as the denominators exceed N. To verify the formula we have only to observe that, among the first N integers

$$1, 2, 3, \ldots, N$$

$N = [N]$ are divisible by 1, so their total contribution to $R(N)$ is $4[N]$; $\left[\frac{N}{3} \right]$ are divisible by 3, so their total contribution to $R(N)$ is $-4 \left[\frac{N}{3} \right]$; and so on. This proves formula (2).

But we have already seen that the averages

$$\frac{r(1) + r(2) + \cdots + r(N)}{N} \to \pi$$

as $N \to \infty$. Therefore

$$R(N) \sim \pi N$$

and hence

$$\frac{1}{4} \pi N \sim \left[\frac{N}{1} \right] - \left[\frac{N}{3} \right] + \left[\frac{N}{5} \right] - \left[\frac{N}{7} \right] + \cdots. \tag{3}$$

Leibniz's formula will follow at once provided we can show that we can remove every square bracket and still retain the asymptotic equality (for then we need only cancel the Ns throughout). To see that this is permissible, let us take $N = k^2$ with k even (just to simplify the notation) and split the right side into two sums:

$$\sum_{1} = \left[\frac{k^2}{1} \right] - \left[\frac{k^2}{3} \right] + \cdots - \left[\frac{k^2}{2k-1} \right]$$

and

$$\sum_2 = \left[\frac{k^2}{2k+1}\right] - \left[\frac{k^2}{2k+3}\right] + \cdots$$

so that $\sum_1$ is the sum of the first k terms and $\sum_2$ the sum of the remaining terms. Now the removal of a single square bracket from $\sum_1$ introduces an error of at most 1; therefore if we remove all of them we commit an error of at most k, so that

$$\sum_1 = k^2\left(1 - \frac{1}{3} + \frac{1}{5} - \cdots - \frac{1}{2k-1}\right) + \theta k$$

where $|\theta| < 1$. On the other hand, the value of $\sum_2$ lies between 0 and its first term, since the terms alternate in sign and decrease in magnitude. Thus $\sum_2 = \lambda k$, where $|\lambda| < 1$. Therefore, when we divide by $N = k^2$, equation (3) becomes

$$\frac{\pi}{4} \sim 1 - \frac{1}{3} + \frac{1}{5} - \cdots - \frac{1}{2k-1} + \epsilon k^{-1}$$

where $|\epsilon| < 2$, and when $k \to \infty$, equation (1) follows.

Thus we have obtained in a purely arithmetical way a formula for the most important geometric constant. Virgil wrote: "*Numero deus impare gaudet*" (God rejoices in the odd numbers).[10]

c. "Snowflakes" and "Cantor Dust": The Simplest Fractals

Nature's features are irregular. The soft lines and delicate symmetries visible from afar give rise on closer scrutiny to a myriad of imperfections. "Clouds are not spheres, mountains are not cones, coastlines are not circles, and bark is not smooth, nor does lightning travel in a straight line." So wrote Benoit Mandelbrot on the very first page of his 1982 book *The Fractal Geometry of Nature* [290]. Recognizing that standard Euclidean geometry is inadequate to deal with such seemingly formless shapes, Mandelbrot introduced a revolutionary new geometry of wondrous and exotic forms to bring order out of the apparent chaos. He called the new shapes *fractals*.

Freeman Dyson has given the following eloquent summary of Mandelbrot's theme:[11]

"Fractal" is a word invented by Mandelbrot to bring together under one heading a large class of objects that have certain structural fea-

---

tures in common although they appear in diverse contexts in astronomy, geography, biology, fluid dynamics, probability theory, and pure mathematics. ... A great revolution of ideas separates the classical mathematics of the 19th century from the modern mathematics of the 20th. Classical mathematics had its roots in the regular geometric structures of Euclid and the continuously evolving dynamics of Newton. Modern mathematics began with Cantor's set theory and Peano's space-filling curve. Historically, the revolution was forced by the discovery of mathematical structures that did not fit the patterns of Euclid and Newton. These new structures were regarded by contemporary mathematicians as "pathological." They were described as a "gallery of monsters," kin to the cubist painting and atonal music that were upsetting established standards of taste in the arts at about the same time. The mathematicians who created the monsters regarded them as important in showing that the world of pure mathematics contains a richness of possibilities going far beyond the simple structures that they saw in nature. Twentieth-century mathematics flowered in the belief that it had transcended completely the limitations imposed by its natural origins.

Now, as Mandelbrot points out with one example after another, we see that nature has played a joke on the mathematicians. The 19th-century mathematicians may have been lacking in imagination, but nature was not. The same pathological structures that the mathematicians invented to break loose from 19th-century naturalism turn out to be inherent in familiar objects all around us in nature.

The Cantor Set. The principle of Eudoxus is far too restrictive. If we wish to "exhaust" a given magnitude it is not necessary to remove more than half of that magnitude and then successively remove more than half of what remains behind. Any fixed proportion will do. A striking illustration in the case in which the given proportion is equal to one-third is provided by a remarkable set of points known as the *Cantor set,* after Georg Cantor (1845–1918), the creator of the modern theory of infinite sets.

The construction of the Cantor set begins with the unit interval $[0, 1]$, represented as a line segment. First we remove the open middle third. That leaves two segments, $[0, 1/3]$ and $[2/3, 1]$, and we remove the open middle third of each. That leaves four segments, and again the open middle third of each of them is to be removed. When this recursive procedure is carried out indefinitely—at each stage we remove the open middle third of every interval

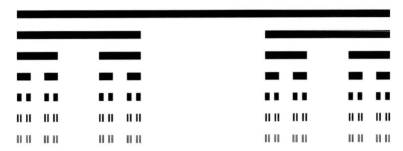

FIGURE 3
The construction of the Cantor set.

that remains from the previous stage—what is left after all the open intervals have been removed is called the Cantor set. We denote it by C.

The construction shows that the Cantor set is exceedingly small. Since one interval of length $1/3$ was removed at the first stage, two intervals, each of length $1/9$, at the second stage, and so on ad infinitum, it follows that the total length of the intervals removed is

$$1/3 + 2(1/3)^2 + 4(1/3)^3 + \cdots = 1.$$

(Reason: Factor out $1/3$ and sum the resulting geometric series.) Thus the process of deleting middle thirds really does "exhaust" the interval $[0, 1]$.

Cantor was led to the set that now bears his name in his efforts to clarify the essential features of a mathematical continuum—and thereby uncover the distinction between a continuous and a discrete set of points. (For the fascinating story of this problem see *Georg Cantor, His Mathematics and Philosophy of the Infinite* [119], pp. 107–110.) A century later, Mandelbrot saw in this strange "dust" of points a perfect model for the propagation of errors within a "noisy" electronic transmission line. Contrary to intuition, there is never a time interval, no matter how small, during which errors are clustered continuously. Transmission noise appears as a Cantor set in time.

But the appearance of simplicity is misleading. Contained within the Cantor set are not only the "visible" endpoints of deleted intervals, $1/3, 2/3, 1/9$, $2/9, 7/9, 8/9, \ldots$, but also all the limits of convergent sequences formed from these endpoints.[12] To see this we have only to view the Cantor set *arithmetically*

---

[12] It is a worthwhile exercise to show, for example, that the number $\frac{1}{4}$ is the limit of a sequence of deleted endpoints and therefore belongs to the Cantor set.

as the set of all real numbers of the form

$$x = \sum_{n=1}^{\infty} \frac{c_n}{3^n} \tag{4}$$

where the coefficients c_n take on only the two values 0 and 2, in other words, as the set of all ternary decimals that do not contain the digit 1. Indeed, at the first stage of the construction, the rejected interval is $(1/3, 2/3)$ and each of its elements has a unique ternary expansion in which the first digit is a 1; at the second stage of the construction, the rejected numbers all have a unique ternary expansion in which the second digit is a 1; and so on. Thus the arithmetic definition agrees with the geometric one.[13]

Paradoxically, while the geometric characterization shows that the Cantor set is very small, the arithmetically equivalent one shows that it is also very large, so large in fact that there is a function

$$f : C \to [0, 1]$$

which maps C *onto* the entire unit interval! Indeed, if x is given by (4), we need only define

$$f(x) = \frac{1}{2} \sum_{n=1}^{\infty} \frac{c_n}{2^n}$$

$$= 0. \left(\frac{c_1}{2}\right) \left(\frac{c_2}{2}\right) \left(\frac{c_3}{2}\right) \cdots \quad \text{(base 2).} \tag{5}$$

Since every real number between 0 and 1 can be expressed as a binary decimal, and since $c_n/2$ takes on either of the values 0 or 1, it follows that f maps C onto $[0, 1]$. Thus the Cantor set is at least as large as the unit interval—in the sense of cardinality—even though its total "length" is zero!

Figure 4 displays two other strange and wondrous shapes obtained by "constructing with holes"—the *Sierpiński carpet* and its three-dimensional analogue, the *Menger sponge*.

---

[13] For the endpoints of deleted intervals, $1/3, 2/3, 1/9, 2/9, 7/9, 8/9, \ldots$, there is an ambiguity: each such number has *two* representations, such as

$$\frac{1}{3} = 0.1000\ldots = 0.0222\ldots \quad \text{(base 3).}$$

Nevertheless, one of these representations is always without a 1.

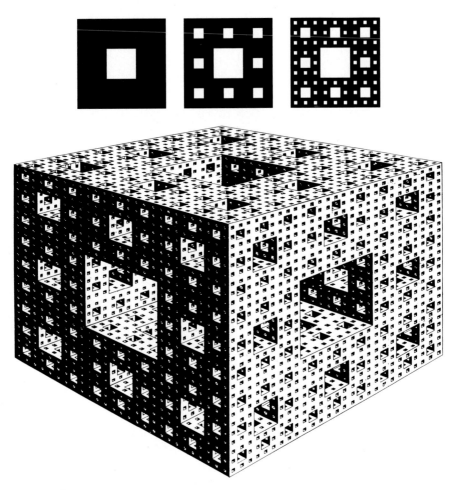

FIGURE 4

Menger sponge: From STUDIES IN GEOMETRY. By Leonard M. Blumenthal
and Karl Menger. Copyright ©1970 by W. H. Freeman and Company. Reprinted
with permission.

A few mathematicians in the early twentieth century conceived monstrous-
seeming objects made by the technique of adding or removing infinitely
many parts. One such shape is the Sierpiński carpet, constructed by cutting the center one-ninth of a square; then cutting out the centers of the
eight smaller squares that remain; and so on. The three-dimensional analogue is the Menger sponge, a solid-looking lattice that has an infinite
surface area, yet zero volume.

—James Gleick [172], p. 101.

The Koch Curve.   The mathematical shape known as the "snowflake" was first described in 1904 by the Swedish mathematician Helge von Koch (1870–1924). Its construction, which is analogous to that of the Cantor set, is based on what might be called *analogy by contrast.* While the Cantor set is formed by a process of repeated deletions, the snowflake is built up through repeated additions. The construction begins with an equilateral triangle as in Figure 5. First take the middle-third of each side and erect an equilateral triangle pointing outward. The resulting figure is a star hexagon or star of David. Next, take each of the twelve sides of the star and repeat the process, erecting a smaller equilateral triangle on each middle-third. The new figure will have 48 sides. Continuing in this fashion, we obtain in the limit a set that resembles an ideal snowflake. The outlines of successive approximations become more and more detailed, just as the successive approximations to the Cantor set become more and more sparse.

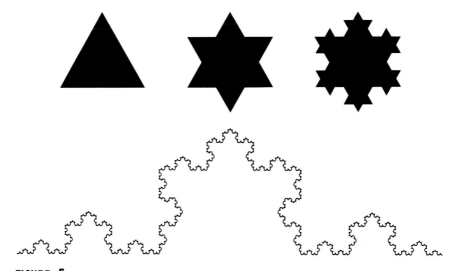

FIGURE 5
The Koch curve: a rough but vigorous model of a coastline.
Courtesy of Benoit B. Mandelbrot, IBM Thomas J. Watson Research Center.

It is not difficult to show that at each stage of the process the new triangles erected on each side never intersect, and that the boundary of the limit—the so-called *Koch curve* or *snowflake curve*—is continuous and yet so "jagged" that it fails to have a tangent at even a single point. The pathological nature of this strange curve is further revealed by the observation that, *while it bounds a finite area, its total length is infinite!*

To prove the last assertion, let P_n denote the perimeter of the polygonal approximation formed at the nth stage. It is at once clear from the construction that

$$P_{n+1} = \frac{4}{3}P_n \qquad n = 1, 2, 3, \ldots$$

since, in passing from the nth stage to the $(n+1)$th stage, we have multiplied the number of sides by 4 while dividing the length of each of them by 3. Therefore

$$P_n = \left(\frac{4}{3}\right)^n P_0$$

where P_0 is the perimeter of the original equilateral triangle. Since the fraction $(4/3)^n$ increases without bound as n tends to infinity, it follows that the snowflake curve has infinite length.

To see that it bounds a finite area, let A_n denote the combined areas of all the equilateral triangles added at the nth stage. Then

$$A_{n+1} = \frac{4}{9}A_n \qquad n = 1, 2, 3, \ldots$$

and it follows that

$$A_1 + A_2 + A_3 + \cdots = A_1 \left(1 + \frac{4}{9} + \left(\frac{4}{9}\right)^2 + \cdots\right)$$

$$= \frac{9}{5}A_1.$$

If A_0 denotes the area of the original equilateral triangle, then $A_1 = \frac{1}{3}A_0$, and therefore

$$\text{Area of the snowflake} = \sum_{n=0}^{\infty} A_n = \frac{8}{5}A_0$$

which is finite.

Expressing his wonder over this remarkable curve, E. Cesàro wrote in 1905:

> This endless imbedding of this shape into itself gives us an idea of what Tennyson describes somewhere as the *inner* infinity, which is after all the only one we could conceive in Nature. Such similarity between the whole and its parts, even its infinitesimal parts, leads us to consider the triadic Koch curve as truly marvelous. Had it been given life, it would not be possible to do away with it without destroying

it altogether for it would rise again and again from the depths of its triangles, as life does in the Universe.[14]

The Cantor set and the Koch curve are only two of a number of curious shapes that began to appear with greater frequency toward the end of the 19th century. In 1872, Weierstrass exhibited a class of functions that are continuous everywhere but differentiable nowhere. In 1890, Peano constructed his remarkable "space-filling" curve, a continuous parametric curve that passes through every point of the unit square—thereby showing that a curve need not be 1-dimensional!

Most mathematicians of the period regarded these strange objects with distrust. They viewed them as artificial, unlikely to be of any value in either science or mathematics. "These new functions, violating laws deemed perfect, were looked upon as signs of anarchy and chaos which mocked the order and harmony previous generations had sought."[15] Poincaré called them a "gallery of monsters" and Hermite wrote of turning away "in fear and horror from this lamentable plague of functions which do not have derivatives."

At the same time, the pathology forced the mathematicians of the 20th century to reexamine and clarify the notions of continuity, dimension, and the structure of the real number system. The result was the development of a higher analysis—topology and the theory of functions of one or several variables—in which such pathology was finally seen to be the rule rather than the exception.

Fractal geometry combines these once-perceived oddities into a powerful tool for describing natural phenomena. In fact, it is precisely the chaotic appearance of fractals that renders them so valuable in modeling such seemingly random behavior as Brownian motion, the turbulent flow of fluids, the fracturing of metals, or the clustering of galaxies. The following pictures (see Color Plates 7 and 8) dramatically illustrate how effective fractals can be in generating realistic images on a computer. Such techniques play an important role in modern cinematography.

---

[14] Quoted from *The Fractal Geometry of Nature* [290], p. 43.

[15] M. Kline, *Mathematical Thought from Ancient to Modern Times* [260], p. 973.

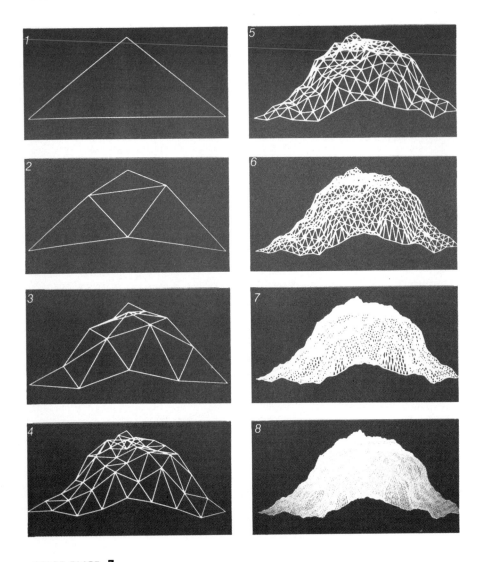

COLOR PLATE 7

Simple technique for generating a realistic image of a mountain is loosely based on the concepts of fractal geometry originally formulated by Benoit B. Mandelbrot of the IBM Thomas J. Watson Research Center. This demonstration of the technique is reproduced through the courtesy of Lucasfilm Ltd. Starting with the single triangle shown in step *1,* the computer program generates step *2* from it by the following procedure. First, break each side of the triangle at its midpoint. Second, displace each midpoint by a distance proportional to the length of the corresponding side. (The factor of proportionality can be generated at random or taken from a table of, say, 100 well-dispersed random numbers.) Third, connect the three new points to one another to form four new triangles. Step *3* is generated from step *2* by applying the same procedure in turn to each of the four new triangles, generating 16 triangles, to each of which the procedure is applied again in step *4,* and so on. Although the subdivision algorithm is simple, it can yield a very complex polygonal surface. The mountainlike surface in step *8* can later be rendered by standard computer-graphics techniques to produce a finished landscape. Figure and caption previously appeared in A. van Dam [423], p. 156.

COLOR PLATE 8
Fractal Landscape: Gaussian Hills That Never Were

Courtesy of Benoit B. Mandelbrot and Richard F. Voss, IBM Thomas J. Watson Research Center.

The name of Carl Friedrich Gauss (1777–1855) appears in nearly every chapter of mathematics and of physics, making him the first (*princeps*) among the mathematicians (including the physicists) of his time. But these imaginary hills being called *Gaussian* is motivated by a probability distribution for which Gauss receives undeserved credit. It is the distribution whose graph is the famous "bell-shaped curve" or "Galton ogive." On [Plate 8], this distribution rules the difference in altitude between any two prescribed points on the map, at least after a suitable transformation.

—Benoit B. Mandelbrot [290], C10

PROBLEMS

1.   The Cissoid of Diocles.   Diocles (circa 180 B.C.) invented the **cissoid** in order to solve the problem of the duplication of the cube. The equation of the cissoid in Cartesian coordinates is

$$y^2 = \frac{x^3}{a - x}$$

where a is a fixed positive number (see Figure 6).

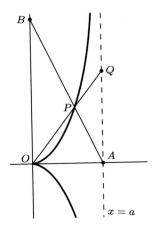

FIGURE 6

Let A be the point where the asymptote of the cissoid, $x = a$, intersects the x-axis and choose B on the positive y-axis so that $OB = 2OA$. Let the line connecting A and B intersect the cissoid at P. Finally, extend OP so that it intersects the asymptote at Q. Show that

$$(AQ)^3 = 2(OA)^3,$$

thereby solving the duplication problem.

2.   Show how the spiral of Archimedes can be used to trisect any given angle. In polar coordinates the equation of the spiral is $r = a\theta$. If the angle is placed so that its initial side lies on the positive x-axis and its vertex at the origin, then its terminal side will intersect the spiral at a point P (see Figure 7). Trisect segment OP at points R and S and draw circles with O as center and OR and

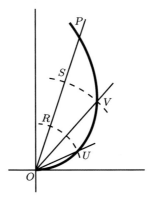

FIGURE 7

OS as radii. Show that if these circles intersect the spiral at points U and V, then the lines OU and OV will trisect the angle.

3.   Show how the spiral of Archimedes can be used to effect the quadrature of the circle.

   Hint. Draw the circle with center at O and radius equal to a, as in Figure 8. If the measure of angle AOB is θ, then OP and the circular arc bounded by the lines OA and OB are both equal to $a\theta$. Now take OB perpendicular to OA.

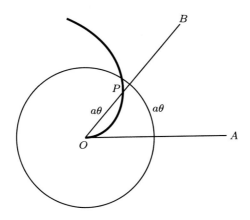

FIGURE 8

4. Calculate the area under the "higher parabola" $y = x^n$ (n a positive integer) between $x = 0$ and $x = a$, arithmetically by means of the following ingenious argument due to Fermat.

a.   Begin by choosing a positive number $r < 1$ and divide the interval $[0, a]$ into an infinite number of unequal subintervals by means of the partition points $a, ar, ar^2, ar^3, \ldots$ (see Figure 9).

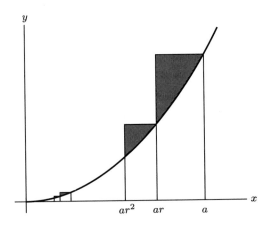

FIGURE 9

b.   Show that the sum of the areas of the circumscribed rectangles is given by the formula

$$A(r) = \frac{a^{n+1}}{1 + r + r^2 + \cdots + r^n}.$$

c.   Conclude that the area under the curve is equal to

$$A = \lim_{r \to 1} A(r) = \frac{a^{n+1}}{n + 1}.$$

(Can the argument be modified when n is not a positive integer?)

5. Derive **Newton's series**

$$1 + \frac{1}{3} - \frac{1}{5} - \frac{1}{7} + \frac{1}{9} + \frac{1}{11} - \cdots = \frac{\pi}{2\sqrt{2}}.$$

Hint. Evaluate the integral $\int_0^1 \frac{1+x^2}{1+x^4}\, dx$ in two different ways.

6. Find the sum of the series

$$1 - \frac{1}{4} + \frac{1}{7} - \frac{1}{10} + \cdots .$$

7. a. Show that a convergent infinite series of the form

$$\gamma_1 + \gamma_1\gamma_2 + \gamma_1\gamma_2\gamma_3 + \cdots$$

is equivalent to the continued fraction

$$\cfrac{\gamma_1}{1 - \cfrac{\gamma_2}{1 + \gamma_2 - \cfrac{\gamma_3}{1 + \gamma_3 - \cdots}}}$$

b. By taking in particular Gregory's series for the arctangent,

$$\tan^{-1} x = x - \frac{x^3}{3} + \frac{x^5}{5} - \frac{x^7}{7} + \cdots ,$$

derive Lord Brouncker's famous expansion for π:

$$\frac{4}{\pi} = 1 + \cfrac{1^2}{2 + \cfrac{3^2}{2 + \cfrac{5^2}{2 + \cdots}}}$$

c. Compare the partial sums of the Leibniz series $1 - \frac{1}{3} + \frac{1}{5} - \frac{1}{7} + \cdots$ with the convergents of the continued fraction in (b).

8. Formulate a conjecture about the value of the continued fraction

$$1 + \cfrac{1}{1 + \cfrac{1}{\frac{1}{2} + \cfrac{1}{\frac{1}{3} + \cfrac{1}{\frac{1}{4} + \cdots}}}}$$

(The proof was discovered by Euler, but it was J. J. Sylvester who first identified the successive convergents.)

9. Observe that

$$\sum_{n=1}^{\infty} \frac{1}{2^n} = 1!, \qquad \sum_{n=1}^{\infty} \frac{n}{2^n} = 2!, \qquad \sum_{n=1}^{\infty} \frac{n^2}{2^n} = 3!.$$

Does the pattern continue?

10.   Establish the relation

$$\frac{\pi}{4} = \tan^{-1}\frac{1}{2} + \tan^{-1}\frac{1}{3}$$

geometrically, by showing that, in the figure consisting of three adjacent squares pictured below, $\gamma = \alpha + \beta$.

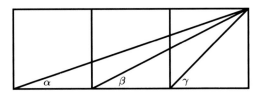

FIGURE 10

11.   **The Cantor Function.**   The function f defined on the Cantor set by formula (5) can be extended in a natural way to all of $[0, 1]$. The graph of the extension has many remarkable properties. Vilenkin called it the *Devil's Staircase*.

   a.   First show that, if a and b are endpoints of one of the open intervals deleted from $[0, 1]$ in the construction of the Cantor set, then $f(a) = f(b)$. For example,

$$f\left(\frac{1}{3}\right) = f(.0222\ldots) = .0111\ldots = \frac{1}{2}$$
$$f\left(\frac{2}{3}\right) = f(.2000\ldots) = .1000\ldots = \frac{1}{2}.$$

(Observe that the numbers $1/3$ and $2/3$ are expressed in the ternary system while their images are in binary.)

   b.   If x belongs to one of the open intervals (a, b) described above, then $f(a) = f(b)$, and we define $f(x)$ to be the common value at the endpoints. The extended function is therefore constant on each interval removed in the construction of the Cantor set. It is called the **Cantor function** and its graph is pictured below (Figure 11). Show that the Cantor function is a continuous nondecreasing function on $[0, 1]$, and that its derivative vanishes everywhere except on the Cantor set.

c. What is the arc length of the graph of the Cantor function? Is the standard integral formula

$$L = \int_0^1 \sqrt{1 + [f'(x)]^2}\, dx$$

applicable?

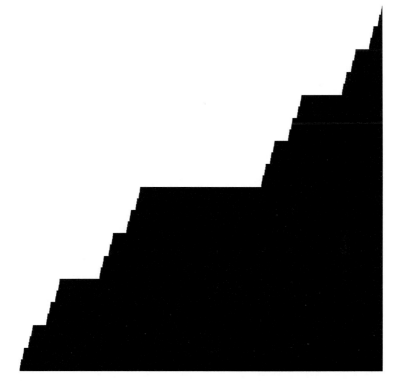

FIGURE 11
The Devil's Staircase

12. Let $\{a_1, a_2, a_3, \ldots\}$ be a sequence of positive numbers that tends monotonically to zero. A well-known result due to Leibniz asserts that the alternating series

$$a_1 - a_2 + a_3 - \cdots \tag{6}$$

is convergent and that the error R_n made in approximating the sum of the series by its nth partial sum is no greater in magnitude than the absolute value of the first term omitted. Show that the bound on the error can be improved considerably, provided that the sequence of differences

$$\{a_1 - a_2, a_2 - a_3, a_3 - a_4, \ldots\}$$

also tends monotonically to zero. Specifically, show that under these conditions

$$\frac{a_{n+1}}{2} \leq |R_n| \leq \frac{a_n}{2}.$$

(The hypotheses are satisfied by a great many important series, among them the alternating harmonic series and the Leibniz series for π.)

Hint. Let S_n be the nth partial sum of the series (6) and define

$$T_n = S_n + (-1)^n \frac{a_{n+1}}{2}.$$

Show that the sequence of closed intervals $[T_2, T_1], [T_4, T_3], [T_6, T_5], \ldots$ is nested.

13. a. The Sierpiński Triangle. The construction of the **Sierpiński triangle**, also known as the **Sierpiński gasket**, begins with an equilateral triangle as pictured below (Figure 12). First we remove the open center triangle. That leaves three closed "corner triangles," and we remove the open center triangle from each. That leaves nine triangles, and again the open center triangle of each of them is to be removed. When this recursive procedure is carried out indefinitely—at each stage we remove the open center triangle from every triangle that remains from the previous stage—what remains after all the open triangles have been removed is called the Sierpiński triangle. What is the area of the Sierpiński triangle?

FIGURE 12

b. Pascal's Triangle Modulo 2. When the entries of Pascal's triangle are reduced modulo 2—so that odd entries are replaced with a 1 and even entries with a 0—the result is a striking fractal-like pattern that resembles the

Sierpiński triangle (see Figure 13). The pattern displays in graphic form a remarkable property of the binomial coefficients that was discovered a century ago by E. Lucas: *The binomial coefficient $\binom{n}{k}$ is odd if and only if for every digit in the binary representation of the number k that is a 1 the corresponding digit in the binary representation of n (reading from right to left) is also a 1.* For example, the binomial coefficient $\binom{14}{6}$ is odd since every 1 in the binary representation of 6 (110) is matched with a 1 in the same position in the binary representation of 14 (1110). (This property of binomial coefficients was used by G. J. Chaitin in his construction of an unsolvable Diophantine equation [91].) Which rows of Pascal's triangle contain only odd entries?

FIGURE 13

    c.   How many odd entries are there in the first 2^n rows of Pascal's triangle. Use your answer to conclude that *almost all binomial coefficients are even.*

14.   The method of computing the area of a planar region by means of approximating rectangles, followed by a passage to the limit, is usually a formidable task. Critique the following method which is based on the observation that a point and not a rectangle is the more elementary unit. Since every geometric figure is made up of points, and since the area of a single point is zero, it follows that the area of every geometric figure is equal to the sum of the series $0 + 0 + 0 + \cdots$, or zero!

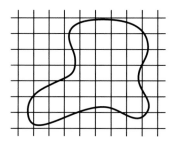 

FIGURE 14

15.   Show that the snowflake curve lies in the smallest circle that contains the original equilateral triangle used in its construction.

16.   a.   Starting with the closed unit square $[0, 1] \times [0, 1]$, remove everything but the four $\frac{1}{3} \times \frac{1}{3}$ "corner squares": $[0, \frac{1}{3}] \times [0, \frac{1}{3}]$, $[0, \frac{1}{3}] \times [\frac{2}{3}, 1]$, $[\frac{2}{3}, 1] \times [0, \frac{1}{3}]$, and $[\frac{2}{3}, 1] \times [\frac{2}{3}, 1]$, as pictured in Figure 15. Then, from each of these, remove everything but the four $\frac{1}{9} \times \frac{1}{9}$ squares located in their respective corners; and so on, ad infinitum. Identify the remaining set of points.
    b.   If S is any nonempty set, its **difference set** $D(S)$ is the set of all differences between elements of S:

$$D(S) = \{x - y : x, y \in S\}.$$

For example, if $S = [0, 1]$, then $D(S) = [-1, 1]$. Show that the difference set of the Cantor set is also $[-1, 1]$.

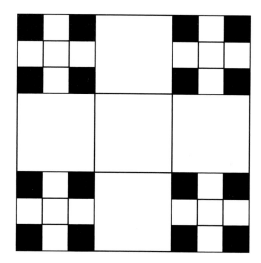

FIGURE 15

17. Show that

$$\frac{1}{1\cdot 2\cdot 3} + \frac{1}{3\cdot 4\cdot 5} + \frac{1}{5\cdot 6\cdot 7} + \cdots = \log 2 - \frac{1}{2}.$$

Hint. First show that

$$\frac{1}{2}\int_0^1 t^{n-1}(1-t)^2\, dt = \frac{1}{n(n+1)(n+2)}.$$

18. Show that

$$1 + \frac{1}{3} + \frac{2!}{3\cdot 5} + \frac{3!}{3\cdot 5\cdot 7} + \cdots = \frac{\pi}{2}.$$

Hint. First show that

$$\int_0^{\pi/2} \cos^{2n+1} x\, dx = \frac{2\cdot 4\cdot 6\cdots 2n}{3\cdot 5\cdot 7\cdots 2n+1}$$

$$= 2^n\frac{n!}{3\cdot 5\cdot 7\cdots 2n+1}.$$

19. Consider the series

$$
\begin{aligned}
1+ x+ &\ x^2 + &\ x^3 + \cdots \\
1+2x+ &\ 3x^2 + &\ 4x^3 + \cdots \\
1+3x+ &\ 6x^2 + &\ 10x^3 + \cdots \\
1+4x+ &\ 10x^2 + &\ 20x^3 + \cdots
\end{aligned}
$$

$$\cdots$$

each one corresponding to one of the columns of Pascal's triangle. Knowing the sum of the first series, find the sum of all the others.

Hint. The series

$$1 + 2x + 3x^2 + 4x^3 + \cdots,$$

for example, can be decomposed into the sum of infinitely many series

$$
\begin{aligned}
1+x +x^2 +x^3 + &\cdots \\
+x +x^2 +x^3 + &\cdots \\
+x^2 +x^3 + &\cdots \\
+x^3 + &\cdots .
\end{aligned}
$$

2. Summing the Reciprocals of the Squares

With the development and growth of the infinitesimal calculus, the allure of the figurate numbers took on an added dimension. Especially intriguing was the problem of finding the sum of the reciprocals of the perfect squares

$$\sum_{n=1}^{\infty} \frac{1}{n^2}.$$

The problem had been posed in 1650 in a book by Pietro Mengoli, a professor of mechanics in Bologna, but, for the next 80 years, it resisted the efforts of all analysts, including Wallis, Leibniz, and the Bernoulli brothers.

That the problem is tantalizing stems in part from the fact that the series bears a close, yet superficial, resemblance to the telescoping series

$$\sum_{n=1}^{\infty} \frac{1}{n(n+1)}$$

whose sum is readily seen to be

$$\sum_{n=1}^{\infty} \left(\frac{1}{n} - \frac{1}{n+1} \right) = 1.$$

The decomposition into partial fractions seems so obvious to us today, but it was not always so, and this simple problem would play an important role in Leibniz's development of the calculus.

a. Figurate Numbers and the Origins of the Calculus

> It is most useful that the true origins of memorable inventions be known, especially of those which were conceived not by accident but by an effort of meditation. The use of this is not merely that history may give everyone his due and others be spurred by the expectation of similar praise, but also that the art of discovery may be promoted and its method become known through brilliant examples. One of the noblest inventions of our time has been a new kind of mathematical analysis, known as the differential calculus; but, while its substance has been adequately explained, its source and original motivation have not been made public. It is almost forty years now that its author invented it . . . .
>
> —Leibniz[16]

It is difficult to chronicle the exact course of scientific discovery or to account for the inspiration that gives rise to great ideas. But in 1714, two years before his death, Leibniz set forth in his essay *Historia et Origo Calculi Differentialis* (*History and Origin of the Differential Calculus*), whose opening lines are quoted above, an account of the development of his own thinking.

In that work and elsewhere, Leibniz always traced his motivation for the calculus back to his early work with infinite series and with sequences of sums and differences.

Soon after his arrival in Paris in 1672—the year Leibniz began his serious study of mathematics—Christian Huygens (1629–1695), perhaps the most renowned scientist on the continent, posed to him that "fateful question" which would ultimately be of profound influence in his mathematical work, namely,

---

[16] *Historia et Origo Calculi Differentialis.* The English translation of this extract of the essay is given by André Weil (Review of J. E. Hofmann's *Leibniz in Paris 1672–1676, Bull. Amer. Math. Soc.* 81 (1975), pp. 676–688). A translation of the complete essay is available in the volume of J. M. Child [93], pp. 22–58.

to find the sum of the reciprocals of the triangular numbers

$$\frac{1}{1} + \frac{1}{3} + \frac{1}{6} + \frac{1}{10} + \cdots + \frac{1}{n(n+1)/2} + \cdots.$$

Leibniz was already familiar with sums and differences of numbers, and he recognized that each term of the series is equal to *twice the difference* between successive terms of the harmonic sequence $1, 1/2, 1/3, 1/4, \ldots$

$$\text{Harmonic sequence:} \quad 1 \quad \frac{1}{2} \quad \frac{1}{3} \quad \frac{1}{4} \quad \frac{1}{5} \cdots$$

$$\text{Differences:} \quad \frac{1}{2} \quad \frac{1}{6} \quad \frac{1}{12} \quad \frac{1}{20} \quad \cdots$$

But the sum of the first n differences is equal to the difference between the first and $(n + 1)$st terms,

$$\left(1 - \frac{1}{2}\right) + \left(\frac{1}{2} - \frac{1}{3}\right) + \left(\frac{1}{3} - \frac{1}{4}\right) + \cdots + \left(\frac{1}{n} - \frac{1}{n+1}\right) = 1 - \frac{1}{n+1}$$

which approaches 1 as $n \to \infty$. Thus the sum required by Huygens is equal to 2.

By generalizing this procedure, Leibniz was led to a remarkable arrangement of numbers which he called the *harmonic triangle* (see Figure 16). The first column of the triangle contains the harmonic progression $1, 1/2, 1/3, \ldots$; the second column contains the successive differences of the first column; the third column the successive differences of the second; and so on in this fashion.

In the harmonic triangle and in the arithmetical triangle there are striking relationships. For example, the arithmetical triangle is founded on the sequence of integers and every element (which is not in the first column) is the sum of the two terms directly above it and to the left; the harmonic triangle is founded on the sequence of reciprocals of the integers, and every element is the sum of the two terms directly below it and to the right. Moreover, in the arithmetical triangle each element (not in the first row or column) is the sum of all the terms in the previous column which lie above it, whereas in the harmonic triangle each element is the sum of all the terms in the next column which lie below it. The second assertion follows from the fact that *in any sequence tending to zero, the first term is equal to the sum of all the differences.* This is precisely the principle that was used earlier to sum the differences of the harmonic progression. Thus the sum of the elements in each column of the harmonic triangle is equal to the leading entry of the preceding column: The elements in the second

Harmonic Triangle

$$\frac{1}{1}$$

$$\frac{1}{2} \quad \frac{1}{2}$$

$$\frac{1}{3} \quad \frac{1}{6} \quad \frac{1}{3}$$

$$\frac{1}{4} \quad \frac{1}{12} \quad \frac{1}{12} \quad \frac{1}{4}$$

$$\frac{1}{5} \quad \frac{1}{20} \quad \frac{1}{30} \quad \frac{1}{20} \quad \frac{1}{5}$$

$$\frac{1}{6} \quad \frac{1}{30} \quad \frac{1}{60} \quad \frac{1}{60} \quad \frac{1}{30} \quad \frac{1}{6}$$

FIGURE 16

Arithmetical Triangle

1

1 1

1 2 1

1 3 3 1

1 4 6 4 1

1 5 10 10 5 1

1 6 15 20 15 6 1

FIGURE 17

column add up to 1; the elements in the third column add up to 1/2; those in the fourth column add up to 1/3; and so on.

Can the reader now identify the elements in each column? The numbers in the second column are, as we have already observed, one-half the reciprocals of the triangular numbers

$$\frac{n(n+1)}{1 \cdot 2}.$$

By subtraction, the numbers in the third column are found to be one-third the reciprocals of the tetrahedral numbers

$$\frac{n(n+1)(n+2)}{1 \cdot 2 \cdot 3}.$$

Similarly, those in the fourth column are one-fourth the reciprocals of the figurate numbers of dimension four

$$\frac{n(n+1)(n+2)(n+3)}{1 \cdot 2 \cdot 3 \cdot 4}.$$

And, by induction, the numbers in the nth column are the reciprocals of the corresponding numbers in the $(n+1)$th column of the arithmetical triangle divided by n. In this way, Leibniz succeeded in summing all the series of the reciprocal figurate numbers: If $F(n, k)$ denotes the nth figurate number of dimension k, then

$$\sum_{n=1}^{\infty} \frac{1}{F(n, k)} = \frac{k}{k-1}.$$

The arithmetical and harmonic triangles reinforced for Leibniz the inverse relation that exists between sums and differences and which would play a dominant role in the discovery and development of the calculus. Here, from one of his early manuscripts (circa 1680), is a statement of the fundamental principle of what we now call the calculus of finite differences. Notice that Leibniz's notation is the same as that used today for the differential calculus.

The fundamental principle of the calculus.

Differences and sums are the inverses of one another, that is to say, the sum of the differences of a series is a term of the series, and the difference of the sums of a series is a term of the series; and I enunciate the former thus, $\int dx = x$, and the latter thus, $d \int x = x$.

Thus, let the differences of a series, the series itself, and the sums of the series, be, let us say,

Diffs.		1	2	3	4	5		$\cdots$	dx
Series	0	1	3	6	10	15		$\cdots$	x
Sums		0	1	4	10	20	35	$\cdots$	$\int x$

Then the terms of the series are the sums of the differences, or $x = \int dx$; thus, $3 = 1 + 2$, $6 = 1 + 2 + 3$, etc.; on the other hand, the differences of the sums of the series are terms of the series, or $d \int x = x$; thus, 3 is the difference between 1 and 4, 6 between 4 and 10.

Also $da = 0$, if it is given that a is a constant quantity, since $a - a = 0$.[17]

But the path to the "higher calculus" led not through arithmetic but geometry. For Leibniz the crucial insight came in 1673, through a single figure that he had run across in Pascal's "Traité des sinus du quart de cercle" (Treatise on the sines of a quadrant of a circle).[18] From this one example "a light suddenly burst upon him" and he was led to his famous *characteristic triangle*. It was then that he realized what Pascal had not—that the determination of the tangent to a curve depended on the *differences* in the ordinates, while the quadrature depended on the *sum* of the ordinates, in the sense of Cavalieri (namely, the sum of infinitely thin rectangles making up the area). Just as the formation of the arithmetical and harmonic triangles is based on the inverse operations of summing and differencing, so too in geometry the problems of quadratures and tangents are inverses of each other.

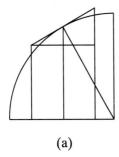
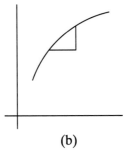

(a)   (b)

FIGURE 18

(a) Pascal's infinitesimal triangle; (b) Leibniz's characteristic triangle

Two decades later, in a letter to L'Hôpital, he summarized his decisive discovery as follows:

---

[17] Quoted from J. M. Child [93], p. 142.

[18] See *A Source Book in Mathematics, 1200–1800* [408], pp. 239–241, for an English translation.

[Using] what I call the "characteristic triangle" ... formed from the elements of the coordinates and the curve, I thus found as it were in the twinkling of an eyelid nearly all the theorems that I afterward found in the works of Barrow and Gregory. Up to that time, I was not sufficiently versed in the calculus of Descartes, and as yet did not make use of equations to express the nature of curved lines; but, on the advice of Huygens, I set to work at it, and I was far from sorry that I did so: for it gave me the means almost immediately of finding my differential calculus. This was as follows. I had for some time previously taken a pleasure in finding the sums of series of numbers, and for this I had made use of the well-known theorem, that, in a series decreasing to infinity, the first term is equal to the sum of all the differences. From this I had obtained what I call the "harmonic triangle," as opposed to the "arithmetical triangle" of Pascal; for M. Pascal had shown how one might obtain the sums of the figurate numbers, which arise when finding sums and sums of sums of the natural scale of arithmetical numbers. I on the other hand found that the fractions having figurate numbers for their denominators are the differences and the differences of the differences, etc., of the natural harmonic scale (that is, the fractions $1/1$, $1/2$, $1/3$, $1/4$, etc.), and that thus one could give the sums of the series of figurate fractions

$$1/1 + 1/3 + 1/6 + 1/10 + \text{etc.,} \qquad 1/1 + 1/4 + 1/10 + 1/20 + \text{etc.}$$

Recognizing from this the great utility of differences and seeing that by the calculus of M. Descartes the ordinates of the curve could be expressed numerically, I saw that to find quadratures or the sums of the ordinates was the same thing as to find an ordinate (that of the quadratrix), of which the difference is proportional to the given ordinate. I also recognized almost immediately that to find tangents is nothing else but to find differences, and that to find quadratures is nothing else but to find sums, provided that one supposes that the differences are incomparably small. I saw also that of necessity the differential magnitudes could be freed from the fraction and the root-symbol, and that thus tangents could be found without getting into difficulties over irrationals and fractions. And there you have the story of the origin of my method ....[19]

---

[19] Quoted from J. M. Child [93], pp. 220–222.

b. What Euler Saw

> And so is satisfied the burning desire of my brother [James] who, re-
> alizing that the investigation of the sum was more difficult than any-
> one would have thought, openly confessed that all his zeal had been
> mocked. If only my brother were alive now.
>
> —John Bernoulli, on learning of Euler's triumph[20]

The problem of summing the reciprocals of the squares occupied Leibniz all his life, but the solution always remained beyond his reach. The problem was already famous when Euler took up the matter around 1730, and little about the series was known. "If someone should succeed," wrote James Bernoulli, "in finding what till now withstood our efforts and communicate it to us, we shall be much obliged to him."

In characteristic fashion, Euler's first efforts were to obtain good approximations to the desired sum

$$\sum_{n=1}^{\infty} \frac{1}{n^2}$$

—no small accomplishment in view of the slow convergence of the series. In fact, it is a simple matter to show that the remainder after n terms lies between $1/n$ and $1/(n+1)$, and therefore no great accuracy can be obtained by adding up the terms of the series one by one.[21]

Euler's first contribution came in 1731 when he was able to show by elementary means that the given series could be transformed into a new series which converges extremely rapidly:

$$\sum_{n=1}^{\infty} \frac{1}{n^2} = (\log 2)^2 + \sum_{n=1}^{\infty} \frac{1}{2^{n-1}n^2}.$$

Thus, by approximating

$$\sum_{n=1}^{\infty} \frac{1}{2^{n-1}n^2} \quad \text{and} \quad \log 2 = -\log\left(1 - \frac{1}{2}\right) = \sum_{n=1}^{\infty} \frac{1}{n2^n},$$

he computed the sum $\sum 1/n^2$ numerically to 6 decimal places (1.644934).

---

[20] *Werke,* vol. 4, p. 22.

[21] Ralph Boas makes the point well: "You may well suggest that all we have to do is to wait for an improved computer. Well, several lines of argument suggest that the shortest useful unit of time is about 10^{-23}-second (the time for light to travel the width of an electron); no one has yet given names to units this small. A machine that added a term in 10^{-23}-second could not yet have added 10^{41} terms if it had started at the beginning of the universe." ([44], p. 153)

The following year, in a paper entitled *Methodus Generalis Summandi Progressiones,* he stated his famous method for approximating sums which is now known as the *Euler–Maclaurin summation formula* and which he used to compute $\sum 1/n^2$ to 20 decimal places,

$$1.64493406684822643647.$$

But this is only an approximation and his goal was to find the exact value. Two years later, in 1734, he did.

Theorem 2 (Euler).   *The sum of the reciprocals of the perfect squares is $\pi^2/6$.*

Euler's original argument was based on daring assumptions which he could not logically support and which would take him ten years to put on a sound basis. But the method is inspired and it opened up whole new areas of mathematical investigation whose importance would far transcend the immediate applications.

His proof begins with a bold extension of a simple algebraic fact: If a polynomial equation

$$P(x) = 0$$

of degree n has n distinct roots

$$r_1, r_2, r_3, \ldots, r_n$$

then $P(x)$ can be represented as a product of n linear factors

$$P(x) = A(x - r_1)(x - r_2) \cdots (x - r_n).$$

Equivalently, if none of the roots is equal to zero, then the decomposition into linear factors can also be written in the form

$$P(x) = B\left(1 - \frac{x}{r_1}\right)\left(1 - \frac{x}{r_2}\right) \cdots \left(1 - \frac{x}{r_n}\right).$$

Euler reasoned that a similar representation ought to be valid for the transcendental function $\frac{\sin x}{x}$, which by virtue of the known power series expansion of the sine, can be viewed as a "polynomial of infinite degree":

$$\frac{\sin x}{x} = 1 - \frac{x^2}{3!} + \frac{x^4}{5!} - \frac{x^6}{7!} + \cdots. \tag{1}$$

Since the roots of the equation

$$\frac{\sin x}{x} = 0$$

are $\pi, -\pi, 2\pi, -2\pi, \ldots$, Euler concluded by analogy that

$$
\begin{aligned}
\frac{\sin x}{x} &= \left(1 - \frac{x}{\pi}\right)\left(1 + \frac{x}{\pi}\right)\left(1 - \frac{x}{2\pi}\right)\left(1 + \frac{x}{2\pi}\right)\cdots \\
&= \left(1 - \frac{x^2}{\pi^2}\right)\left(1 - \frac{x^2}{4\pi^2}\right)\left(1 - \frac{x^2}{9\pi^2}\right)\cdots .
\end{aligned}
\tag{2}
$$

This infinite product, which converges for all values of x, is one of the most beautiful formulas in mathematics. If we formally expand the right-hand side in a power series and then compare corresponding terms in (1), we obtain in the simplest case, by comparing second-degree terms,

$$
\frac{1}{\pi^2} + \frac{1}{4\pi^2} + \frac{1}{9\pi^2} + \cdots = \frac{1}{3!}
$$

or

$$
1 + \frac{1}{4} + \frac{1}{9} + \cdots = \frac{\pi^2}{6}.
$$

This remarkable discovery at once established Euler as a mathematician of the first rank.

His method, of course, was open to serious objections, as he himself realized. Even if one admits a priori that $\frac{\sin x}{x}$ can be represented as a product of linear factors corresponding to the nonzero roots of the equation $\sin x = 0$, and even if it is permissible to calculate formally with that infinite product, there is still the question of whether all the roots of $\sin x = 0$ are *real*. If there were, for example, in addition to the "visible" real roots other imaginary roots, then the entire calculation would be invalidated. Euler recognized the difficulties and over the next ten years never ceased in his efforts to put the method on a sound basis.

Yet even without the final rigor there were still good reasons for Euler to trust his discovery. To begin with, the numerical evidence was compelling—his approximation for $\sum 1/n^2$ agreed with $\frac{\pi^2}{6}$ to the last decimal place. Furthermore, by comparing other coefficients in the product expansion of $\frac{\sin x}{x}$, he discovered the sum of other remarkable series, among them the sum of the reciprocals of the fourth powers,

$$
\sum_{n=1}^{\infty} \frac{1}{n^4} = \frac{\pi^4}{90}.
$$

Again the numerical evidence supported the theoretical results.

And, finally, the method leads to other known results. If, for example, in formula (2) we take $x = \pi/2$, then the resulting equation is

$$\frac{2}{\pi} = \left(1 - \frac{1}{4}\right)\left(1 - \frac{1}{16}\right)\left(1 - \frac{1}{36}\right)\left(1 - \frac{1}{64}\right)\cdots$$

$$= \left(\frac{1\,3}{2\,2}\right)\left(\frac{3\,5}{4\,4}\right)\left(\frac{5\,7}{6\,6}\right)\left(\frac{7\,9}{8\,8}\right)\cdots$$

which is Wallis's celebrated formula for π.

Commenting on the method, Pólya writes:

> Euler's decisive step was daring. In strict logic, it was an outright fallacy: he applied a rule to a case for which the rule was not made, a rule about algebraic equations to an equation which is not algebraic. In strict logic, Euler's step was not justified. Yet it was justified by analogy, by the analogy of the most successful achievements of a rising science that he called himself a few years later the "Analysis of the Infinite." Other mathematicians, before Euler, passed from finite differences to infinitely small differences, from sums with a finite number of terms to sums with an infinity of terms, from finite products to infinite products. And so Euler passed from equations of finite degree (algebraic equations) to equations of infinite degree, applying the rules made for the finite to the infinite.[22]

The method was bold and perhaps even reckless but mathematics was certainly the richer for it.[23]

---

[22] G. Pólya, *Induction and Analogy in Mathematics* [338], p. 21.

[23] A great many proofs of Euler's formula are now known, some of them quite elementary. Perhaps the simplest proof, based only on elementary algebra and trigonometry, was discovered by A. M. Yaglom and I. M. Yaglom [457], volume II, pp. 131–133 (and independently by Papadimitriou [329]). Other proofs, of varying degrees of difficulty, are given in [12], [36], [71], [262], [296], and [387]. For a survey of elementary methods, see T. M. Apostol, *American Mathematical Monthly,* 80 (1973) 425–431. A lengthy bibliography is given by E. L. Stark in *Mathematics Magazine,* 47 (1974) 197–202.

PROBLEMS

1. Summing by Differences. Establish the following relations by representing each series in the form of a telescoping sum:

(a)
$$\sum_{n=1}^{\infty} \tan^{-1} \frac{1}{2n^2} = \frac{\pi}{4}.$$

(b)
$$\sum_{n=1}^{\infty} \tan^{-1} \frac{2}{n^2} = \frac{3\pi}{4}.$$

Hint. In each case, use the identity

$$\tan^{-1} p - \tan^{-1} q = \tan^{-1} \frac{p-q}{1+pq}.$$

2. Show that the series

$$\frac{1}{2\log 2} + \frac{1}{3\log 3} + \cdots + \frac{1}{n\log n} + \cdots$$

is divergent by proving that

$$\log\log(n+1) < \log\log n + \frac{1}{n\log n}.$$

3. Show that the reciprocal of every integer greater than 1 is the sum of a finite number of consecutive terms of the infinite series

$$\sum_{n=1}^{\infty} \frac{1}{n(n+1)}.$$

4. Prove that, for every natural number m,

$$\sum_{n=1}^{\infty} \frac{1}{n(n+m)} = \frac{1 + \frac{1}{2} + \frac{1}{3} + \cdots + \frac{1}{m}}{m}.$$

5. Prove that, for every natural number m,

$$\sum_{n=1}^{\infty} \frac{1}{n(n+1)(n+2)\cdots(n+m)} = \frac{1}{m\cdot m!}.$$

Hint. Find an explicit formula for the partial sums.

6. Show that if we approximate the sum of the series $\sum 1/n^2$ by its nth partial sum, we commit an error that lies between $1/n$ and $1/(n+1)$.

7. The function

$$f(x) = \sum_{n=1}^{\infty} \frac{x^n}{n^2},$$

which is defined and continuous for $-1 \le x \le 1$, is called the **dilogarithm**. Euler (*Opera Omnia*, ser. 1, vol. 14, pp. 40–41) used the formula

$$\sum_{n=1}^{\infty} \frac{1}{n^2} = \log x \cdot \log(1-x) + f(x) + f(1-x),$$

valid for $0 < x < 1$, to approximate the sum of the series on the left-hand side. Prove the formula.

Hint. Show by taking derivatives that the right-hand side is a constant. Evaluate the constant by letting $x \to 0$.

8. It is often possible to find the exact value of an improper integral by identifying it with an infinite series having a known sum. Show that the following integrals can be evaluated in this way:

(a)
$$\int_0^1 \frac{\log(1+x)}{x} \, dx$$

(b)
$$\int_0^{\infty} \frac{x}{e^x - 1} \, dx.$$

Hint. For the first integral, express the integrand as a power series and integrate term-by-term. For the second, express the integrand as a geometric series and do the same.

9. Try to find a third derivation for the sum of the reciprocals of the squares, knowing that

$$\arcsin x = x + \frac{1}{2} \cdot \frac{x^3}{3} + \frac{1 \cdot 3}{2 \cdot 4} \cdot \frac{x^5}{5} + \frac{1 \cdot 3 \cdot 5}{2 \cdot 4 \cdot 6} \cdot \frac{x^7}{7} + \cdots.$$

Hint. Put $x = \sin t$ and integrate term-by-term between $t = 0$ and $t = \frac{\pi}{2}$.

10. Try to find a fourth derivation for the sum of the reciprocals of the squares, knowing that

$$(\arcsin x)^2 = x^2 + \frac{2}{3} \cdot \frac{x^4}{2} + \frac{2 \cdot 4}{3 \cdot 5} \cdot \frac{x^6}{3} + \frac{2 \cdot 4 \cdot 6}{3 \cdot 5 \cdot 7} \cdot \frac{x^8}{4} + \cdots.$$

Hint. Put $x = \sin t$ and integrate term-by-term between $t = 0$ and $t = \frac{\pi}{2}$.

11. Show that

$$\sum_{n=1}^{\infty} \frac{1}{n^2 \binom{2n}{n}} = \frac{\pi^2}{18}.$$

12. Show that

$$\left(\sum_{-\infty}^{\infty} \frac{(-1)^n}{2n + 1} \right)^2 = \sum_{-\infty}^{\infty} \frac{1}{(2n + 1)^2}.$$

The square of a sum is equal to a sum of squares!

13. Show that all squares beyond the first can be packed into the first without overlapping.

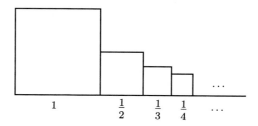

$$1 \qquad \frac{1}{2} \qquad \frac{1}{3} \quad \frac{1}{4}$$

FIGURE 19

14. By comparing further coefficients in Euler's expansion of $\sin x$ as an infinite product, show that

$$\sum_{n=1}^{\infty} \frac{1}{n^4} = \frac{\pi^4}{90}.$$

15. a. Euler's method leads to the conjecture

$$1 - \sin x = \left(1 - \frac{2x}{\pi} \right)^2 \left(1 + \frac{2x}{3\pi} \right)^2 \left(1 - \frac{2x}{5\pi} \right)^2 \left(1 + \frac{2x}{7\pi} \right)^2 \cdots.$$

Show that this is not only analogous to, but a consequence of, the product representation of $\sin x$. Why does each linear factor appear twice? (For numerous other consequences of Euler's product formula for $\sin x$, see [338], pp. 30–34.)

b.   Use part (a) to give another derivation of the Leibniz series:

$$\frac{\pi}{4} = 1 - \frac{1}{3} + \frac{1}{5} - \frac{1}{7} + \cdots .$$

16.   Using the infinite product expansion for $\sin x$, derive the partial fractions decomposition

$$\cot x = \cdots + \frac{1}{x + 2\pi} + \frac{1}{x + \pi} + \frac{1}{x} + \frac{1}{x - \pi} + \frac{1}{x - 2\pi} + \cdots .$$

17.   Using the result of problem 16, show that

$$\cot x = \frac{1}{x} - \frac{2x}{\pi^2} \left(\sum_{1}^{\infty} \frac{1}{n^2} \right)$$

$$- \frac{2x^3}{\pi^4} \left(\sum_{1}^{\infty} \frac{1}{n^4} \right) \tag{3}$$

$$- \frac{2x^5}{\pi^6} \left(\sum_{1}^{\infty} \frac{1}{n^6} \right)$$

$$- \cdots .$$

18.   Summing the Reciprocals of the Even Powers.   It was shown in chapter 3 (§3, problem 21) that the power series expansion for $x \cot x$ is

$$x \cot x = \sum_{n=0}^{\infty} (-4)^n \frac{B_{2n}}{(2n)!} x^{2n} \tag{4}$$

where B_n is the nth Bernoulli number. Discover from (3) and (4), as Euler did, the remarkable result: *For each even integer n,*

$$\zeta(n) = 1 + \frac{1}{2^n} + \frac{1}{3^n} + \frac{1}{4^n} + \cdots = \frac{2^{n-1} \pi^n |B_n|}{n!} . \tag{5}$$

In particular,[24]

---

[24] The function

$$\zeta(s) = \sum_{n=1}^{\infty} \frac{1}{n^s} = 1 + \frac{1}{2^s} + \frac{1}{3^s} + \cdots$$

has come to be known as the **Riemann zeta function**, although Euler began investigating its properties around 1730, more than 100 years before Riemann. (For a fascinating account of Euler's work, see R. Ayoub, "Euler and the zeta function" [19].) The zeta function plays a fundamental

$$\zeta(2) = \frac{\pi^2}{6}, \quad \zeta(4) = \frac{\pi^4}{90}, \quad \zeta(6) = \frac{\pi^6}{945}, \quad \zeta(8) = \frac{\pi^8}{9450}.$$

19. The derivation of formula (5) shows that the Bernoulli numbers B_{2n} are of alternating sign and that $(-1)^{n-1}B_{2n}$ is positive. Show, furthermore, that $|B_{2n}|$ increases with extreme rapidity, so fast in fact that

$$\left| \frac{B_{2n+2}}{B_{2n}} \right| \to \infty.$$

Hint. The value of $\sum_{k=1}^{\infty} 1/k^{2n}$ lies between 1 and 2 for every n.

20. Prove the following well-known result of number theory:

Theorem (E. Cesàro, 1881). *If u and v are positive integers chosen at random, the probability that $\gcd(u, v) = 1$ is $6/\pi^2$.*

A precise formulation of the theorem, which carefully defines what is meant here by "chosen at random," can be found in *The Art of Computer Programming* [263], volume 2 (p. 314, exercise 10). Here, the reader is asked to explore two heuristic arguments each of which renders the result plausible.

First Proof. Begin by assuming that there exists a well-defined probability P that $\gcd(u, v) = 1$. Show that, for any positive integer n, the probability that $\gcd(u, v) = n$ is equal to P/n^2. Conclude that the sum of these probabilities

---

role in the study of the distribution of the prime numbers. The number-theoretic properties of $\zeta(s)$ derive from Euler's remarkable identity

$$\zeta(s) = \prod_{p} \frac{1}{1 - \frac{1}{p^s}}$$

in which the primes $p = 2, 3, 5, 7, \ldots$ appear explicitly on the right-hand side of the formula but not on the left. (For a formal proof, expand each factor on the right in the form of a geometric series and then multiply out. For a rigorous proof, see [234], p. 76.) Thus, for example, when $s = 2$, we find

$$\prod_{p} \left(1 - \frac{1}{p^2}\right) = \frac{6}{\pi^2}$$

since $\zeta(2) = \pi^2/6$. After 250 years little about the arithmetical properties of $\zeta(s)$ is known. No one has found the exact value of $\zeta(2n + 1)$, and it is only recently that $\zeta(3)$ was shown to be irrational (see [425]).

over all possible values of n equals unity, so that

$$\frac{1}{\mathbf{P}} = \sum_{n=1}^{\infty} \frac{1}{n^2}.$$

Second Proof. If two integers are chosen at random, the probability that at least one of them is not divisible by the prime number p is equal to $1 - 1/p^2$. Therefore, the required probability $\mathbf{P}$ is given by the formula

$$\mathbf{P} = \prod_{p} \left(1 - \frac{1}{p^2}\right)$$

where the product is taken over *all* prime numbers p. Now apply Euler's formula for the zeta function (see footnote 24).

21. Euler's Summation Formula

> Euler calculated without any apparent effort, just as men breathe, as eagles sustain themselves in the air.
>
> —Arago

In 1732, Euler discovered an ingenious method for calculating the value of a finite sum

$$f(0) + f(1) + \cdots + f(n-1)$$

based on a novel application of the calculus. After two hundred and sixty years, it remains one of the most important tools of the numerical analyst. Discover, as Euler did, the famous *summation formula* by following these steps.

a.   Under the assumption that $f(x)$ is an analytic function, the goal is to find another analytic function $F(x)$ such that

$$f(x) = F(x+1) - F(x).$$

Using Taylor's theorem, show that

$$f(x) = F'(x) + \frac{F''(x)}{2!} + \frac{F'''(x)}{3!} + \cdots. \tag{6}$$

b.   Equation (6) bears a close formal resemblance to the equation $y = x + a_2 x^2 + a_3 x^3 + \cdots$, except in place of x and its higher powers we have $F'(x)$ and its higher derivatives. Now, it is a simple matter to "invert" the power series and, at least formally, solve for x in terms of y. The solution takes the form of another power series $x = y + b_2 y^2 + b_3 y^3 + \cdots$, where the b_n can be found recursively by the method of undetermined coefficients (see, for example, [341], volume I, p. 95). Euler reasoned that, in equation (6), $F'(x)$ could be expressed similarly in terms of $f(x)$ and its successive derivatives. Thus:

$$F'(x) = f(x) + b_1 f'(x) + b_2 f''(x) + \cdots. \tag{7}$$

Assuming that such a representation is valid, show that the coefficients b_n are given by the formula

$$b_n = \frac{B_n}{n!} \qquad \text{for } n = 1, 2, 3, \ldots,$$

where B_n is the nth Bernoulli number.

c. Show, by virtue of the fundamental theorem of calculus, that the required sum $f(0) + \cdots + f(n-1)$ can now be found by integration:

$$f(0) + f(1) + \cdots + f(n-1) = F(n) - F(0)$$
$$= \int_0^n F'(x)\, dx.$$

This is one version of Euler's summation formula. For a rigorous proof, see [42]. A symbolic proof, based on the formal properties of the difference operator Δ ($\Delta F(x) = F(x+1) - F(x)$), which is the discrete analogue of the differential operator $\frac{d}{dx}$, can be found in [71], pp. 304–305.

22. Divergent Series

> The divergent series are the invention of the devil, and it is a shame to base on them any demonstration whatsoever.
>
> —Abel

> The series is divergent; therefore we may be able to do something with it.
>
> —Heaviside

The Bernoulli numbers B_{2n} grow so rapidly in magnitude that, for most functions that occur in applications, the series expansion (7) for $F'(x)$ is divergent. Nevertheless, for such functions, Euler's formula can often be used to great advantage. As an illustration, let us try to approximate

$$\zeta(3) = 1 + \frac{1}{2^3} + \cdots + \frac{1}{(n-1)^3} + \sum_{k=n}^{\infty} \frac{1}{k^3} \tag{8}$$

by adding up the first $n - 1$ terms of the series and then using the summation formula to approximate the remainder.

a.   Begin by showing (at least formally) that for any analytic function $f(x)$,

$$\sum_{k=n}^{\infty} f(k) = \int_n^{\infty} F'(x)\, dx \qquad n = 1, 2, 3, \ldots.$$

b.   Next, by taking $f(x) = 1/x^3$ and integrating the series for $F'(x)$ term-by-term, show that the remainder in (8) is given by the formula

$$R_n = \frac{1}{2n^2} + \frac{1}{2n^3} + \frac{3B_2}{2n^4} + \frac{5B_4}{2n^6} + \frac{7B_6}{2n^8} + \cdots.$$

The series on the right is divergent for every value of n, yet it can be used to give excellent numerical approximations. The reason lies in the fact that *the difference between R_n and any partial sum is less in magnitude than the magnitude of the first term omitted* ([262], pp. 531ff.). Thus the series behaves like a convergent alternating series. At first the terms decrease in magnitude very rapidly, even for relatively small values of n. Only later do they begin to increase. Thus Euler could correctly prescribe that the series be summed "until it begins to diverge."

c.   By taking $n = 10$ and only 6 Bernoulli numbers, $B_2, B_4, B_6, B_8, B_{10}, B_{12}$, calculate $\zeta(3)$ with an error not exceeding 10^{-15}. Show also that to achieve this much accuracy by adding the terms of the series one by one would require the addition of more than twenty million terms!

The logical foundations of the theory of divergent series—and the full understanding of those that are useful in the representation and evaluation of functions—were finally secured in the late nineteenth century, independently by Poincaré and Stieltjes. For an excellent account of the history of this rich and fruitful branch of analysis, see *Mathematical Thought from Ancient to Modern Times* [260], chapter 47.

3. The Pentagonal Number Theorem

> It appears to me that if one wants to make progress in mathematics,
> one should study the masters and not the pupils.
>
> —Niels Henrik Abel
>
> Read Euler, read Euler, he is the master of us all.
>
> —Pierre-Simon Laplace

It is fitting that we conclude these essays by following the advice given by Laplace: "Read Euler." Extraordinarily versatile in both of the main currents of mathematics, the continuous and the discrete, Euler stands out among the masters for the clarity and elegance of his exposition, his honesty in acknowledging what he believes but cannot prove, and his willingness to reveal the origins of his discoveries. George Pólya sums it up well when he writes:

> A master of inductive research in mathematics, he made important discoveries (on infinite series, in the Theory of Numbers, and in other branches of mathematics) by induction, that is, by observation, daring guess, and shrewd verification. In this respect, however, Euler is not unique; other mathematicians, great and small, used induction extensively in their work.
>
> Yet Euler seems to me almost unique in one respect: he takes pains to present the relevant inductive evidence carefully, in detail, in good order. He presents it convincingly but honestly, as a genuine scientist should do. His presentation is "the candid exposition of the ideas that led him to those discoveries" and has a distinctive charm. Naturally enough, as any other author, he tries to impress his readers, but, as a really good author, he tries to impress his readers only by such things as have genuinely impressed himself.[25]

Here, then, is a sample of Euler's writing, an account of one of his most profound discoveries, a "most extraordinary law" concerning the positive integers which has come to be known as the pentagonal number theorem.[26]

---

[25] From *Induction and Analogy in Mathematics* [338], p. 90.

[26] The English translation of Euler's memoir given here, with a few unessential alterations, is due to Pólya [338], pp. 91–98. In footnote 1, Pólya writes: "The original is in French; see Euler's *Opera Omnia,* ser. 1, vol. 2, p. 241–253. The alterations consist in a different notation [footnote 27], in the arrangement of a table [explained in footnote 28], in slight changes affecting a few formulas, and in dropping a repetition of former arguments in the last No. 13 of the memoir. The reader may consult the easily available original." In "Euler's pentagonal number theorem" [7], G. E. Andrews gives a full modern exposition of Euler's proof.

DISCOVERY OF A MOST
EXTRAORDINARY LAW OF THE NUMBERS
CONCERNING THE SUM OF THEIR DIVISORS

1. Till now the mathematicians tried in vain to discover some order in the sequence of the prime numbers and we have every reason to believe that there is some mystery which the human mind shall never penetrate. To convince oneself, one has only to glance at the tables of the primes, which some people took the trouble to compute beyond a hundred thousand, and one perceives that there is no order and no rule. This is so much more surprising as the arithmetic gives us definite rules with the help of which we can continue the sequence of the primes as far as we please, without noticing, however, the least trace of order. I am myself certainly far from this goal, but I just happened to discover an extremely strange law governing the sums of the divisors of the integers which, at the first glance, appear just as irregular as the sequence of the primes, and which, in a certain sense, comprise even the latter. This law, which I shall explain in a moment, is, in my opinion, so much more remarkable as it is of such a nature that we can be assured of its truth without giving it a perfect demonstration. Nevertheless, I shall present such evidence for it as might be regarded as almost equivalent to a rigorous demonstration.

2. A prime number has no divisors except unity and itself, and this distinguishes the primes from the other numbers. Thus 7 is a prime, for it is divisible only by 1 and itself. Any other number which has, besides unity and itself, further divisors, is called composite, as for instance, the number 15, which has, besides 1 and 15, the divisors 3 and 5. Therefore, generally, if the number p is prime, it will be divisible only by 1 and p; but if p was composite, it would have, besides 1 and p, further divisors. Therefore, in the first case, the sum of its divisors will be $1 + p$, but in the latter it would exceed $1 + p$. As I shall have to consider the sum of divisors of various numbers, I shall use[27] the sign $\sigma(n)$ to denote the sum of the divisors of the number n. Thus, $\sigma(12)$ means the sum of all the divisors of 12, which are 1, 2, 3, 4, 6, and 12; therefore, $\sigma(12) = 28$. In the same way, one can see that $\sigma(60) = 168$ and $\sigma(100) = 217$. Yet, since unity is only divisible by itself, $\sigma(1) = 1$. Now, 0 (zero) is divisible by all numbers. Therefore, $\sigma(0)$ should be properly infinite. (However, I shall assign to it later a finite value, different in different cases, and this will turn out serviceable.)

---

[27] Euler was the first to introduce a symbol for the sum of the divisors; he used $\int n$, not the modern $\sigma(n)$ of the text.

3. Having defined the meaning of the symbol $\sigma(n)$, as above, we see clearly that if p is a prime $\sigma(p) = 1 + p$. Yet $\sigma(1) = 1$ (and not $1 + 1$); hence we see that 1 should be excluded from the sequence of the primes; 1 is the beginning of the integers, neither prime nor composite. If, however, n is composite, $\sigma(n)$ is greater than $1 + n$.

In this case we can easily find $\sigma(n)$ from the factors of n. If $a, b, c, d, \ldots$ are different primes, we see easily that

$$\sigma(ab) = 1 + a + b + ab = (1 + a)(1 + b) = \sigma(a)\sigma(b),$$

$$\sigma(abc) = (1 + a)(1 + b)(1 + c) = \sigma(a)\sigma(b)\sigma(c),$$

$$\sigma(abcd) = \sigma(a)\sigma(b)\sigma(c)\sigma(d)$$

and so on. We need particular rules for the powers of primes, as

$$\sigma(a^2) = 1 + a + a^2 = \frac{a^3 - 1}{a - 1}$$

$$\sigma(a^3) = 1 + a + a^2 + a^3 = \frac{a^4 - 1}{a - 1}$$

and, generally,

$$\sigma(a^n) = \frac{a^{n+1} - 1}{a - 1}.$$

Using this, we can find the sum of the divisors of any number, composite in any way whatever. This we see from the formulas

$$\sigma(a^2 b) = \sigma(a^2)\sigma(b)$$

$$\sigma(a^3 b^2) = \sigma(a^3)\sigma(b^2)$$

$$\sigma(a^3 b^4 c) = \sigma(a^3)\sigma(b^4)\sigma(c)$$

and, generally,

$$\sigma(a^\alpha b^\beta c^\gamma d^\delta e^\epsilon) = \sigma(a^\alpha)\sigma(b^\beta)\sigma(c^\gamma)\sigma(d^\delta)\sigma(e^\epsilon).$$

For instance, to find $\sigma(360)$ we set, since 360 factorized is $2^3 \cdot 3^2 \cdot 5$,

$$\sigma(360) = \sigma(2^3)\sigma(3^2)\sigma(5) = 15 \cdot 13 \cdot 6 = 1170.$$

4. In order to show the sequence of the sums of the divisors, I add the following table[28] containing the sums of the divisors of all integers from 1 up to 99.

n	0	1	2	3	4	5	6	7	8	9
0	—	1	3	4	7	6	12	8	15	13
10	18	12	28	14	24	24	31	18	39	20
20	42	32	36	24	60	31	42	40	56	30
30	72	32	63	48	54	48	91	38	60	56
40	90	42	96	44	84	78	72	48	124	57
50	93	72	98	54	120	72	120	80	90	60
60	168	62	96	104	127	84	144	68	126	96
70	144	72	195	74	114	124	140	96	168	80
80	186	121	126	84	224	108	132	120	180	90
90	234	112	168	128	144	120	252	98	171	156

If we examine a little the sequence of these numbers, we are almost driven to despair. We cannot hope to discover the least order. The irregularity of the primes is so deeply involved in it that we must think it impossible to disentangle any law governing this sequence, unless we know the law governing the sequence of the primes itself. It could appear even that the sequence before us is still more mysterious than the sequence of the primes.

5. Nevertheless, I observed that this sequence is subject to a completely definite law and could even be regarded as a *recurring* sequence. This mathematical expression means that each term can be computed from the foregoing terms, according to an invariable rule. In fact, if we let $\sigma(n)$ denote any term of this sequence, and $\sigma(n-1)$, $\sigma(n-2)$, $\sigma(n-3)$, $\sigma(n-4)$, $\sigma(n-5)$, ... the preceding terms, I say that the value of $\sigma(n)$ can always be combined from some of the

---

[28] The number in the intersection of the row marked 60 and the column marked 7, that is, 68, is $\sigma(67)$. If p is prime, $\sigma(p)$ is in heavy print. This arrangement of the table is a little more concise than the arrangement in the original.

preceding as prescribed by the following formula:

$$\sigma(n) = \sigma(n-1) + \sigma(n-2) - \sigma(n-5) - \sigma(n-7)$$
$$+ \sigma(n-12) + \sigma(n-15) - \sigma(n-22) - \sigma(n-26)$$
$$+ \sigma(n-35) + \sigma(n-40) - \sigma(n-51) - \sigma(n-57)$$
$$+ \sigma(n-70) + \sigma(n-77) - \sigma(n-92) - \sigma(n-100)$$
$$+ \cdots$$

On this formula we must make the following remarks.

I.   In the sequence of the signs $+$ and $-$, each arises twice in succession.

II.   The law of the numbers $1, 2, 5, 7, 12, 15, \ldots$ which we have to subtract from the proposed number n, will become clear if we take their differences:

Nrs. 1, 2, 5, 7, 12, 15, 22, 26, 35, 40, 51, 57, 70, 77, 92, 100, ...

Diff. 1, 3, 2, 5, 3, 7, 4, 9, 5, 11, 6, 13, 7, 15, 8, ...

In fact, we have here, alternately, all the integers $1, 2, 3, 4, 5, 6, \ldots$ and the odd numbers $3, 5, 7, 9, 11, \ldots$, and hence we can continue the sequence of these numbers as far as we please.

III.   Although this sequence goes to infinity, we must take, in each case, only those terms for which the numbers under the sign σ are still positive and omit the σ for negative values.

IV.   If the sign $\sigma(0)$ turns up in the formula, we must, as its value in itself is indeterminate, substitute for $\sigma(0)$ the number n proposed.

6.   After these remarks it is not difficult to apply the formula to any given particular case, and so anybody can satisfy himself of its truth by as many examples as he may wish to develop. And since I must admit that I am not in a position to give it a rigorous demonstration, I will justify it by a sufficiently large number of examples.

$$\sigma(1) \ =\sigma(0) \qquad\qquad\qquad\qquad = 1 \qquad\qquad = 1$$

$$\sigma(2) \ =\sigma(1) \ +\sigma(0) \qquad\qquad\qquad = 1 + \ 2 \qquad\qquad = 3$$

$$\sigma(3) \ =\sigma(2) \ +\sigma(1) \qquad\qquad\qquad = 3 + \ 1 \qquad\qquad = 4$$

$$\sigma(4) \ =\sigma(3) \ +\sigma(2) \qquad\qquad\qquad = 4 + \ 3 \qquad\qquad = 7$$

$$\sigma(5) \ =\sigma(4) \ +\sigma(3) \ -\sigma(0) \qquad\qquad = 7 + \ 4 - \ 5 \qquad\quad = 6$$

$$\sigma(6) \ =\sigma(5) \ +\sigma(4) \ -\sigma(1) \qquad\qquad = 6 + \ 7 - \ 1 \qquad\quad = 12$$

$$\sigma(7) \ =\sigma(6) \ +\sigma(5) \ -\sigma(2) \ -\sigma(0) \qquad = 12 + \ 6 - \ 3 - \ 7 \qquad = 8$$

$$\sigma(8) \ =\sigma(7) \ +\sigma(6) \ -\sigma(3) \ -\sigma(1) \qquad = 8 + 12 - \ 4 - \ 1 \qquad = 15$$

$$\sigma(9) \ =\sigma(8) \ +\sigma(7) \ -\sigma(4) \ -\sigma(2) \qquad = 15 + \ 8 - \ 7 - \ 3 \qquad = 13$$

$$\sigma(10) =\sigma(9) \ +\sigma(8) \ -\sigma(5) \ -\sigma(3) \qquad = 13 + 15 - \ 6 - \ 4 \qquad = 18$$

$$\sigma(11) =\sigma(10) +\sigma(9) \ -\sigma(6) \ -\sigma(4) \qquad = 18 + 13 - 12 - \ 7 \qquad = 12$$

$$\sigma(12) =\sigma(11) +\sigma(10) -\sigma(7) \ -\sigma(5) \ +\sigma(0) \quad = 12 + 18 - \ 8 - \ 6 + 12 \quad = 28$$

$$\sigma(13) =\sigma(12) +\sigma(11) -\sigma(8) \ -\sigma(6) \ +\sigma(1) \quad = 28 + 12 - 15 - 12 + \ 1 \quad = 14$$

$$\sigma(14) =\sigma(13) +\sigma(12) -\sigma(9) \ -\sigma(7) \ +\sigma(2) \quad = 14 + 28 - 13 - \ 8 + \ 3 \quad = 24$$

$$\sigma(15) =\sigma(14) +\sigma(13) -\sigma(10) -\sigma(8) \ +\sigma(3) \ +\sigma(0) = 24 + 14 - 18 - 15 + \ 4 + 15 = 24$$

$$\sigma(16) =\sigma(15) +\sigma(14) -\sigma(11) -\sigma(9) \ +\sigma(4) \ +\sigma(1) = 24 + 24 - 12 - 13 + \ 7 + \ 1 = 31$$

$$\sigma(17) =\sigma(16) +\sigma(15) -\sigma(12) -\sigma(10) +\sigma(5) \ +\sigma(2) = 31 + 24 - 28 - 18 + \ 6 + \ 3 = 18$$

$$\sigma(18) =\sigma(17) +\sigma(16) -\sigma(13) -\sigma(11) +\sigma(6) \ +\sigma(3) = 18 + 31 - 14 - 12 + 12 + \ 4 = 39$$

$$\sigma(19) =\sigma(18) +\sigma(17) -\sigma(14) -\sigma(12) +\sigma(7) \ +\sigma(4) = 39 + 18 - 24 - 28 + \ 8 + \ 7 = 20$$

$$\sigma(20) =\sigma(19) +\sigma(18) -\sigma(15) -\sigma(13) +\sigma(8) \ +\sigma(5) = 20 + 39 - 24 - 14 + 15 + \ 6 = 42$$

I think these examples are sufficient to discourage anyone from imagining that it is by mere chance that my rule is in agreement with the truth.

7. Yet somebody could still doubt whether the law of the numbers $1, 2, 5, 7,$ $12, 15, \ldots$ which we have to subtract is precisely that one which I have indicated, since the examples given imply only the first six of these numbers. Thus, the law could still appear as insufficiently established and, therefore, I will give some examples with larger numbers.

    I.   Given the number 101, find the sum of its divisors. We have

$$\sigma(101) = \sigma(100) + \sigma(99) - \sigma(96) - \sigma(94)$$
$$+ \ \sigma(89) + \sigma(86) - \sigma(79) - \sigma(75)$$
$$+ \ \sigma(66) + \sigma(61) - \sigma(50) - \sigma(44)$$
$$+ \ \sigma(31) + \sigma(24) - \ \sigma(9) - \ \sigma(1)$$
$$= \quad 217 + \quad 156 - \quad 252 - \quad 144$$
$$+ \quad\ \ 90 + \quad 132 - \quad\ \ 80 - \quad 124$$
$$+ \quad 144 + \quad\ \ 62 - \quad\ \ 93 - \quad\ \ 84$$
$$+ \quad\ \ 32 + \quad\ \ 60 - \quad\ \ 13 - \quad\ \ 1$$
$$= \quad 893 - \quad 791$$
$$= \quad 102$$

and hence we could conclude, if we would not have known it before, that 101 is a prime number.

II.   Given the number 301, find the sum of its divisors. We have

diff.   1   3   2   5
$$\sigma(301) = \sigma(300) + \sigma(299) - \sigma(296) - \sigma(294) +$$
   3   7   4   9
$$+\sigma(289) + \sigma(286) - \sigma(279) - \sigma(275) +$$
   5   11   6   13
$$+\sigma(266) + \sigma(261) - \sigma(250) - \sigma(244) +$$
   7   15   8   17
$$+\sigma(231) + \sigma(224) - \sigma(209) - \sigma(201) +$$
   9   19   10   21
$$+\sigma(184) + \sigma(175) - \sigma(156) - \sigma(146) +$$
  11   23   12   25
$$+\sigma(125) + \sigma(114) - \ \sigma(91) - \ \sigma(79) +$$
  13   27   14
$$+\sigma(54) \ + \ \sigma(41) - \ \sigma(14) - \ \sigma(0).$$

We see by this example how we can, using the differences, continue the formula as far as is necessary in each case. Performing the computations, we find

$$\sigma(301) = 4939 - 4587 = 352.$$

We see hence that 301 is not a prime. In fact, $301 = 7 \cdot 43$ and we obtain

$$\sigma(301) = \sigma(7)\sigma(43) = 8 \cdot 44 = 352$$

as the rule has shown.

8.　The examples that I have just developed will undoubtedly dispel any qualms which we might have had about the truth of my formula. Now, this beautiful property of the numbers is so much more surprising as we do not perceive any intelligible connection between the structure of my formula and the nature of the divisors with the sum of which we are here concerned. The sequence of the numbers $1, 2, 5, 7, 12, 15, \ldots$ does not seem to have any relation to the matter in hand. Moreover, as the law of these numbers is "interrupted" and they are in fact a mixture of two sequences with a regular law, of $1, 5, 12, 22, 35, 51, \ldots$ and $2, 7, 15, 26, 40, 57, \ldots$, we would not expect that such an irregularity can turn up in Analysis. The lack of demonstration must increase the surprise still more, since it seems wholly impossible to succeed in discovering such a property without being guided by some reliable method which could take the place of a perfect proof. I confess that I did not hit on this discovery by mere chance, but another proposition opened the path to this beautiful property—another proposition of the same nature which must be accepted as true although I am unable to prove it. And although we consider here the nature of integers to which the Infinitesimal Calculus does not seem to apply, nevertheless I reached my conclusion by differentiations and other devices. I wish that somebody would find a shorter and more natural way, in which the consideration of the path that I followed might be of some help, perhaps.

9.　In considering the partitions of numbers, I examined, a long time ago, the expression

$$(1 - x)(1 - x^2)(1 - x^3)(1 - x^4)(1 - x^5)(1 - x^6)(1 - x^7)(1 - x^8) \ldots,$$

in which the product is assumed to be infinite. In order to see what kind of series will result, I multiplied actually a great number of factors and found

$$1 - x - x^2 + x^5 + x^7 - x^{12} - x^{15} + x^{22} + x^{26} - x^{35} - x^{40} + \cdots.$$

The exponents of x are the same which enter into the above formula; also the signs $+$ and $-$ arise twice in succession. It suffices to undertake this multiplication and to continue it as far as it is deemed proper to become convinced of the truth of this series. Yet I have no other evidence for this, except a long induction which I have carried out so far that I cannot in any way doubt the law governing

the formation of these terms and their exponents. I have long searched in vain
for a rigorous demonstration of the equation between the series and the above
infinite product $(1-x)(1-x^2)(1-x^3)\ldots$, and I have proposed the same question to some of my friends with whose ability in these matters I am familiar, but
all have agreed with me on the truth of this transformation of the product into
a series, without being able to unearth any clue of a demonstration. Thus, it
will be a known truth, but not yet demonstrated, that if we put

$$s = (1-x)(1-x^2)(1-x^3)(1-x^4)(1-x^5)(1-x^6)\ldots$$

the same quantity s can also be expressed as follows:

$$s = 1 - x - x^2 + x^5 + x^7 - x^{12} - x^{15} + x^{22} + x^{26} - x^{35} - x^{40} + \cdots.$$

For each of us can convince himself of this truth by performing the multiplication as far as he may wish; and it seems impossible that the law which has
been discovered to hold for 20 terms, for example, would not be observed in
the terms that follow.

10.   As we have thus discovered that those two infinite expressions are equal
even though it has not been possible to demonstrate their equality, all the conclusions which may be deduced from it will be of the same nature, that is, true
but not demonstrated. Or, if one of these conclusions could be demonstrated,
one could reciprocally obtain a clue to the demonstration of that equation; and
it was with this purpose in mind that I maneuvered those two expressions in
many ways, and so I was led among other discoveries to that which I explained
above; its truth, therefore, must be as certain as that of the equation between
the two infinite expressions. I proceeded as follows. Being given that the two
expressions

I. $s = (1-x)(1-x^2)(1-x^3)(1-x^4)(1-x^5)(1-x^6)(1-x^7)\cdots$

II. $s = 1 - x - x^2 + x^5 + x^7 - x^{12} - x^{15} + x^{22} + x^{26} - x^{35} - x^{40} + \cdots$

are equal, I got rid of the factors in the first by taking logarithms

$$\log s = \log(1-x) + \log(1-x^2) + \log(1-x^3) + \log(1-x^4) + \cdots.$$

In order to get rid of the logarithms, I differentiate and obtain the equation

$$\frac{1}{s}\frac{ds}{dx} = -\frac{1}{1-x} - \frac{2x}{1-x^2} - \frac{3x^2}{1-x^3} - \frac{4x^3}{1-x^4} - \frac{5x^4}{1-x^5} - \cdots$$

or

$$-\frac{x}{s}\frac{ds}{dx} = \frac{x}{1-x} + \frac{2x^2}{1-x^2} + \frac{3x^3}{1-x^3} + \frac{4x^4}{1-x^4} + \frac{5x^5}{1-x^5} + \cdots .$$

From the second expression for s, as infinite series, we obtain another value for the same quantity

$$-\frac{x}{s}\frac{ds}{dx} = \frac{x + 2x^2 - 5x^5 - 7x^7 + 12x^{12} + 15x^{15} - 22x^{22} - 26x^{26} + \cdots}{1 - x - x^2 + x^5 + x^7 - x^{12} - x^{15} + x^{22} + x^{26} - \cdots}.$$

11.   Let us put

$$-\frac{x}{s}\frac{ds}{dx} = t.$$

We have above two expressions for the quantity t. In the first expression, I expand each term into a geometric series and obtain

$$\begin{aligned}
t = x + \;& x^2 + \; x^3 + \; x^4 + \; x^5 + \; x^6 + \; x^7 + \; x^8 + \cdots \\
+ \;& 2x^2 \qquad\quad +2x^4 \qquad\quad +2x^6 \qquad\quad +2x^8 + \cdots \\
& \qquad\quad +3x^3 \qquad\qquad\quad +3x^6 \qquad\qquad\quad + \cdots \\
& \qquad\qquad\quad +4x^4 \qquad\qquad\qquad\quad +4x^8 + \cdots \\
& \qquad\qquad\qquad\quad +5x^5 \qquad\qquad\qquad\quad + \cdots \\
& \qquad\qquad\qquad\qquad\quad +6x^6 \qquad\qquad\qquad + \cdots \\
& \qquad\qquad\qquad\qquad\qquad\quad +7x^7 \qquad\quad + \cdots \\
& \qquad\qquad\qquad\qquad\qquad\qquad\quad +8x^8 + \cdots .
\end{aligned}$$

Here we see easily that each power of x arises as many times as its exponent has divisors, and that each divisor arises as a coefficient of the same power of x. Therefore, if we collect the terms with like powers, the coefficient of each power of x will be the sum of the divisors of its exponent. And, therefore, using the above notation $\sigma(n)$ for the sum of the divisors of n, I obtain

$$t = \sigma(1)x + \sigma(2)x^2 + \sigma(3)x^3 + \sigma(4)x^4 + \sigma(5)x^5 + \cdots .$$

The law of the series is manifest. And, although it might appear that some induction was involved in the determination of the coefficients, we can easily satisfy ourselves that this law is a necessary consequence.

12.   By virtue of the definition of t, the last formula of No. 10 can be written as follows:

$$\begin{aligned}
&t(1 - x - x^2 + x^5 + x^7 - x^{12} - x^{15} + x^{22} + x^{26} - \cdots) \\
&- x - 2x^2 + 5x^5 + 7x^7 - 12x^{12} - 15x^{15} + 22x^{22} + 26x^{26} - \cdots = 0.
\end{aligned}$$

Substituting for t the value obtained at the end of No. 11, we find

$$0 = \sigma(1)x + \sigma(2)x^2 + \sigma(3)x^3 + \sigma(4)x^4 + \sigma(5)x^5 + \sigma(6)x^6 + \cdots$$
$$-x - \sigma(1)x^2 - \sigma(2)x^3 - \sigma(3)x^4 - \sigma(4)x^5 - \sigma(5)x^6 - \cdots$$
$$- \quad 2x^2 - \sigma(1)x^3 - \sigma(2)x^4 - \sigma(3)x^5 - \sigma(4)x^6 - \cdots$$
$$+ \quad 5x^5 + \sigma(1)x^6 + \cdots .$$

Collecting the terms, we find the coefficient for any given power of x. This coefficient consists of several terms. First comes the sum of the divisors of the exponent of x, and then sums of divisors of some preceding numbers, obtained from that exponent by subtracting successively $1, 2, 5, 7, 12, 15, 22, 26, \ldots$. Finally, if it belongs to this sequence, the exponent itself arises. We need not explain again the signs assigned to the terms just listed. Therefore, generally, the coefficient of x^n is

$$\sigma(n) - \sigma(n-1) - \sigma(n-2) + \sigma(n-5) + \sigma(n-7)$$
$$- \sigma(n-12) - \sigma(n-15) + \cdots .$$

This is continued as long as the numbers under the sign σ are not negative. Yet, if the term $\sigma(0)$ arises, we must substitute n for it.

13. Since the sum of the infinite series considered in the foregoing No. 12 is 0, whatever the value of x may be, the coefficient of each single power of x must necessarily be 0. Hence we obtain the law that I explained above in No. 5; I mean the law that governs the sum of the divisors and enables us to compute it recursively for all numbers. In the foregoing development, we may perceive some reason for the signs, some reason for the sequence of the numbers

$$1, 2, 5, 7, 12, 15, 22, 26, 35, 40, 51, 57, 70, 77, \ldots$$

and, especially, a reason why we should substitute for $\sigma(0)$ the number n itself, which could have appeared the strangest feature of my rule. This reasoning, although still very far from a perfect demonstration, will certainly lift some doubts about the most extraordinary law that I explained here.

We keep returning to the positive integers, with new methods, new insights, and a new appreciation for the elegant interplay that exists between the continuous and the discrete. In the end, it is perhaps not surprising for in the nineteenth century, when analysis was finally placed on a firm footing, it was arithmetic that served as the foundation.

Euler's work reminds us of what we should strive for in the teaching of our students or ourselves—to observe, to experiment, to search for patterns, and, above all, to rejoice not just in the final rigor, but in that wondrous process of discovery that must always precede it. Perhaps then we can feel justified in saying:

> We shall not cease from exploration,
> And the end of all our exploring
> Will be to arrive where we started
> And know the place for the first time.
>
> —T. S. Eliot, *Four Quartets*

APPENDIX:
THE CONGRUENCE NOTATION

In his celebrated treatise on numbers, *Disquisitiones arithmeticae,* published in 1801, Gauss introduced a remarkable notation which clarifies and simplifies many problems concerning the divisibility of integers. In so doing he created a new branch of number theory called the theory of congruences.

Let m be a fixed positive integer. Two integers a and b whose difference $a - b$ is divisible by m are said to be **congruent with respect to the modulus** m, or simply **congruent** modulo m. To designate the congruence, Gauss proposed the notation

$$a \equiv b \pmod{m}.$$

Thus, for example,

$$21 \equiv 0 \pmod{3}, \qquad 2^{11} \equiv 1 \pmod{23}$$

are valid congruences since 21 is divisible by 3 and $2^{11} - 1$ is divisible by 23.

It is clear that congruences are of great practical importance. For example, the hands of a clock indicate the hour modulo 12, and the statement "today is Sunday" is a congruence property, modulo 7, of the number of days that have passed since some fixed date. As Hardy said, "The absolute values of numbers are comparatively unimportant; we want to know what time it is, not how many minutes have passed since the creation."[1]

The symbol $\equiv$ reinforces the analogy between congruence and equality by stressing that congruent numbers are to be thought of as equivalent. Thus, for

---

[1] "An Introduction to the Theory of Numbers" [207], p. 70.

example, if we are interested only in the last digit of a number, then numbers which differ by a multiple of 10 are effectively the same. In fact congruence modulo m means precisely "equality except for the addition of some multiple of m."

It is an immediate consequence of the definition that congruent numbers, when divided by the modulus, leave the same remainders (or "residues"); and, conversely, numbers with the same remainders are congruent. In particular, every integer is congruent modulo m to exactly one of the numbers

$$0, 1, 2, \ldots, m - 1.$$

Thus modular arithmetic is a finite arithmetic, and every problem in it can be solved, at least in principle, by trial. For example, we find by trial that the congruence

$$x^2 \equiv 1 \pmod 8$$

has just four solutions, namely,

$$x = 1, 3, 5, 7,$$

while the congruence

$$x^2 \equiv 2 \pmod 8$$

has none.

The power of Gauss's notation derives from the fact that in modular arithmetic many of the laws of ordinary arithmetic have close analogues. Two congruences with the same modulus can be added, subtracted, and multiplied in just the same way as two equations. If

$$a \equiv b \pmod m \qquad \text{and} \qquad c \equiv d \pmod m, \tag{1}$$

then

$$a + c \equiv b + d \pmod m$$
$$a - c \equiv b - d \pmod m$$
$$ac \equiv bd \pmod m.$$

The first two statements are obvious. For the third, we need only multiply the first congruence in (1) by c and the second by b to see that $ac \equiv bc \equiv bd$ (mod m). Alternatively, we can express the difference $ac - bd$ in the form $(a - b)c + (c - d)b$, showing that it is divisible by m.

It is obvious that the aforementioned rules may be applied to several congruences. In particular, a congruence may be multiplied by itself any number of times so that

$$a \equiv b \pmod{m}$$

implies

$$a^n \equiv b^n \pmod{m}$$

for any natural number n.

Finally, it should be pointed out that whereas a congruence may be multiplied throughout by a constant, division by a nonzero constant is not always permissible. For example,

$$6 \equiv 0 \pmod{6}$$

but if we divide both sides by 2 we obtain the incorrect result $3 \equiv 0 \pmod{6}$. Dividing by a common factor is legitimate provided the factor is relatively prime to the modulus. Thus, from the relation

$$ak \equiv bk \pmod{m},$$

with k relatively prime to m, we conclude first that $(a - b)k$ is divisible by m and then, by the fundamental theorem of arithmetic, that $a - b$ is divisible by m.

Obeying so many of the laws of ordinary arithmetic, the congruence notation does far more than just simplify the exposition—it gives rise to new and important problems, which otherwise might not have been posed, and it suggests new methods of proof. In this way, it is just as significant and useful in number theory as Leibniz's differential notation is in the infinitesimal calculus.

At first the arithmetic may appear foreign, but the reader who gains confidence in its use will have acquired a powerful number-theoretic tool. For now we content ourselves with a few simple applications.

Example 1. "Casting Out Nines"

As a first illustration of the usefulness of congruences, we derive the well-known rules for the divisibility of a number by 3 or 9 or 11. In decimal form any natural number n has a representation

$$n = a + 10b + 100c + \cdots$$

where $a, b, c, \ldots$ are the digits of the number, read from right to left, so that a is the number of units, b the number of tens, and so on. Since $10 \equiv 1 \pmod{9}$, it follows that $10^n \equiv 1 \pmod 9$ for every n. Accordingly we conclude from the representation of n above that

$$n \equiv a + b + c + \cdots \pmod 9. \qquad (2)$$

In other words, any number n differs from the sum of its digits by a multiple of 9, and in particular, n *is divisible by 9 if and only if the sum of its digits is divisible by 9*. This is the rule of "casting out nines."

It is obvious that if two numbers are congruent for a particular modulus m, they remain congruent for any modulus that divides m. Therefore, (2) remains valid when 9 is replaced by 3, and we conclude that a number is divisible by 3 if and only if the sum of its digits is divisible by 3.

The rule for 11 is based on the fact that $10 \equiv -1 \pmod{11}$, so that

$$10^2 \equiv 1, \ 10^3 \equiv -1, \ 10^4 \equiv 1, \ \ldots \pmod{11}.$$

Therefore,

$$n \equiv a - b + c - \cdots \pmod{11},$$

and n is divisible by 11 if and only if $a - b + c - \cdots$ is divisible by 11. For example, to test the divisibility of 96415 by 11, we form $5 - 1 + 4 - 6 + 9 = 11$. Since this is divisible by 11, so is 96415.

Example 2. Modular Exponentiation

The application of modular arithmetic often facilitates computations that would be impractical—if not impossible—to carry out in ordinary arithmetic. As an illustration let us find the remainder obtained by dividing

$$3^{100}$$

by 101, without ever calculating 3^{100}. If we express the exponent in binary form—that is to say, as a sum of powers of 2—then the computation can be simplified considerably. In our case

$$100 = 64 + 32 + 4$$

so that

$$3^{100} = 3^{64} \cdot 3^{32} \cdot 3^4.$$

Now $3^4 = 81 \equiv -20 \pmod{101}$, and we find by successive squarings and reductions modulo 101 that

$$3^8 \equiv 20^2 \equiv -4, \quad 3^{16} \equiv 16,$$

$$3^{32} \equiv 16^2 \equiv -47, \quad 3^{64} \equiv 47^2 \equiv -13 \pmod{101}.$$

By multiplying the congruences for the fourth, thirty-second, and sixty-fourth powers of 3 we obtain

$$3^{100} \equiv (-13)(-47)(-20) \equiv (-13)(31) \equiv 1 \pmod{101}$$

so the required remainder is 1.

This procedure is very efficient, especially in a binary computer, where the exponent is already stored in binary form.

Example 3. The Chinese Remainder Theorem

It is often necessary, in both theory and practice, to find a number that has prescribed remainders when divided by two or more moduli. Such problems arose in ancient Chinese puzzles and their solution has come to be known as the Chinese remainder theorem. Typical of these puzzles is the "basket of eggs" problem. When the eggs were removed 2, 3, 4, 5, and 6 at a time, there was always one left; but when they were removed 7 at a time, the basket was empty. In mathematical terms, we are to find all values of x for which

$$x \equiv 1 \pmod{2, 3, 4, 5, 6}, \qquad x \equiv 0 \pmod{7},$$

x being the number of eggs.

The simultaneous solution of a system of several linear congruences relies ultimately on the solution of a single linear congruence in one unknown. Such a congruence takes the form

$$ax \equiv b \pmod{m}$$

and it is to this problem that we turn our attention first.

As an example, consider the congruence

$$2x \equiv 1 \pmod{7}.$$

Since every integer is congruent modulo 7 to one of the numbers 0, 1, 2, 3, 4, 5, 6, we may let x range over these values and we find that the unique solution is $x \equiv 4$. In similar fashion, we find that the congruence $2x \equiv 1 \pmod{12}$ has no solutions. An event that occurs every other day must ultimately occur on Sunday, while one that occurs every other month need never occur in January. The difference is accounted for by the fact that 2 and 7 are relatively prime while 2 and 12 are not.

Lemma.   *The linear congruence* $ax \equiv b \pmod{m}$ *is solvable provided that a and m are relatively prime. If this is the case, then there is a unique solution modulo m.*

The simplest proof is based on a general logical principle which is so elementary that it may be thought too obvious to be worth mentioning: *If m objects are put into m boxes, with no two objects in the same box, then there must be one in every box.* Let us take as "objects" the m distinct numbers

$$0, a, 2a, 3a, \ldots, (m-1)a.$$

No two of them can be congruent $\pmod{m}$, for the relation

$$ax \equiv ay \pmod{m}$$

implies, by the cancellation law, that $x \equiv y \pmod{m}$. Therefore, by the principle just enunciated, they must be congruent in some order to the numbers

$$0, 1, 2, \ldots, m-1.$$

In particular, there is exactly one x between 0 and $m-1$ for which $ax \equiv b$ $\pmod{m}$.

By virtue of the lemma we can resolve a problem alluded to earlier: When is division possible in modular arithmetic? We see now that an integer a has an "inverse" modulo m—that is, the congruence

$$ax \equiv 1 \pmod{m}$$

is solvable—if and only if a and m are relatively prime.

The Chinese remainder theorem is a simple consequence of this fact.

Chinese Remainder Theorem.   *Every system of linear congruences*

$$x \equiv c_1 \pmod{m_1}, \ldots, x \equiv c_k \pmod{m_k},$$

in which the moduli are relatively prime in pairs, is solvable. Moreover, the solution is unique modulo the product of the moduli.

Proof.   The uniqueness of the solution—provided of course that one exists—is trivial. For if x and y are two solutions, then $x \equiv y \pmod{m_i}$ for every i, and therefore (since the m_i are relatively prime), $x \equiv y \pmod{m_1 m_2 \cdots m_k}$.

We complete the proof by actually exhibiting a solution in a very convenient form. For each $i = 1, 2, \ldots k$, let M_i be the product of all the moduli except m_i. Then M_i and m_i are relatively prime and hence the linear congruence

$$a_i M_i \equiv 1 \pmod{m_i}$$

is solvable. Let

$$x = a_1 M_1 c_1 + a_2 M_2 c_2 + \cdots + a_k M_k c_k.$$

Since $M_j \equiv 0 \pmod{m_i}$ whenever $j \neq i$, we see that

$$x \equiv a_i M_i c_i \equiv c_i \pmod{m_i}$$

for every i, so that x satisfies the given system.

The theorem was apparently first stated and proved in its full generality by Euler in 1734. The method, however, is ancient, the first known source being the *Arithmetic* of the Chinese mathematician Sun Tzu (+3rd century), who called his rule the "great generalization."

As an illustration let us solve the "basket of eggs" problem that was posed earlier. We are required to find a number x, divisible by 7, which leaves the remainder 1 when divided by 2, 3, 4, 5, or 6. By the latter condition, x exceeds 1 by a multiple of 60, and therefore the problem reduces to

$$x \equiv 1 \pmod{60}, \qquad x \equiv 0 \pmod{7}.$$

The solution then takes the form

$$x \equiv a_1 M_1 c_1 \pmod{420}$$

where $c_1 = 1$, $M_1 = 7$, and a_1 is a solution of the congruence $7a_1 \equiv 1 \pmod{60}$. We find at length the solution $a_1 = 43$, whence

$$x \equiv 301 \pmod{420}.$$

The smallest number of eggs the basket could have contained is therefore $x = 301$.

PROBLEMS

1. There is a famous puzzle concerning an island inhabited by five men and their pet monkey. One day the men gathered a large pile of coconuts, which they proposed to divide equally among themselves in the morning. During the night one of the men arose and decided to take his share. After giving one coconut to the monkey, he divided the remaining pile into five equal parts, took his one-fifth share, and left the remaining coconuts in a single pile. The other four men did likewise. Each in turn went to the pile and, in ignorance of what had happened previously, followed the same procedure. In the morning all five men went to the diminished pile, gave one coconut to the monkey and divided the remaining coconuts equally among themselves. What is the least number of coconuts the original pile could have contained?

2. Determine all positive integers n for which $2^n + 1$ is divisible by 3.

3. Find the last two digits of the number 2^{1000}.

4. Find the last two digits of the number $9^{9^{9^9}}$. Gauss is said to have called this number "a measurable infinity"; it has more than a quarter of a million digits ([385], p. 202).

5. Prove that all odd numbers satisfy the congruences

$$x^2 \equiv 1 \ (\text{mod } 8), \ x^4 \equiv 1 \ (\text{mod } 16), \ x^8 \equiv 1 \ (\text{mod } 32), \ldots.$$

6. Prove that, for any positive integer n,

$$1^n + 2^n + 3^n + 4^n$$

is divisible by 5 if and only if n is not divisible by 4.

7. Prove that $2222^{5555} + 5555^{2222}$ is divisible by 7.

8. Let a, b, c, d be integers. Prove that the product of the six differences

$$b - a, \ c - a, \ c - b, \ d - a, \ d - b, \ d - c$$

is divisible by 12.

9. The following method can be used to compute x^n for any positive integer n. First write n in the binary system. Then replace each 0 by the letter S and each 1, except for the first, by the pair of letters SX. The result provides us with a rule for computing x^n: simply interpret S as the operation of squaring and interpret X as the operation of multiplying by x. For example, if $n = 21$, its binary representation is 10101, so we form the sequence SSXSSX. This tells us that we should "square, square, multiply by x, square, square, and multiply by x"; in other words, we should compute successively x^2, x^4, x^5, x^{10}, x^{20}, x^{21}. Justify the method.

10. How many zeros are there at the end of the number

$$1000! = 1 \times 2 \times 3 \times \cdots \times 1000?$$

11. Prove that, for any set of n positive integers, there is a subset of them whose sum is divisible by n.

REFERENCES

1. J. C. Abbott, ed., *The Chauvenet Papers*, Mathematical Association of America, Washington, D.C., 1978.
2. D. J. Albers and G. L. Alexanderson, eds., *Mathematical People: Profiles and Interviews*, Birkhäuser, Boston, 1985.
3. D. J. Albers, G. L. Alexanderson, and C. Reid, eds., *More Mathematical People*, Harcourt Brace Jovanovich, Boston, 1990.
4. G. L. Alexanderson, L. F. Klosinski, and L. C. Larson, eds., *The William Lowell Putnam Mathematical Competition*: *Problems and Solutions, 1965–1984*, Mathematical Association of America, Washington, D.C., 1985.
5. G. Almkvist and B. Berndt, Gauss, Landen, Ramanujan, the arithmetic-geometric mean, ellipses, π, and the *Ladies Diary*, *Amer. Math. Monthly*, 95 (1988) 585–608.
6. G. E. Andrews, *The Theory of Partitions*, Addison-Wesley, Reading, 1976.
7. ——, Euler's pentagonal number theorem, *Math. Mag.*, 56 (1983) 279–284.
8. T. M. Apostol, chairman of the editorial committee, *Selected Papers on Calculus*, Mathematical Association of America, Washington, D.C., 1969.
9. ——, *Mathematical Analysis*, 2nd ed., Addison-Wesley, Reading, 1974.
10. ——, *Introduction to Analytic Number Theory*, Springer-Verlag, New York, 1976.
11. ——, chairman of the editorial committee, *Selected Papers on Precalculus*, Mathematical Association of America, Buffalo, 1977.
12. ——, A proof that Euler missed: evaluating $\zeta(2)$ the easy way, *Math. Intell.*, 5, no. 3 (1983) 59–60.
13. R. C. Archibald, "Golden section" and "A Fibonacci series", *Amer. Math. Monthly*, 25 (1918) 232–238.
14. ——, Outline of the history of mathematics, published as a supplement to The Second Herbert Ellsworth Slaught Memorial Paper, *Amer. Math. Monthly*, January 1949.
15. Archimedes, *The Works of Archimedes*, edited by T. L. Heath, Dover, New York, 1953.
16. R. B. Ash, *Information Theory*, Dover, New York, 1990.

17. W. Aspray and P. Kitcher, eds., *History and Philosophy of Modern Mathematics*, University of Minnesota Press, Minneapolis, 1988.

18. R. Ayoub, *An Introduction to the Analytic Theory of Numbers*, American Mathematical Society, Providence, 1963.

19. ——, Euler and the zeta function, *Amer. Math. Monthly*, 81 (1974) 1067–1086.

20. ——, The lemniscate and Fagnano's contributions to elliptic integrals, *Archive for History of Exact Sciences*, 29 (1984) 131–149.

21. W. W. Rouse Ball and H. S. M. Coxeter, *Mathematical Recreations and Essays*, 12th ed., University of Toronto Press, Toronto, 1974. (A revision of Ball's *Mathematical Recreations and Problems of Past and Present Times,* first published by Macmillan, 1892. Reprinted by Dover, New York, 1987.)

22. E. J. Barbeau, *Polynomials*, Springer-Verlag, New York, 1989.

23. P. Barberini, *Raffaello e la Sezione Aurea*, Bora, Bologna, 1984.

24. M. F. Barnsley, *Fractals Everywhere*, Academic Press, Boston, 1988.

25. M. E. Baron, *The Origins of the Infinitesimal Calculus*, Dover, New York, 1987.

26. E. Batschelet, *Introduction to Mathematics for Life Scientists*, 3rd ed., Springer-Verlag, Berlin, 1979.

27. E. F. Beckenbach, *Inequalities*, Springer-Verlag, Berlin, 1965.

28. E. F. Beckenbach and R. Bellman, *An Introduction to Inequalities*, Mathematical Association of America, Washington, D.C., 1961.

29. P. Beckmann, *A History of Pi*, 4th ed., Golem Press, Boulder, 1977.

30. N. G. W. H. Beeger, On even numbers m dividing $2^m - 2$, *Amer. Math. Monthly*, 58 (1951) 553–555.

31. A. H. Beiler, *Recreations in the Theory of Numbers*, 2nd ed., Dover, New York, 1966.

32. E. T. Bell, *Men of Mathematics*, Simon and Schuster, New York, 1937.

33. ——, *The Last Problem*, Mathematical Association of America, Washington, D.C., 1990.

34. R. E. Bellman, *Dynamic Programming*, Princeton University Press, Princeton, 1957.

35. P. Benacerraf and H. Putnam, eds., *Philosophy of Mathematics: Selected Readings*, 2nd ed., Cambridge University Press, Cambridge, 1983.

36. B. Berndt, Elementary evaluation of $\zeta(2n)$, *Math. Mag.*, 48 (1975) 148–154.

37. Jakob Bernoulli, *Ars conjectandi*, Basel, 1713. (Reprinted in *Die Werke von Jakob Bernoulli*, volume 3, pp. 107–286, Birkhäuser, Basel, 1975.)

38. P. Biler and A. Witkowski, *Problems in Mathematical Analysis*, Marcel Dekker, New York, 1990.

39. G. Birkhoff, ed., *A Source Book in Classical Analysis*, Harvard University Press, Cambridge, 1973.

40. S. Bloch, The proof of the Mordell conjecture, *Math. Intell.*, 6, no. 2 (1984) 41-47.

41. R. P. Boas, Jr., *Entire Functions*, Academic Press, New York, 1954.

42. ——, Partial sums of infinite series, and how they grow, *Amer. Math. Monthly*, 84 (1977) 237–258.

43. ——, "Anomalous Cancellation," in *Mathematical Plums*, Ross Honsberger, ed., Mathematical Association of America, Washington, D.C., 1979.

44. ——, "Convergence, Divergence, and the Computer," in *Mathematical Plums*, Ross Honsberger, ed., Mathematical Association of America, Washington, D.C., 1979.

45. ——, "Some Remarkable Sequences of Integers," in *Mathematical Plums*, Ross Honsberger, ed., Mathematical Association of America, Washington, D.C., 1979

46. ——, *A Primer of Real Functions*, 3rd ed., Mathematical Association of America, Washington, D.C., 1981.

47. R. P. Boas, Jr. and J. W. Wrench, Jr., Partial sums of the harmonic series, *Amer. Math. Monthly*, 78 (1971) 864–870.

48. G. Boole, *Calculus of Finite Differences*, 5th ed., Chelsea, New York, 1970.

49. E. Borel, *Probability and Certainty*, Walker, New York, 1963.

50. W. Borho and H. Hoffmann, Breeding amicable numbers in abundance, *Math. of Comp.*, 46 (1986) 281–293.

51. J. M. Borwein and G. de Barra, Nested radicals, *Amer. Math. Monthly*, 98 (1991) 735–739.

52. J. M. Borwein and P. B. Borwein, The arithmetic-geometric mean and fast computation of elementary functions, *SIAM Review* 26 (1984) 351–366.

53. ——, *Pi and the AGM*, Wiley, New York, 1987.

54. ——, On the complexity of familiar functions and numbers, *SIAM Review* 30 (1988) 589–601.

55. ——, Ramanujan and pi, *Sci. Amer.* (February 1988) 112–117.

56. ——, *A Dictionary of Real Numbers*, Wadsworth, Belmont, 1990.

57. J. M. Borwein, P. B. Borwein, and D. H. Bailey, Ramanujan, modular equations, and approximations to pi or how to compute one billion digits of pi, *Amer. Math. Monthly*, 96 (1989) 201–219.

58. U. Bottazzini, *The Higher Calculus: A History of Real and Complex Analysis from Euler to Weierstrass*, Springer-Verlag, New York, 1986.

59. O. Bottema, R. Z. Djordjević, R. Z. Janić, D. S. Mitrinović, and P. M. Vasić, *Geometric Inequalities*, Wolters-Noordhoff, Groningen, 1969.

60. C. B. Boyer, Pascal's formula for the sums of powers of the integers, *Scripta Math.*, 9 (1943) 237–244.

61. ——, *The History of the Calculus and its Conceptual Development*, Dover, New York, 1959.

62. ——, *A History of Mathematics*, 2nd ed., revised by U. C. Merzbach, Wiley, New York, 1991.

63. R. P. Brent, Fast multiple-precision evaluation of elementary functions, *J. of the Assoc. for Comp. Mach.*, 23 (1976) 242–251.

64. ——, Computation of the regular continued fraction for Euler's constant, *Math. of Comp.*, 31 (1977) 771–777.

65. R. P. Brent and G. L. Cohen, A new lower bound for odd perfect numbers, *Math. of Comp.*, 53 (1989) 431–437 and S-7 to S-24.

66. R. P. Brent, G. L. Cohen, and H. J. J. te Riele, Improved techniques for lower bounds for odd perfect numbers, *Math. of Comp.*, 57 (1991) 857–868.

67. R. P. Brent and E. M. McMillan, Some new algorithms for high-precision computation of Euler's constant, *Math. of Comp.*, 34 (1980) 305–312.

68. D. M. Bressoud, *Factorization and Primality Testing*, Springer-Verlag, New York, 1989.

69. J. Brillhart, D. H. Lehmer, J. L. Selfridge, B. Tuckerman, and S. S. Wagstaff, Jr., *Factorizations of $b^n \pm 1$, $b = 2, 3, 5, 6, 7, 10, 11, 12$ up to high powers*, 2nd ed., American Mathematical Society, Providence, 1988.

70. J. Brillhart, P. L. Montgomery, and R. D. Silverman, Tables of Fibonacci and Lucas factorizations, and Supplement, *Math. of Comp.*, 50 (1988) 251–260 and S-1 to S-15.

71. T. J. Bromwich, *An Introduction to the Theory of Infinite Series*, 2nd ed., Macmillan, New York, 1965.

72. F. E. Browder, ed., *Mathematical Developments Arising from Hilbert Problems*, Proceedings of Symposia in Pure Mathematics, Volume XXVIII, American Mathematical Society, Providence, 1976.

73. R. C. Buck, Prime-representing functions, *Amer. Math. Monthly*, 53 (1946) 265.

74. ——, Mathematical induction and recursive definitions, *Amer. Math. Monthly*, 70 (1963) 128–135.

75. ——, *Advanced Calculus*, 3rd ed., McGraw-Hill, New York, 1978.

76. W. K. Bühler, *Gauss: A Biographical Study*, Springer-Verlag, Berlin, 1981.

77. W. Burnside, *Theory of Probability*, Dover, New York, 1959.

78. D. M. Burton, *The History of Mathematics*, Allyn and Bacon, Boston, 1985.

79. ——, *Elementary Number Theory*, 2nd ed., William C. Brown, Dubuque, 1989.

80. W. H. Bussey, Fermat's method of infinite descent, *Amer. Math. Monthly*, 25 (1918) 333–337.

81. F. Cajori, *A History of Mathematics*, 4th ed., Chelsea, New York, 1985.

82. P. Calabrese, A note on alternating series, *Amer. Math. Monthly*, 69 (1962) 215–217. (Reprinted in *Selected Papers on Calculus* [8], pp. 352–353.)

83. R. Calinger, ed., *Classics of Mathematics*, Moore Publishing Co., Oak Park, 1982.

84. Georg Cantor, Ueber unendliche, lineare Punktmannichfaltigkeiten, Nr. 5, *Math. Ann.*, 21 (1883) 545–591. (Reprinted in his *Gesammelte Abhandlungen*, pp. 165–209, Verlag von Julius Springer, Berlin, 1932.)

85. L. Carlitz, Note on irregular primes, *Proc. of the Amer. Math. Soc.*, 5 (1954) 329–331.

86. B. C. Carlson, Algorithms involving arithmetic and geometric means, *Amer. Math. Monthly*, 78 (1971) 496–505.

87. R. D. Carmichael, *The Theory of Numbers, and Diophantine Analysis*, Dover, New York, 1959.

88. J. D. Cassini, Une nouvelle progression de nombres, *Histoire de l'Académie Royale des Sciences, Paris*, 1 (1733) 201.

89. D. Castellanos, The ubiquitous π, *Math. Mag.*, Part I, 61 (1988) 67–98; Part II, 61 (1988) 148–163.

90. Augustin-Louis Cauchy, *Cours d'Analyse de l'École Royale Polytechnique*, Imprimerie Royale, Paris, 1821.

91. G. J. Chaitin, Randomness in arithmetic, *Sci. Amer.* (July 1988) 80–85.

92. E. W. Cheney, *Introduction to Approximation Theory*, 2nd ed., Chelsea, New York, 1982.

93. J. M. Child, *The Early Mathematical Manuscripts of Leibniz*, Open Court, Chicago, 1920.

94. D. V. Chudnovsky and G. V. Chudnovsky, The computation of classical constants, *Proc. of the National Academy of Sciences USA*, 86 (1989) 8178–8182.

95. D. I. A. Cohen, An explanation of the first digit phenomenon, *J. of Comb. Thy., Series A*, 20 (1976) 367–370.

96. A. J. Coleman, A simple proof of Stirling's formula, *Amer. Math. Monthly*, 58 (1951) 334–336. (Reprinted in *Selected Papers on Calculus* [8], pp. 325–327.)

97. ——, The probability integral, *Amer. Math. Monthly*, 61 (1954) 710–711.

98. S. D. Collingwood, ed., *The Lewis Carroll Picture Book*, T. Fisher Unwin, London, 1899. (Reprinted by Dover, 1961, under the new title *Diversions and Digressions of Lewis Carroll*.)

99. W. N. Colquitt and L. Welsh, Jr., A new Mersenne prime, *Math. of Comp.*, 56 (1991) 867–870.

100. L. Comtet, *Advanced Combinatorics: The Art of Finite and Infinite Expansions*, Reidel, Boston, 1974.

101. S. D. Conte and C. de Boor, *Elementary Numerical Analysis: An Algorithmic Approach*, 3rd ed., McGraw-Hill, New York, 1980.

102. T. A. Cook, *The Curves of Life*, Dover, New York, 1979.

103. J. L. Coolidge, Two geometrical applications of the method of least squares, *Amer. Math. Monthly*, 20 (1913) 187–190.

104. ——, The story of the binomial theorem, *Amer. Math. Monthly*, 56 (1949) 147–157.

105. ——, The number e, *Amer. Math. Monthly*, 57 (1950) 591–602. (Reprinted in *Selected Papers on Calculus* [8], pp. 8–10.)

106. ——, *An Introduction to Mathematical Probability*, Dover, New York, 1962.

107. ——, *The Mathematics of Great Amateurs*, 2nd ed., Oxford University Press, Oxford, 1990.

108. E. T. Copson, *Asymptotic Expansions*, Cambridge University Press, Cambridge, 1965.

109. R. Courant and D. Hilbert, *Methods of Mathematical Physics*, Interscience, New York, 1962.

110. R. Courant and F. John, *Introduction to Calculus and Analysis*, Interscience, New York, 1974.

111. R. Courant and H. Robbins, *What is Mathematics?*, Oxford University Press, London, 1953.

112. D. A. Cox, The arithmetic-geometric mean of Gauss, *L'Enseign. Math.*, 30 (1984) 275–330.

113. ——, Gauss and the arithmetic-geometric mean, *Notices of the Amer. Math. Soc.*, 32 (1985) 147–151.

114. H. S. M. Coxeter, The golden section, phyllotaxis, and Wythoff's game, *Scripta Math.*, 19 (1953) 135–143.

115. ——, *Contributions of Geometry to the Mainstream of Mathematics*, National Science Foundation Summer Mathematics Institute Notes, The Oklahoma Agricultural and Mechanical College, 1955.

116. ——, *Introduction to Geometry*, Wiley, New York, 1961.

117. R. E. Crandall, On the "$3x + 1$" problem, *Math. of Comp.*, 32 (1978) 1281–1292.

118. H. M. Cundy and A. D. Rollet, *Mathematical Models*, 3rd ed., Tarquin Publications, Norfolk, 1981.

119. J. W. Dauben, *Georg Cantor*, Princeton University Press, Princeton, 1990.

120. H. Davenport, *The Higher Arithmetic*, 5th ed., Cambridge University Press, Cambridge, 1982. (Reprinted by Dover, New York, 1983.)

121. M. Davis, Hilbert's tenth problem is unsolvable, *Amer. Math. Monthly*, 80 (1973) 233–269.

122. M. Davis and R. Hersh, Hilbert's 10th problem, *Sci. Amer.* (November 1973) 84–91. (Reprinted in *The Chauvenet Papers* [1], Volume II, pp. 555–571.)

123. P. J. Davis, Leonhard Euler's integral: A historical profile of the gamma function, *Amer. Math. Monthly*, 66 (1959) 849–869. (Reprinted in *The Chauvenet Papers* [1], Volume II, pp. 332–351.)

124. ——, *Interpolation and Approximation*, Blaisdell, New York, 1963. (Reprinted by Dover, New York, 1975.)

125. ——, Are there coincidences in mathematics?, *Amer. Math. Monthly*, 88 (1981) 311–320.

126. P. J. Davis and R. Hersh, *The Mathematical Experience*, Birkhäuser, Boston, 1981.

127. P. J. Davis and P. Rabinowitz, *Numerical Integration*, Blaisdell, Waltham, 1967.

128. J.-P. Delahaye, Chaitin's equation: An extension of Gödel's theorem, *Notices of the Amer. Math. Soc.*, 36 (1989) 984–987.

129. Augustus De Morgan, *The Differential and Integral Calculus*, Baldwin and Cradock, London, 1842.

130. ——, *A Budget of Paradoxes*, 2nd ed., Open Court, La Salle, 1974.

131. R. L. Devaney, *An Introdution to Chaotic Dynamical Systems*, 2nd ed., Addison-Wesley, Redwood City, 1989.

132. ——, *Chaos, Fractals, and Dynamics: Computer Experiments in Mathematics*, Addison-Wesley, Menlo Park, 1990.

133. R. L. Devaney and L. Keen, eds., *Chaos and Fractals: The Mathematics Behind the Computer Graphics*, American Mathematical Society, Providence, 1989.

134. K. J. Devlin, *Mathematics: The New Golden Age*, Penguin Books, New York, 1988.

135. L. E. Dickson, *History of the Theory of Numbers*, volume 1: Divisibility and Primality, volume 2: Diophantine Analysis, Chelsea, New York, 1971.

136. J. D. Dixon, Factorization and primality tests, *Amer. Math. Monthly*, 91 (1984) 333–352.

137. H. Dörrie, *100 Great Problems of Elementary Mathematics*, Dover, New York, 1965.

138. U. Dudley, Formulas for primes, *Math. Mag.*, 56 (1983) 17–22.

139. ——, *A Budget of Trisections*, Springer-Verlag, New York, 1987.

140. W. Dunham, *Journey through Genius*, Wiley, New York, 1990.

141. J. Dutka, On square roots and their representations, *Archive for History of Exact Sciences*, 36 (1986) 21–39.

142. ——, On the St. Petersburg paradox, *Archive for History of Exact Sciences*, 39 (1988) 13–39.

143. F. Dyson, Characterizing irregularity, *Science*, 200 (1978) 677–678.

144. A. W. F. Edwards, *Pascal's Arithmetical Triangle*, Oxford University Press, New York, 1987.

145. C. H. Edwards, Jr., *The Historical Development of the Calculus*, Springer-Verlag, New York, 1979.

146. H. M. Edwards, *Fermat's Last Theorem*, Springer-Verlag, New York, 1977.

147. N. D. Elkies, On $A^4 + B^4 + C^4 = D^4$, *Math. of Comp.*, 51 (1988) 825–835.

148. P. Erdős and R. L. Graham, *Old and New Problems and Results in Combinatorial Number Theory*, L'Enseign. Math., Université de Genève, Genève, 1980.

149. Leonhard Euler, *Introduction to Analysis of the Infinite*, 2 vols., Springer-Verlag, New York, 1988–1990.

150. H. W. Eves, *Great Moments in Mathematics (Before 1650)*, Mathematical Association of America, Washington, D.C., 1980.

151. ——, *Great Moments in Mathematics (After 1650)*, Mathematical Association of America, Washington, D.C., 1981.

152. ——, *An Introduction to the History of Mathematics*, 6th ed., Saunders, Philadelphia, 1990.

153. H. Eves and E. P. Starke, eds., The Otto Dunkel Memorial Problem Book, *Amer. Math. Monthly*, 64 (1957), Supplement to Number 7. (Contains an Index of Solutions to *Monthly* Problems, 1918–1950.)

154. R. P. Feinerman and D. J. Newman, *Polynomial Approximation*, Williams & Wilkins, Baltimore, 1974.

155. W. Feller, A direct proof of Stirling's formula, *Amer. Math. Monthly*, 74 (1967) 1223–1225.

156. ——, *An Introduction to Probability Theory and Its Applications*, Volume I, 3rd ed., 1970; Volume II, 2nd ed., 1968, Wiley, New York.

157. Pierre de Fermat, *Oeuvres de Fermat*, edited by Paul Tannery and Charles Henry, 5 vols., Gauthier-Villars, Paris, 1891–1922.

158. J. D. Foley and A. van Dam, *Fundamentals of Interactive Computer Graphics*, Addison-Wesley, Reading, 1983.

159. A. A. Fraenkel, *Abstract Set Theory*, 4th ed., American Elsevier, New York, 1976.

160. D. Freedman, R. Pisani, and R. Purves, *Statistics*, Norton, New York, 1978.

161. M. Gardner, *The Unexpected Hanging*, Simon and Schuster, New York, 1969.

162. ——, *Mathematical Magic Show*, Alfred A. Knopf, New York, 1977.

163. ——, *Wheels, Life and Other Mathematical Amusements*, Freeman, New York, 1983.

164. ——, About phi, an irrational number that has some remarkable geometrical expressions, *Sci. Amer.* (August 1959) 128–134. (Reprinted with additions in his book *The 2nd Scientific American Book of Mathematical Puzzles & Diversions*, University of Chicago Press, Chicago, 1987, pp. 89–103.)

165. Carl Friedrich Gauss, *Theory of the Motion of the Heavenly Bodies Moving about the Sun in Conic Sections,* Dover, New York, 1963.

166. ——, *Disquisitiones Arithmeticae*, rev. ed., Springer-Verlag, New York, 1986.

167. L. Gehér, I. Kovács, L. Pintér, G. Szász, eds., *Contests in Higher Mathematics*, Akadémiai Kiadó, Budapest, 1968.

168. B. R. Gelbaum and J. M. H. Olmsted, *Counterexamples in Analysis*, Holden-Day, Oakland, 1964.

169. S. I. Gelfand, M. L. Gerver, A. A. Kirilov, N. N. Konstantinov, and A. G. Kushnirenko, *Sequences, Combinations, Limits*, MIT Press, Cambridge, 1969.

170. A. O. Gelfond, *Calculus of Finite Differences*, Hindustan Publishing Corp., Delhi, 1971.

171. A. M. Gleason, R. E. Greenwood, and L. M. Kelly, eds., *The William Lowell Putnam Mathematical Competition: Problems and Solutions, 1938–1964*, Mathematical Association of America, Washington, D.C., 1980.

172. J. Gleick, *Chaos: Making a New Science*, Penguin Books, New York, 1988.

173. B. V. Gnedenko and A. Ya. Khinchin, *An Elementary Introduction to the Theory of Probability*, Dover, New York, 1962.

174. C. Goffman and G. Pedrick, *First Course in Functional Analysis*, 2nd ed., Chelsea, New York, 1983.

175. R. R. Goldberg, *Methods of Real Analysis*, 2nd ed., Wiley, New York, 1976.

176. S. Goldberg, *Introduction to Difference Equations*, Wiley, New York, 1958. (Reprinted by Dover, New York, 1986.)

177. ——, *Probability: An Introduction*, Prentice-Hall, Englewood Cliffs, 1962. (Reprinted by Dover, New York, 1987.)

178. H. H. Goldstine, *History of Numerical Analysis from the 16th through the 19th Century*, Springer-Verlag, New York, 1977.

179. S. W. Golomb, Combinatorial proof of Fermat's "little" theorem, *Amer. Math. Monthly*, 63 (1956) 718.

180. L. I. Golovina and I. M. Yaglom, *Induction in Geometry*, D. C. Heath and Company, Boston, 1963.

181. E. Goursat, *A Course in Mathematical Analysis*, translated by E. R. Hedrick, Ginn and Company, Boston, 1904.

182. J. V. Grabiner, Is mathematical truth time-dependent?, *Amer. Math. Monthly*, 81 (1974) 354–365.

183. ——, *The Origin of Cauchy's Rigorous Calculus*, MIT Press, Cambridge, 1981.

184. ——, Who gave you the epsilon? Cauchy and the origins of rigorous calculus, *Amer. Math. Monthly*, 90 (1983) 185–194.

185. R. L. Graham, On finite sums of unit fractions, *Proc. of the London Math. Soc.*, 14 (1964) 193–207.

186. R. L. Graham, D. E. Knuth, and O. Patashnik, *Concrete Mathematics*, Addison-Wesley, Reading, 1990.

187. A. Granville, The set of exponents, for which Fermat's last theorem is true, has density one, *Comptes Rendus/Mathematical Reports, Academy of Science, Canada*, 7 (1985) 55–60.

188. I. Grattan-Guinness, *The Development of the Foundations of Mathematical Analysis from Euler to Riemann*, MIT Press, Cambridge, 1970.

189. ——, ed., *From the Calculus to Set Theory, 1630–1910*, Duckworth, London, 1980.

190. E. Grosswald, *Representations of Integers as Sums of Squares*, Springer-Verlag, New York, 1985.

191. R. K. Guy, *Unsolved Problems in Number Theory*, Springer-Verlag, New York, 1981.

192. ——, The strong law of small numbers, *Amer. Math. Monthly*, 95 (1988) 697–712.

193. ——, The second strong law of small numbers, *Math. Mag.*, 63 (1990) 3–20.

194. Jacques Hadamard, *An Essay on the Psychology of Invention in the Mathematical Field*, Dover, New York, 1954.

195. T. Hall, *Carl Friedrich Gauss*, MIT Press, Cambridge, 1970.

196. P. R. Halmos, How to write mathematics, *L'Enseign. Math.*, 16 (1970) 123–152.

197. ——, Has progress in mathematics slowed down?, *Amer. Math. Monthly*, 97 (1990) 561–588.

198. ——, *Problems for Mathematicians, Young and Old*, Mathematical Association of America, Washington, D.C., 1991.

199. J. Hambidge, "Notes on the Logarithmic Spiral, Golden Section and the Fibonacci Series," in *Dynamic Symmetry*, Yale University Press, New Haven, 1920.

200. R. W. Hamming, An elementary discussion of the transcendental nature of the elementary transcendental functions, *Amer. Math. Monthly*, 77 (1970) 294–297.

201. H. Hancock, *Elliptic Integrals*, Dover, New York, 1958.

202. ——, *Lectures on the Theory of Maxima and Minima of Functions of Several Variables*, Dover, New York, 1960.

203. E. Hansen, *A Table of Series and Products*, Prentice-Hall, Englewood Cliffs, 1975.

204. G. H. Hardy, *Divergent Series*, Oxford University Press, Oxford, 1949.

205. ——, *The Integration of Functions of a Single Variable*, 2nd ed., Hafner, New York, 1971.

206. ——, *A Course of Pure Mathematics*, 10th ed., Cambridge University Press, Cambridge, 1975.

207. ——, An Introduction to the Theory of Numbers, in *The Chauvenet Papers* [1], Volume I, pp. 67–94. (The sixth Josiah Willard Gibbs Lecture, read at New York City, December 28, 1928. Previously published in the *Bull. of the Amer. Math. Soc.*, 35 (1929) 778–818.)

208. ——, *Ramanujan: Twelve Lectures on Subjects Suggested by His Life and Work*, 3rd ed., Chelsea, New York, 1978.

209. ——, *A Mathematician's Apology*, Cambridge University Press, Cambridge, 1979.

210. G. H. Hardy, J. E. Littlewood, and G. Pólya, *Inequalities*, 2nd ed., Cambridge University Press, Cambridge, 1952.

211. G. H. Hardy and E. M. Wright, *An Introduction to the Theory of Numbers*, 5th ed., Oxford University Press, Oxford, 1988.

212. F. Hartt, *Art: A History of Painting, Sculpture, and Architecture*, 3rd ed., Prentice-Hall, Englewood Cliffs, 1989.

213. T. L. Heath, *Diophantus of Alexandria*, 2nd ed., Dover, New York, 1964.

214. ——, *A History of Greek Mathematics*, Volume I: *From Thales to Euclid*, Volume II: *From Aristarchus to Diophantus*, Dover, New York, 1981.

215. D. R. Heath-Brown, Fermat's last theorem for "almost all" exponents, *Bull. of the London Math. Soc.*, 17 (1985) 15–16.

216. ——, The first case of Fermat's Last Theorem, *Math. Intell.*, 7, no. 4 (1985) 40–47, 55.

217. J. M. Henle and E. M. Kleinberg, *Infinitesimal Calculus*, MIT Press, Cambridge, 1979.

218. R. Herz-Fischler, *A Mathematical History of Division in Extreme and Mean Ratio*, Wilfrid Laurier University Press, Waterloo, 1987.

219. S. Hildebrandt and A. Tromba, *Mathematics and Optimal Form*, Scientific American Library, New York, 1985.

220. P. Hilton and J. Pedersen, Looking into Pascal's triangle: combinatorics, arithmetic, and geometry, *Math. Mag.*, 60 (1987) 305–316.

221. ——, Catalan numbers, their generalization, and their uses, *Math. Intell.*, 13, no. 2 (1991) 64–75.

222. A. M. Hinz, The tower of Hanoi, *L'Enseign. Math.*, 35 (1989) 289–321.

223. E. W. Hobson, *Squaring the Circle and Other Monographs*, Chelsea, New York, 1969.

224. J. E. Hofmann, *Leibniz in Paris 1672–1676*, Cambridge University Press, New York, 1974.

225. V. E. Hoggatt, *Fibonacci and Lucas Numbers*, Houghton-Mifflin, Boston, 1969.

226. R. Honsberger, *Ingenuity in Mathematics*, Mathematical Association of America, Washington, D.C., 1970.

227. ——, *Mathematical Gems*, Mathematical Association of America, Washington, D.C., 1973.

228. ——, *Mathematical Gems II*, Mathematical Association of America, Washington, D.C., 1976.

229. ——, *Mathematical Morsels*, Mathematical Association of America, Washington, D.C., 1978.

230. ——, ed., *Mathematical Plums*, Mathematical Association of America, Washington, D.C., 1979.

231. ——, *Mathematical Gems III*, Mathematical Association of America, Washington, D.C., 1985.

232. ——, *More Mathematical Morsels*, Mathematical Association of America, Washington, D.C., 1991.

233. G. Horton, A note on the calculation of Euler's constant, *Amer. Math. Monthly*, 23 (1916) 73. (Reprinted in *Selected Papers on Calculus* [8], pp. 388–389.)

234. L.-K. Hua, *Introduction to Number Theory*, Springer-Verlag, Berlin, 1982.

235. H. E. Huntley, *The Divine Proportion*, Dover, New York, 1970.

236. D. Jackson, A proof of Weierstrass's theorem, *Amer. Math. Monthly*, 41 (1934) 309–312. (Reprinted in *Selected Papers on Calculus* [8], pp. 227–231.)

237. L. B. W. Jolley, *Summation of Series*, 2nd ed., Dover, New York, 1961.

238. J. P. Jones and Y. V. Matijasevič, Proof of recursive unsolvability of Hilbert's tenth problem, *Amer. Math. Monthly*, 98 (1991) 689–709.

239. J. P. Jones, D. Sato, H. Wada, and D. Wiens, Diophantine representation of the set of prime numbers, *Amer. Math. Monthly*, 83 (1976) 449–464.

240. K. Jordan, *Calculus of Finite Differences*, 3rd ed., Chelsea, New York, 1965.

241. M. Kac, *Statistical Independence in Probability, Analysis and Number Theory*, Mathematical Association of America, Washington, D.C., 1959.

242. ——, *Enigmas of Chance*, University of California Press, Berkeley, 1987.

243. R. Kanigel, *The Man Who Knew Infinity*, Charles Scribner's Sons, New York, 1991.

244. E. Kasner and J. Newman, *Mathematics and the Imagination*, Simon and Schuster, New York, 1940.

245. N. D. Kazarinoff, *Geometric Inequalities*, Mathematical Association of America, Washington, D.C., 1961.

246. ——, *Analytic Inequalities*, Holt, Rinehart and Winston, New York, 1964.

247. H. J. Keisler, *Foundations of Infinitesimal Calculus*, Prindle, Weber & Schmidt, Boston, 1976.

248. ——, *Elementary Calculus*, 2nd ed., Prindle, Weber & Schmidt, Boston, 1986.

249. J. G. Kemeny, J. L. Snell, and G. L. Thompson, *Introduction to Finite Mathematics*, 3rd ed., Prentice-Hall, Englewood Cliffs, 1974.

250. J. Kiefer, Sequential minimax search for a maximum, *Proc. of the Amer. Math. Soc.*, 4 (1953) 502–506.

251. P. Kitcher, *The Nature of Mathematical Knowledge*, Oxford University Press, New York, 1984.

252. G. Klambauer, *Problems and Propositions in Analysis*, Marcel Dekker, New York, 1979.

253. ——, *Aspects of Calculus*, Springer-Verlag, New York, 1986.

254. M. S. Klamkin, *International Mathematical Olympiads, 1979–1985 and Forty Supplementary Problems*, Mathematical Association of America, Washington, D.C., 1986.

255. V. Klee and S. Wagon, *Old and New Unsolved Problems in Plane Geometry and Number Theory*, Mathematical Association of America, Washington, D.C., 1991.

256. F. Klein, *Elementary Mathematics from an Advanced Standpoint: Arithmetic, Algebra, Analysis*, Dover, New York, 1953.

257. ——, *Famous Problems of Elementary Geometry*, 2nd ed., Dover, New York, 1956.

258. M. Kline, *Mathematics, The Loss of Certainty*, Oxford University Press, New York, 1980.

259. ——, Euler and infinite series, *Math. Mag.*, 56 (1983) 307–314.

260. ——, *Mathematical Thought from Ancient to Modern Times*, Oxford University Press, New York, 1972.

261. R. A. Knoebel, Exponentials reiterated, *Amer. Math. Monthly*, 88 (1981), 235–252.

262. K. Knopp, *Theory and Application of Infinite Series*, 2nd ed., Dover, New York, 1990.

263. D. E. Knuth, *The Art of Computer Programming*, Volume 1: *Fundamental Algorithms,* Volume 2: *Seminumerical Algorithms,* Addison-Wesley, Reading, 1969.

264. N. Koblitz, *A Course in Number Theory and Cryptography*, Springer-Verlag, New York, 1987.

265. B. A. Kordemsky, *The Moscow Puzzles*, Scribner's, New York, 1972.

266. P. P. Korovkin, *Inequalities*, MIR Publishers, Moscow, 1987.

267. J. Kramer, The Fibonacci series in twentieth-century music, *J. of Music Theory*, 17 (1973) 110–149.

268. E. Kranakis, *Primality and Cryptography*, Wiley, New York, 1986.

269. L. Krüger, L. J. Daston, and M. Heidelberger, eds., *The Probabilistic Revolution,* Volume 1: *Ideas in History,* MIT Press, Cambridge, 1987.

270. J. Kürschák, ed., *Hungarian Problem Book I*, Mathematical Association of America, Washington, D.C., 1963.

271. ——, ed., *Hungarian Problem Book II*, Mathematical Association of America, Washington, D.C., 1963.

272. J. C. Lagarias, The $3x + 1$ problem and its generalizations, *Amer. Math. Monthly*, 92 (1985) 3–23.

273. I. Lakatos, *Proofs and Refutations: The Logic of Mathematical Discovery*, Cambridge University Press, New York, 1987.

274. L. C. Larson, *Problem-Solving through Problems*, Springer-Verlag, New York, 1983.

275. E. J. Lee and J. S. Madachy, The history and discovery of amicable numbers, *J. of Rec. Math.*, 5 (1972), Part I, 77–93; Part II, 153–173; Part III, 231–249; Errata, 6 (1973) 164, 229.

276. D. H. Lehmer, Interesting series involving the central binomial coefficient, *Amer. Math. Monthly*, 92 (1985) 449–457.

277. D. N. Lehmer, Hunting big game in the theory of numbers, *Scripta Math.*, 1 (1932) 229–235.

278. F. Le Lionnais, ed., *Great Currents of Mathematical Thought,* Volume I: *Mathematics: Concepts and Development,* Volume II: *Mathematics in the Arts and Sciences,* Dover, New York, 1971.

279. ——, *Les Nombres Remarquables*, Hermann, Paris, 1983.

280. W. J. LeVeque, *Topics in Number Theory*, Addison-Wesley, Reading, 1956.

281. ——, *Fundamentals of Number Theory*, Addison-Wesley, Reading, 1977.

282. J. E. Littlewood, *Littlewood's Miscellany*, edited by B. Bollobás, Cambridge University Press, Cambridge, 1986. (Formerly titled: *A Mathematician's Miscellany*.)

283. H. Lockwood, *A Book of Curves*, Cambridge University Press, Cambridge, 1971.

284. E. Lucas, *Récréations Mathématiques*, 4 vols., Gauthier-Villars, Paris, 1891–1896.

285. ——, *L'Arithmétique Amusante*, Gauthier-Villars, Paris, 1895.

286. R. D. Luce, R. R. Bush, and E. Galanter, eds., *Handbook of Mathematical Psychology*, Volume I, Wiley, New York, 1963.

287. A. Madeleine-Perdrillat, *Seurat*, Skira/Rizzoli, New York, 1990.

288. M. S. Mahoney, *The Mathematical Career of Pierre de Fermat (1601–1665)*, Princeton University Press, Princeton, 1973.

289. L. E. Maistrov, *Probability Theory: A Historical Sketch*, Academic Press, New York, 1974.

290. B. B. Mandelbrot, *The Fractal Geometry of Nature*, Freeman, New York, 1982.

291. H. B. Mann and D. Shanks, A necessary and sufficient condition for primality, and its source, *J. of Comb. Thy., Series A,* 13 (1972) 131–134.

292. H. B. Mann and W. A. Webb, A short proof of Fermat's theorem for $n = 3$, *Math. Student*, 46 (1978) 103–104.

293. E. Maor, *To Infinity and Beyond: A Cultural History of the Infinite*, Birkhäuser, Boston, 1987.

294. G. Markowsky, Misconceptions about the golden ratio, *Coll. Math. J.*, 23 (1992) 2–19.

295. A. I. Markushevich, *Recursion Sequences*, Mir Publishers, Moscow, 1975.

296. Y. Matsuoka, An elementary proof of the formula $\sum_{k=1}^{\infty} 1/k^2 = \pi^2/6$, *Amer. Math. Monthly*, 68 (1961) 485–487. (Reprinted in *Selected Papers on Calculus* [8], pp. 372–373.)

297. R. D. Mauldin, ed., *The Scottish Book*, Birkhäuser, Boston, 1981.

298. K. O. May, *Index of the American Mathematical Monthly, Volumes 1 through 80 (1894–1973)*, Mathematical Association of America, Washington, D.C., 1977.

299. N. S. Mendelsohn, An application of a famous inequality, *Amer. Math. Monthly*, 58 (1951) 563. (Reprinted in *Selected Papers on Calculus* [8], pp. 380–381.)

300. H. Meschkowski, *Ways of Thought of Great Mathematicians*, Holden-Day, San Francisco, 1964.

301. G. Miel, Of calculations past and present: the Archimedean algorithm, *Amer. Math. Monthly*, 90 (1983) 17–35.

302. W. H. Mills, A prime-representing function, *Bull. of the Amer. Math. Soc.*, 53 (1947) 604.

303. G. J. Mitchison, Phyllotaxis and the Fibonacci series, *Science*, 196 (1977) 270–275.

304. D. S. Mitrinović, *Elementary Inequalities*, P. Noordhoff, Groningen, 1964.

305. ——, *Analytic Inequalities*, Springer-Verlag, New York, 1970.

306. C. N. Moore, Summability of series, *Amer. Math. Monthly*, 39 (1932) 62–71. (Reprinted in *Selected Papers on Calculus* [8], pp. 333–341.)

307. R. E. M. Moore, Mosaic units: patterns in ancient mosaics, *Fibonacci Quarterly*, 8 (1970) 281–310.

308. R. E. Moritz, *On Mathematics*, Dover, New York, 1958. (Formerly titled: *Memorabilia Mathematica or the Philomath's Quotation-Book*.)

309. R. K. Morley, The remainder in computing by series, *Amer. Math. Monthly*, 57 (1950), 550–551. (Reprinted in *Selected Papers on Calculus* [8], pp. 324–325.)

310. F. Mosteller, *Fifty Challenging Problems in Probability with Solutions*, Dover, New York, 1987.

311. E. Nagel and J. R. Newman, *Gödel's Proof*, New York University Press, New York, 1958.

312. T. Nagell, *Introduction to Number Theory*, 2nd ed., Chelsea, New York, 1981.

313. J. Needham, *Science and Civilisation in China, Volume 3—Mathematics and the Sciences of the Heavens and the Earth*, Cambridge University Press, Cambridge, 1959.

314. D. J. Newman, Simple analytic proof of the prime number theorem, *Amer. Math. Monthly*, 87 (1980) 693–696.

315. ——, A simplified version of the fast algorithms of Brent and Salamin, *Math. of Comp.*, 44 (1985) 207–210.

316. J. R. Newman, ed., *The World of Mathematics*, 4 vols., Simon and Schuster, New York, 1956.

317. I. M. Niven, A simple proof that π is irrational, *Bull. of the Amer. Math. Soc.*, 53 (1947) 509.

318. ——, *Mathematics of Choice, or How to Count without Counting*, Mathematical Association of America, Washington, D.C., 1965.

319. ——, Formal power series, *Amer. Math. Monthly*, 76 (1969) 871–889.

320. ——, *Maxima and Minima without Calculus*, Mathematical Association of America, Washington, D.C., 1981.

321. I. M. Niven, H. S. Zuckerman, and H. L. Montgomery, *An Introduction to the Theory of Numbers*, 5th ed., Wiley, New York, 1991.

322. C. S. Ogilvy and J. T. Anderson, *Excursions in Number Theory*, Oxford University Press, New York, 1966. (Reprinted by Dover, New York, 1988.)

323. C. D. Olds, *Continued Fractions*, Mathematical Association of America, Washington, D.C., 1975.

324. O. Ore, On the averages of the divisors of a number, *Amer. Math. Monthly*, 55 (1948) 615–619.

325. ——, *Niels Henrik Abel, Mathematician Extraordinary*, Chelsea, New York, 1974.

326. ——, *Number Theory and Its History*, McGraw-Hill, New York, 1948. (Reprinted by Dover, New York, 1988.)

327. J. C. Oxtoby, *Measure and Category*, 2nd ed., Springer-Verlag, New York, 1980.

328. L. Pacioli, *Divina proportione*, Venice, 1509 (French translation by Librarie du Compagnonnage, Paris, 1980).

329. I. Papadimitriou, A simple proof of the formula $\sum_{k=1}^{\infty} k^{-2} = \pi^2/6$, *Amer. Math. Monthly*, 80 (1973) 424–425.

330. Blaise Pascal, *Traité du Triangle Arithmétique avec quelques autres petits traitez sur la mesme matiere*, Paris, 1665. (Reprinted in *Œuvres de Blaise Pascal*, 3, 445–503.)

331. ——, *Œuvres de Blaise Pascal*, edited by Léon Brunschvicg and Pierre Boutroux, Hachette, Paris, 1904–1914 (Kraus reprint, Nendeln/Lichtenstein, 1976–1978.)

332. H.-O. Peitgen, H. Jurgens, and D. Saupe, *Fractals for the Classroom*, Springer-Verlag, New York, 1991.

333. H.-O. Peitgen and P. H. Richter, *The Beauty of Fractals*, Springer-Verlag, Berlin, 1986.

334. H.-O. Peitgen and D. Saupe, eds., *The Science of Fractal Images*, Springer-Verlag, New York, 1988.

335. W. Penney, Problem 95: Penney-Ante, *J. of Rec. Math.*, 7 (1974) 321.

336. I. Peterson, *The Mathematical Tourist: Snapshots of Modern Mathematics*, Freeman, New York, 1988.

337. G. M. Phillips, Archimedes the numerical analyst, *Amer. Math. Monthly*, 88 (1981) 165–169.

338. G. Pólya, *Induction and Analogy in Mathematics*, Volume I of *Mathematics and Plausible Reasoning,* Princeton University Press, Princeton, 1954.

339. ——, *Patterns of Plausible Inference*, Volume II of *Mathematics and Plausible Reasoning,* Princeton University Press, Princeton, 1954.

340. ——, On picture-writing, *Amer. Math. Monthly*, 63 (1956) 689–697.

341. ——, *Mathematical Discovery*, 2 vols., Wiley, New York, 1962.

342. ——, *Collected Papers, Volume IV : Probability, Combinatorics, Teaching and Learning in Mathematics*, MIT Press, Cambridge, 1984.

343. ——, *How to Solve It*, 2nd ed., Princeton University Press, Princeton, 1988.

344. G. Pólya and J. Kilpatrick, *The Stanford Mathematics Problem Book*, Teachers College Press, New York, 1974.

345. G. Pólya and G. Szegö, *Problems and Theorems in Analysis*, 2 vols., Springer-Verlag, New York, 1972–1976.

346. C. Pomerance, Recent developments in primality testing, *Math. Intell.*, 3, no. 3 (1981), 97–105.

347. ——, The search for prime numbers, *Sci. Amer.* (December 1982) 136–147.

348. C. Pomerance, J. L. Selfridge, and S. S. Wagstaff, Jr., The pseudoprimes to $25 \cdot 10^9$, *Math. of Comp.*, 35 (1980) 1003–1026.

349. R. Preston, Profiles: The mountains of pi, *The New Yorker* (March 2, 1992) 36–67.

350. J. D. Pryce, *Basic Methods of Linear Functional Analysis*, Hutchinson University Library, London, 1973.

351. H. Rademacher, *Lectures on Elementary Number Theory*, Blaisdell, New York, 1964.

352. ——, *Higher Mathematics from an Elementary Point of View*, Birkhäuser, Boston, 1983.

353. H. Rademacher and O. Toeplitz, *The Enjoyment of Mathematics*, Princeton University Press, Princeton, 1964. (Reprinted by Dover, New York, 1990.)

354. C. H. Raifaizen, A simpler proof of Heron's formula, *Math. Mag.*, 44 (1971) 27–28.

355. R. A. Raimi, The first digit problem, *Amer. Math. Monthly*, 83 (1976) 521–538.

356. R. A. Reiff, *Geschichte der unendlichen Reihen*, H. Lauppsche Buchhandlung, Tübingen, 1889. (Reprinted by Martin Sändig, Wiesbaden, 1969.)

357. A. Rényi, *A Diary on Information Theory*, Wiley, New York, 1987.

358. P. Ribenboim, *13 Lectures on Fermat's Last Theorem*, Springer-Verlag, New York, 1979.

359. ——, "1093", *Math. Intell.*, 5, no. 2 (1983) 28–34.

360. ——, Recent results about Fermat's last theorem, *Expositiones Mathematicae*, 5, (1987) 75–90.

361. ——, Euler's famous prime generating polynomial and the class number of imaginary quadratic fields, *L'Enseign. Math.*, 34 (1988) 23–42.

362. ——, *The Book of Prime Number Records*, 2nd ed., Springer-Verlag, New York, 1989.

363. ——, *The Little Book of Big Primes*, Springer-Verlag, New York, 1991.

364. H. J. J. te Riele, On generating new amicable pairs from given amicable pairs, *Math. of Comp.*, 42 (1984) 219–223.

365. ——, Computation of all the amicable pairs below 10^{10}, *Math. of Comp.*, 47 (1986) 361–368.

366. H. Riesel, *Prime Numbers and Computer Methods for Factorization*, Birkhäuser, Boston, 1985.

367. K. H. Rosen, *Elementary Number Theory and Its Applications*, 2nd ed., Addison-Wesley, Reading, 1988.

368. M. I. Rosen, A proof of the Lucas-Lehmer test, *Amer. Math. Monthly*, 95 (1988) 855–856.

369. R. Roy, The discovery of the series formula for π by Leibniz, Gregory and Nilakantha, *Math. Mag.*, 63 (1990) 291–306.

370. E. Salamin, Computation of π using arithmetic-geometric mean, *Math. of Comp.*, 30 (1976) 565–570.

371. E. Schrödinger, *Science and Humanism*, Cambridge University Press, Cambridge, 1951.

372. M. R. Schroeder, Where is the next Mersenne prime hiding?, *Math. Intell.*, 5, no. 3 (1983) 31–33.

373. ——, *Number Theory in Science and Communication*, 2nd ed., Springer-Verlag, New York, 1990.

374. D. Shanks, *Solved and Unsolved Problems in Number Theory*, 3rd ed., Chelsea, New York, 1985.

375. C. E. Shannon and W. Weaver, *The Mathematical Theory of Communication*, University of Illinois Press, Urbana, 1962.

376. H. N. Shapiro, *Introduction to the Theory of Numbers*, Wiley, New York, 1982.

377. H. S. Shapiro, *Smoothing and Approximation of Functions*, Van Nostrand Reinhold, New York, 1969.

378. Yu. A. Shashkin, *Fixed Points*, American Mathematical Society, Providence, 1991.

379. D. O. Shklarsky, N. N. Chentzov, and I. M. Yaglom, *The USSR Olympiad Problem Book*, Freeman, San Francisco, 1962.

380. J. A. Shohat, On a certain transformation of infinite series, *Amer. Math. Monthly*, 40 (1933) 226–229. (Reprinted in *Selected Papers on Calculus* [8], pp. 342–345.)

381. W. Sierpiński, Sur une courbe cantorienne qui contient une image biunivoque et continue de toute courbe donnée, *Comptes Rendus de L'Académie des Sciences, Paris*, 162 (1916) 629–632.

382. ——, *Pythagorean Triangles*, Yeshiva University, New York, 1962.

383. ——, *A Selection of Problems in the Theory of Numbers*, Macmillan, New York, 1964.

384. ——, *250 Problems in Elementary Number Theory*, American Elsevier, New York, 1970.

385. ——, *Elementary Theory of Numbers*, 2nd English ed., North-Holland, New York, 1988.

386. R. D. Silverman, A perspective on computational number theory, *Notices of the Amer. Math. Soc.*, 38 (1991) 562–568.

387. G. F. Simmons, *Calculus with Analytic Geometry*, McGraw-Hill, New York, 1985.

388. P. Singh, The so-called Fibonacci numbers in ancient and medieval India, *Hist. Math.*, 12 (1985) 229–244.
389. N. J. A. Sloane, *A Handbook of Integer Sequences*, Academic Press, New York, 1973.
390. ——, The persistence of a number, *J. of Rec. Math.*, 6 (1973) 97–98.
391. L. L. Smail, *History and Synopsis of the Theory of Summable Infinite Processes*, University of Oregon, Eugene, 1925.
392. D. E. Smith, *History of Mathematics*, 2 vols., Dover, New York, 1958.
393. ——, *A Source Book in Mathematics*, 2 vols., Dover, New York, 1959.
394. I. S. Sominskii, *The Method of Mathematical Induction*, Heath, Boston, 1963.
395. M. R. Spiegel, An elementary method for evaluating an infinite integral, *Amer. Math. Monthly*, 58 (1951) 555–558.
396. ——, Remarks concerning the probability integral, *Amer. Math. Monthly*, 63 (1956) 35–37.
397. M. Spivak, *Calculus*, 2nd ed., Publish or Perish, Wilmington, 1980.
398. ——, *Answer Book for Calculus*, 2nd ed., Publish or Perish, Houston, 1984.
399. H. Steinhaus, *One Hundred Problems in Elementary Mathematics*, Dover, New York, 1979.
400. I. Stewart, *The Problems of Mathematics*, Oxford University Press, 1987.
401. ——, Mathematical recreations, *Sci. Amer.* (December 1990) 128–131.
402. T. J. Stieltjes, *Oeuvres Complètes, Volume II*, Wolters-Noordhoff, Groningen, 1918.
403. S. M. Stigler, *The History of Statistics: The Measurement of Uncertainty before 1900*, Harvard University Press, Cambridge, 1986.
404. J. Stillwell, *Mathematics and Its History*, Springer-Verlag, New York, 1989.
405. S. Straszewicz, *Mathematical Problems and Puzzles from the Polish Mathematical Olympiads*, Pergamon Press, New York, 1965.
406. K. D. Stroyan and W. A. J. Luxemburg, *Introduction to the Theory of Infinitesimals*, Academic Press, New York, 1976.
407. D. J. Struik, *A Concise History of Mathematics*, 3rd ed., Dover, New York, 1967.
408. ——, ed., *A Source Book in Mathematics, 1200–1800*, Harvard University Press, Cambridge, 1969.
409. J. J. Sylvester, *The Collected Mathematical Papers of James Joseph Sylvester*, 4 vols., Cambridge University Press, Cambridge, 1904–1912.
410. B. Sz.-Nagy, *Introduction to Real Functions and Orthogonal Expansions*, Oxford University Press, New York, 1965.
411. J. W. Tanner and S. S. Wagstaff, Jr., New congruences for the Bernoulli numbers, *Math. of Comp.*, 48 (1987) 341–350.
412. ——, New bound for the first case of Fermat's last theorem, *Math. of Comp.*, 53 (1989) 743–750.
413. A. E. Taylor and W. R. Mann, *Advanced Calculus*, 3rd ed., Wiley, New York, 1983.
414. D. W. Thompson, *On Growth and Form*, 2nd ed., Cambridge University Press, Cambridge, 1952.
415. V. M. Tikhomirov, *Stories about Maxima and Minima*, American Mathematical Society and Mathematical Association of America, Washington, D.C., 1990.
416. I. Todhunter, *A History of the Mathematical Theory of Probability from the Time of Pascal to that of Laplace*, Chelsea, New York, 1965.

417. O. Toeplitz, *The Calculus: A Genetic Approach*, University of Chicago Press, Chicago, 1981.
418. A. Tucker, *Applied Combinatorics*, 2nd ed., Wiley, New York, 1984.
419. V. A. Uspenskii, *Pascal's Triangle*, MIR Publishers, Moscow, 1976.
420. J. V. Uspensky, *Introduction to Mathematical Probability*, McGraw-Hill, New York, 1937.
421. J. V. Uspensky and M. A. Heaslet, *Elementary Number Theory*, McGraw-Hill, New York, 1937.
422. S. Vajda, *Fibonacci & Lucas Numbers and the Golden Section*, Halsted Press, New York, 1989.
423. A. van Dam, Computer software for graphics, *Sci. Amer.* (September 1984) 146–159.
424. B. Van der Pol and H. Bremmer, *Operational Calculus Based on the Two-Sided Laplace Integral*, 3rd ed., Chelsea, New York, 1987.
425. A. Van der Poorten, A proof that Euler missed . . . Apéry's proof of the irrationality of $\zeta(3)$, *Math. Intell.*, 1, no. 4 (1979) 195–203.
426. B. L. Van der Waerden, *Science Awakening*, 3rd ed., Wolters-Noordhofff, Groningen, 1971.
427. ——, *Geometry and Algebra in Ancient Civilizations*, Springer-Verlag, New York, 1983.
428. H. S. Vandiver, Is there an infinity of regular primes?, *Scripta Math.*, 21 (1955) 306–309.
429. N. Ya. Vilenkin, *Stories about Sets*, Academic Press, New York, 1968.
430. ——, *Combinatorics*, Academic Press, New York, 1971.
431. ——, *Method of Successive Approximations*, MIR Publishers, Moscow, 1979.
432. H. Von Koch, Sur une courbe continue sans tangente, obtenue par une construction géométrique élémentaire, *Arkiv för mathematik, astronomi och fysik*, 1 (1904) 681–704. (Reproduced in Une méthode géométrique élémentaire pour l'étude de certaines questions de la théorie des courbes planes, *Acta Math.*, 30 (1906) 145–174.)
433. J. Von Neumann, *Collected Works*, Volume V, Pergamon Press, Oxford, 1963.
434. N. N. Vorobyov, *The Fibonacci Numbers*, Heath, Boston, 1963.
435. S. Wagon, The evidence: The Collatz problem, *Math. Intell.*, 7, no. 1 (1985) 72–76.
436. ——, The evidence: Odd perfect numbers, *Math. Intell.*, 7, no. 2 (1985) 66–68.
437. ——, Is π normal?, *Math. Intell.*, 7, no. 3 (1985) 65–67.
438. ——, The evidence: Fermat's last theorem, *Math. Intell.*, 8, no. 1 (1986) 59–61.
439. ——, The evidence: Primality testing, *Math. Intell.*, 8, no. 3 (1986) 58–61.
440. ——, Editor's Corner: The Euclidean algorithm strikes again, *Amer. Math. Monthly*, 97 (1990) 125–129.
441. C. M. Walsh, Fermat's note XLV, *Ann. of Math.*, 29 (1928) 412–432.
442. W. Weaver, Lewis Carroll and a geometrical paradox, *Amer. Math. Monthly*, 45 (1938) 234–236.
443. A. Weil, Two lectures on number theory, past and present, *L'Enseign. Math.*, 20 (1974) 87–110.
444. ——, *Number Theory: An Approach through History from Hammurapi to Legendre*, Birkhäuser, Boston, 1984.

445. D. Wells, *The Penguin Dictionary of Curious and Interesting Numbers*, Penguin Books, Middlesex, 1986.

446. H. Weyl, *Philosophy of Mathematics and Natural Science*, Princeton University Press, Princeton, 1959.

447. D. T. Whiteside, Patterns of mathematical thought in the later seventeenth century, *Archive for History of Exact Sciences*, 1 (1960–1962) 179–388.

448. E. T. Whittaker and G. Robinson, *The Calculus of Observations*, Blackie & Son, London, 1924.

449. E. T. Whittaker and G. N. Watson, *A Course of Modern Analysis*, Cambridge University Press, Cambridge, 1943.

450. W. A. Whitworth, The equiangular spiral, its chief properties proved geometrically, *Messenger of Mathematics*, I (1862) 5–13.

451. R. L. Wilder, *Introduction to the Foundations of Mathematics*, 2nd ed., Wiley, New York, 1983.

452. H. S. Wilf, *Generatingfunctionology*, Academic Press, Boston, 1990.

453. C. P. Willans, On formulae for the nth prime, *Math. Gaz.*, 48 (1964) 413–415.

454. H. C. Williams, Factoring on a computer, *Math. Intell.*, 6, no. 3 (1984) 29–36.

455. H. C. Williams and H. Dubner, The primality of $R1031$, *Math. of Comp.*, 47 (1986) 703–711.

456. D. Wood, The towers of Brahma and Hanoi revisited, *J. of Rec. Math.*, 14 (1981) 17–24.

457. A. M. Yaglom and I. M. Yaglom, *Challenging Mathematical Problems with Elementary Solutions,* 2 vols., Holden-Day, San Francisco, 1964–1967. (Reprinted by Dover, New York, 1987.)

458. D. Yarden, A bibliography of the Fibonacci sequence, *Riveon Lematematika*, 2 (1948) 36–45.

459. R. M. Young, *Excursions in Calculus*, Mathematical Association of America, Washington, D.C., 1992.

460. G. U. Yule and M. G. Kendall, *An Introduction to the Theory of Statistics*, 14th ed., Hafner, New York, 1968.

461. D. Zagier, The first 50 million prime numbers, *Math. Intell.*, 0 (1977) 7–19.

462. ——, A one-sentence proof that every prime $p \equiv 1 \pmod 4$ is a sum of two squares, *Amer. Math. Monthly*, 97 (1990) 144.

463. E. Zeckendorf, Représentation des nombres naturels par une somme de nombres de Fibonacci ou de nombres de Lucas, *Bull. de la Soc. Royale des Sci. de Liège,* 41 (1972) 179–182.

SOURCES FOR SOLUTIONS

Chapter 1

1.1 *American Mathematical Monthly* 79 (1972) 67–69
   Excursions in Calculus [459] 5
1.2 *The Art of Computer Programming,* Volume 1 [263] 19
1.3 *Scientific American*, August 1970, 110
1.4 *American Mathematical Monthly* 74 (1967) 773
1.5 *Mathematical Plums* [230] 113–129
1.6 *Science Awakening* [426] 96
1.7 *Science Awakening* [426] 126
1.8 *Mathematical Gazette* 33 (1949) 41
1.9 *The Stanford Mathematics Problem Book* [344] 7
1.10 *Induction and Analogy in Mathematics* [338] 116
1.11 *Elementary Theory of Numbers* [385] 125
1.12 *Hungarian Problem Book II* [271] 15
1.13 *The USSR Olympiad Problem Book* [379] 17
1.14 *Elementary Theory of Numbers* [385] 118
1.15 *Excursions in Calculus* [459] 78
1.16 *American Mathematical Monthly* 89 (1982) 498
1.17 *Problems and Propositions in Analysis* [252] 66
1.18 *American Mathematical Monthly* 95 (1988) 699
1.19 *American Mathematical Monthly* 95 (1988) 699
1.20 *Mathematical Morsels* [229] 114
1.21 *American Mathematical Monthly* 92 (1985) 3–23
1.22 *The Stanford Mathematics Problem Book* [344] 8
1.23 *Induction and Analogy in Mathematics* [338] 43f
1.24 *Mathematics Magazine* 23 (1949) 109
1.25 Unsolved

1.26 *Elementary Theory of Numbers* [385] 411
1.27 *The USSR Olympiad Problem Book* [379] 13
1.28 *American Mathematical Monthly* 95 (1988) 704
1.29 *The Penguin Dictionary of Curious and Interesting Numbers* [445] 35
1.30 *American Mathematical Monthly* 40 (1933) 607

2.1 (a) *Mathematical Discovery,* Volume I [341] 62
(b) *Mathematical Gazette* 42 (1958) 116–117
2.2 *Elementary Theory of Numbers* [385] 84
2.3 *American Mathematical Monthly* 69 (1962) 168–169
Mathematical Gazette 55 (1971) 36
2.4 *Mathematical Discovery,* Volume I [341] 62
2.5 *Mathematical Discovery,* Volume I [341] 81
2.6 *The USSR Olympiad Problem Book* [379] 30
2.7 *The Method of Mathematical Induction* [394] 15
2.8 *Excursions in Number Theory* [322] 91
2.9 *Problems and Propositions in Analysis* [252] 4
2.10 *Method of Successive Approximations* [431] 20f
2.11 *Mathematical Discovery,* Volume I [341] 70f
2.12 *Mathematics Magazine* 60 (1987) 306
2.13 *American Mathematical Monthly* 58 (1951) 566
2.14 *American Mathematical Monthly* 71 (1964) 912–913
2.15 *The USSR Olympiad Problem Book* [379] 32
2.16 *Induction in Geometry* [180] 7–11
2.17 *The USSR Olympiad Problem Book* [379] 32
2.18 *Problems and Propositions in Analysis* [252] 107
2.19 *The USSR Olympiad Problem Book* [379] 24
Problems and Theorems in Analysis, Volume II [345] 112
American Mathematical Monthly 71 (1964) 1115
2.20 *Induction and Analogy in Mathematics* [338] 120

3.1 *The Higher Arithmetic* [120] 20
3.2 (a) *An Introduction to the Theory of Numbers* [211] 14
(b) *Solved and Unsolved Problems in Number Theory* [374] 13
(c) *An Introduction to the Theory of Numbers* [211] 14
3.3 *American Mathematical Monthly* 69 (1962) 567
3.4 *The USSR Olympiad Problem Book* [379] 21
3.5 *The Art of Computer Programming,* Volume 2 [263] 316
3.6 *Solved and Unsolved Problems in Number Theory* [374] 9f
3.7 *Excursions in Calculus* [459] 128
3.8 *Fibonacci & Lucas Numbers and the Golden Section* [422] 83
3.9 *Introduction to Analytic Number Theory* [10] 62
3.10 *Prime Numbers and Computer Methods for Factorization* [366] 11
3.11 *Elementary Number Theory* [421] 107f
3.12 *Journal of Recreational Mathematics* 15 (1982–83) 85–87
3.13 *The Art of Computer Programming,* Volume 2 [263] 316
3.14 (a) *Combinatorics* [430] 52f

(b) *Combinatorics* [430] 52f

(c) *American Mathematical Monthly* 52 (1945) 96

3.15 *Combinatorics* [430] 50

4.1 *Elementary Theory of Numbers* [385] 41

4.2 *Scripta Mathematica* 11 (1945) 188

4.3 *The USSR Olympiad Problem Book* [379] 29

4.4 *Excursions in Number Theory* [322] 68f

4.5 *Elementary Theory of Numbers* [385] 40

4.6 *Hungarian Problem Book II* [271] 10

4.7 *Elementary Theory of Numbers* [385] 109

4.8 *Elementary Theory of Numbers* [385] 39

4.9 *Elementary Theory of Numbers* [385] 125

4.10 *The USSR Olympiad Problem Book* [379] 29

4.11 *Induction and Analogy in Mathematics* [338] 59f

4.12 *Elementary Theory of Numbers* [385] 62

4.13 *The USSR Olympiad Problem Book* [379] 28

4.14 *The USSR Olympiad Problem Book* [379] 28

4.15 *The Penguin Dictionary of Curious and Interesting Numbers* [445] 122

4.16 *Elementary Theory of Numbers* [385] 51

4.17 *Mathematical Gazette* 65 (1981) 277

4.18 *Elementary Theory of Numbers* [385] 34

4.19 Unsolved

4.20 *A Selection of Problems in the Theory of Numbers* [383] 17f

4.21 *A Selection of Problems in the Theory of Numbers* [383] 18

4.22 *A Selection of Problems in the Theory of Numbers* [383] 18

4.23 *A Selection of Problems in the Theory of Numbers* [383] 18f

4.24 *The USSR Olympiad Problem Book* [379] 29

   Mathematical Gazette 28 (1944) 76

   American Mathematical Monthly 52 (1945) 278

4.25 *Mathematics and Its History* [404] 145f

Chapter 2

1.1 *Concrete Mathematics* [186] 108

1.2 *An Introduction to the Theory of Numbers* [211] 13

1.3 *Elementary Theory of Numbers* [385] 116

1.4 *Elementary Theory of Numbers* [385] 117

1.5 *Elementary Theory of Numbers* [385] 375

1.6 *An Introduction to the Theory of Numbers* [211] 15

1.7 *The USSR Olympiad Problem Book* [379] 18

1.8 Putnam Competition, 1971 [4] 15

1.9 *Elementary Theory of Numbers* [385] 372

1.10 *Elementary Theory of Numbers* [385] 375

1.11 *Unsolved Problems in Number Theory* [191] 6

1.12 Unsolved

1.13 *Concrete Mathematics* [186] 145
1.14 *Elementary Theory of Numbers* [385] 131
1.15 *American Mathematical Monthly* 71 (1964) 795
1.16 (a) Putnam Competition, 1962 [171] 64
       (b) *Concrete Mathematics* [186] 48
       (c) *Concrete Mathematics* [186] 50
1.17 *Concrete Mathematics* [186] 243
1.18 *Concrete Mathematics* [186] (a,b) 187f, (c) 231

2.1   *Elementary Number Theory* [421] 9f
2.2   *History of the Theory of Numbers,* Volume II [135] 3
2.3   *History of the Theory of Numbers,* Volume II [135] 2
2.4   Putnam Competition, 1958 [171] 52
2.5   *Elementary Theory of Numbers* [385] 87
2.6   *Elementary Theory of Numbers* [385] 87
2.7   Putnam Competition, 1958 [171] 54
2.8   *Elementary Theory of Numbers* [385] 42f
2.9   *Mathematical Gazette* 65 (1981) 87-92
2.10 *Mathematical Discovery,* Volume I [341] 79
2.11 *Mathematical Discovery,* Volume I [341] 80
2.12 *Mathematical Discovery,* Volume I [341] 80
2.13 *Concrete Mathematics* [186] 159
2.14 *Concrete Mathematics* [186] 160
2.15 (a) *Mathematical Discovery,* Volume I [341] 179
       (b) *Problems and Theorems in Analysis,* Volume I [345] 6
2.16 *Problems and Propositions in Analysis* [252] 51
2.17 *Induction and Analogy in Mathematics* [338] 118
2.18 *Concrete Mathematics* [186] 186f
2.19 *Fermat's Last Theorem* [146] 233
2.20 *Fermat's Last Theorem* [146] 234
2.21 *An Introduction to the Theory of Numbers* [211] 91

3.1   (a) *Introduction to Number Theory* [234] 14
       (b) *Elementary Number Theory* [79] 256
3.2   *Excursions in Number Theory* [322] 87
3.3   *Introduction to the Theory of Numbers* [376] 50
3.4   *American Mathematical Monthly* 62 (1955) 257
3.5   *Elementary Theory of Numbers* [385] 177
3.6   *Elementary Theory of Numbers* [385] 226
3.7   (a) *Concrete Mathematics* [186] 131
       (b) *Fermat's Last Theorem* [146] 24
3.8   *Number Theory: An Approach through History* [444] 57
3.9   (a) *Elementary Theory of Numbers* [385] 229
       (b) *Elementary Theory of Numbers* [385] 230
       (c) *Solved and Unsolved Problems in Number Theory* [374] 116
3.10 *Elementary Theory of Numbers* [385] 232f
3.11 *Elementary Theory of Numbers* [385] 218

Chapter 3

1.26 *Fibonacci & Lucas Numbers and the Golden Section* [422] 73
1.27 *Fibonacci & Lucas Numbers and the Golden Section* [422] 79
1.28 *Fibonacci & Lucas Numbers and the Golden Section* [422] 84
1.29 *Concrete Mathematics* [186] 301
1.30 *Fibonacci & Lucas Numbers and the Golden Section* [422] 90
1.31 *American Mathematical Monthly* 56 (1949) 409
1.32 *Fibonacci & Lucas Numbers and the Golden Section* [422] 34f
1.33 *Concrete Mathematics* [186] 281
1.34 *Fibonacci & Lucas Numbers and the Golden Section* [422] 115f

2.1 *Fibonacci & Lucas Numbers and the Golden Section* [422] 52
2.2 *Fibonacci & Lucas Numbers and the Golden Section* [422] 103
2.3 *Mathematical Gems III* [231] 135
2.4 *Mathematical Gems III* [231] 135
2.5 (a) *Fibonacci & Lucas Numbers and the Golden Section* [422] 145
     (b) *Fibonacci & Lucas Numbers and the Golden Section* [422] 145
     (c) *The Divine Proportion* [235] 30
2.6 *The Divine Proportion* [235] 45
2.7 *Fibonacci & Lucas Numbers and the Golden Section* [422] 150
2.8 *Scripta Mathematica* 19 (1953) 142–143
     Putnam Competition, 1959 [171] 57
2.9 *Continued Fractions* [323] 29
2.10 *Measure and Category* [327] 1
2.11 *The Scottish Book* [297] 113–117
2.12 *Calculus* [397] 128
     Answer Book for Calculus [398] 89
2.13 *A Primer of Real Functions* [46] 135f
2.14 *Calculus* [397] 128
     Answer Book for Calculus [398] 89
2.15 *Calculus* [397] 118
     Answer Book for Calculus [398] 82f
2.16 *A Primer of Real Functions* [46] 92f
2.17 *Calculus* [397] 126
     Answer Book for Calculus [398] 86f
2.18 *A Primer of Real Functions* [46] 125
2.19 *Calculus with Analytic Geometry* [387] 37f

3.1 *Generatingfunctionology* [452] 5
3.2 *Generatingfunctionology* [452] 59
3.3 *Generatingfunctionology* [452] 61
3.4 *Concrete Mathematics* [186] 197
3.5 *Generatingfunctionology* [452] 35
3.6 *Concrete Mathematics* [186] 362
3.7 *Concrete Mathematics* [186] 339
3.8 *Concrete Mathematics* [186] 169, 198
3.9 *Generatingfunctionology* [452] 38f
3.10 *Generatingfunctionology* [452] 62

3.11 *Generatingfunctionology* [452] 42
3.12 *Concrete Mathematics* [186] 243f
3.13 *Generatingfunctionology* [452] 22f, 39
3.14 *Introduction to Mathematical Probability* [420] 89
3.15 *American Mathematical Monthly* 63 (1956) 689–697
3.16 *American Mathematical Monthly* 63 (1956) 689–697
3.17 *Applied Combinatorics* [418] 242
3.18 *An Introduction to the Theory of Numbers* [211] 274
3.19 *Applied Combinatorics* [418] 241
3.20 *An Introduction to the Theory of Numbers* [211] 277
3.21 *Concrete Mathematics* [186] 271f
3.22 *Concrete Mathematics* [186] 273

4.1 *Method of Successive Approximations* [431] 16f
4.2 *A Primer of Real Functions* [46] 94
4.3 (a) *Calculus* [397] 433
    Answer Book for Calculus [398] 340
    (b) *Answer Book for Calculus* [398] 340
    (c) *Method of Successive Approximations* [431] 36
4.4 *Method of Successive Approximations* [431] 37
4.5 *Littlewood's Miscellany* [282] 55
4.6 *An Introduction to Chaotic Dynamical Systems* [131] 24f
4.7 *An Introduction to Chaotic Dynamical Systems* [131] (a) 32, (b) 28, (c) 60f
4.8 Putnam Competition, 1966 [4] 5
4.9 *Contests in Higher Mathematics* [167] 16
4.10 *American Mathematical Monthly* 94 (1987) 789–793
4.11 Putnam Competition, 1952 [171] 37
4.12 *American Mathematical Monthly* 68 (1961) 507–508
4.13 Putnam Competition, 1953 [171] 39

Chapter 4

1.1 *What is Mathematics?* [111] 363
1.2 *Inequalities* [210] 19
1.3 *Maxima and Minima without Calculus* [320] 21f
1.4 *Problems and Propositions in Analysis* [252] 156
1.5 *Maxima and Minima without Calculus* [320] 48
1.6 *Induction and Analogy in Mathematics* [338] 131
1.7 *The USSR Olympiad Problem Book* [379] 66
1.8 *Induction and Analogy in Mathematics* [338] 141
1.9 *Induction and Analogy in Mathematics* [338] 141
1.10 *Induction and Analogy in Mathematics* [338] 133
1.11 (a) *Hungarian Problem Book II* [271] 73f
    (b) *Calculus* [397] 213
    Answer Book for Calculus [398] 155
1.12 *A Primer of Real Functions* [46] 169

2.10 *The Calculus of Observations* [448] 217f

3.1 *Theory and Application of Infinite Series* [262] 84
3.2 *Theory and Application of Infinite Series* [262] 72
3.3 *Theory and Application of Infinite Series* [262] 73
3.4 *Theory and Application of Infinite Series* [262] 73
3.5 *Theory and Application of Infinite Series* [262] 73
3.6 *Problems and Theorems in Analysis,* Volume I [345] 16
3.7 *Calculus* [397] 262
3.8 (a) *American Mathematical Monthly* 63 (1956) 343
(b) *Calculus* [397] 429
Answer Book for Calculus [398] 335
(c) *Problems and Propositions in Analysis* [252] 303
(d) Putnam Competition, 1941 [171] 16
(e) *Problems and Propositions in Analysis* [252] 304
3.9 *Calculus* [397] 437
Answer Book for Calculus [398] 347f
3.10 *Methods of Real Analysis* [175] 65
3.11 *An Introduction to the Theory of Numbers* [211] 271
3.12 *Canadian Mathematical Bulletin* 6 (1963) 159-161
3.13 *Pythagorean Triangles* [382] 29
3.14 *Introduction to Finite Mathematics* [249] 134
3.15 *Fibonacci & Lucas Numbers and the Golden Section* [422] 55f
3.16 *Concrete Mathematics* [186] 394f
3.17 *Concrete Mathematics* [186] 379
3.18 *An Introduction to Probability Theory and Its Applications,* Volume 1 [156] 251f
3.19 *Introduction to Mathematical Probability* [420] 113f, 251f
3.20 *Introduction to Mathematical Probability* [420] 258
3.21 *Introduction to Mathematical Probability* [420] 255f
3.22 Putnam Competition, 1958 [171] 51
Mathematical Gazette 74 (1990) 167–169
3.23 *Aspects of Calculus* [253] 497
3.24 *Mathematical Discovery,* Volume I [341] 98
3.25 *One Hundred Problems in Elementary Mathematics* [399] 17
3.26 *Challenging Mathematical Problems with Elementary Solutions,* Volume I [457] 34f
3.27 Putnam Competition, 1958 [171] 52
3.28 *American Mathematical Monthly* 61 (1954) 647–648
3.29 *An Introduction to the Theory of Numbers* [211] 263f
3.30 *Number Theory in Science and Communication* [373] 32f

Chapter 5

1.1 *The Historical Development of the Calculus* [145] 114
1.2 *The Historical Development of the Calculus* [145] 178f
1.3 *Calculus with Analytic Geometry* [387] 446
1.4 (a) *Problems and Propositions in Analysis* [252] 259

(b) *Applied Combinatorics* [418] 242
(c) Putnam Competition, 1977 [4] 30
1.5 *Problem-Solving through Problems* [274] 187
1.6 *Induction and Analogy in Mathematics* [338] 102
1.7 *100 Great Problems of Elementary Mathematics* [137] 23
1.8 *An Introduction to Probability Theory and Its Applications,* Volume I [156] 73f, 311f
1.9 *Problems and Propositions in Analysis* [252] 427
1.10 *Induction and Analogy in Mathematics* [338] 86
1.11 *Pi and the AGM* [53] 5f
1.12 Putnam Competition, 1968 [4] 9
1.13 *Selected Papers on Calculus* [8] 325–327
1.14 *Theory and Application of Infinite Series* [262] 528

2.1 Stieltjes: *Oeuvres Complètes,* Volume II [402] 263f
2.2 *Mathematical Gazette* 75 (1991) 88–89
2.3 *Mathematical Gazette* 75 (1991) 88–89
2.4 Putnam Competition, 1953 [171] 38
2.5 *A Course in Mathematical Analysis,* Volume I [181] 204

3.1 *Problems and Propositions in Analysis* [252] 368
3.2 *Methods of Real Analysis* [175] 286
3.3 *Fundamentals of Interactive Computer Graphics* [158] 519f
3.4 *Interpolation and Approximation* [124] 33
3.5 *Problems for Mathematicians, Young and Old* [198] 73
3.6 Putnam Competition, 1958 [171] 52
   A Primer of Real Functions [46] 123
3.7 *First Course in Functional Analysis* [174] 181
3.8 *Interpolation and Approximation* [124] 121f

4.1 *Introduction to Number Theory* [234] 76f
   Problems and Propositions in Analysis [252] 266f
4.2 *Theory and Application of Infinite Series* [262] 219
4.3 *Mathematics and the Imagination* [244] 312
4.4 *American Mathematical Monthly* 61 (1954) 261–262
4.5 *An Introduction to the Theory of Numbers* [321] 475
4.6 *Mathematical Plums* [230] 56
4.7 *Entire Functions* [41] 5
4.8 Putnam Competition, 1969 [4] 12
4.9 Putnam Competition, 1954 [171] 43
   American Mathematical Monthly 69 (1962) 435–436
4.10 *Mathematical Plums* [230] 42
   American Mathematical Monthly 62 (1955) 123–124
4.11 *Mathematical Plums* [230] 42f
4.12 *Problems and Propositions in Analysis* [252] 270

5.1 *Elementary Theory of Numbers* [385] 137
5.2 *Elementary Theory of Numbers* [385] 164

Appendix

INDEX